ELECTRONICS ESSENTIALS
with Audio Emphasis

Gilbert K.D. Hu, P.E.

ELECTRONICS ESSENTIALS with AUDIO EMPHASIS © 2017

Copyright © 2017 Gilbert K.D. Hu
All rights reserved.

ISBN:
1982042028

ISBN-13:
978-1982042028

Contents

Listed in order of appearance in the book
Chapters from 1 to 7, shown aligned to the left
Supplements from A to Y, shown aligned to the right

1 Getting Ready

Electronics is a Language — A
The Language of Relationship: Graphs & Math — B
SI Units & Prefixes — C
Counterproductive Math Syndromes — D
Individualism vs. Teamwork in Electronics Industry — E
Essentials to Memorize — F
How this Book Relates to Other Audio Courses — G
"Take Apart" Project — H

2 Hello, World!

Three Way, Two Throw, One Pole? — I
Determining Component Values — J
Soldering — K
BOM Project — L
Pricing Electronics — M

3 DC and Network Analysis

Order Out of Chaos - From Matter to Current Electricity — N
More Order - From Energy to Voltage — O
V or E for Voltage? — P
XLR-TRS Cable Project — Q
Impedance Matching, Bridging, and Amplifier Damping Factor — R
How to Really Turn Off Power — S

4   Waves, Time & Frequency Domains, dB

                Project Notebook & Report T

5   AC Analysis

             R with AC; Loudspeakers in Series & Parallel U
         How to Visually Identify Common Passive Filters Quickly V
                     Duality Principle W
          Why Different Formulas for Impedance and Reactance X

6   Amplifiers

            Project Tips: Order of Soldering, Height, etc. Y

7   Op-Amps & Active Circuits

Chapter 1 GETTING READY

> This introduces not just the book, but the concept and ideas behind how to make the most out of learning electronics, with an emphasis in audio. Overcoming fear and adequate allocation of time and planning is essential to student success. Learning electronics is like learning a language; it requires perseverance and immersion.
>
> *Expected Learning Outcome*: The student will understand the overall flow of the book, the implication of electronics as language, the benefit of complexity as enabler for simplicity.

CONTENTS

What is Electronics?	1-3
Instructional Objectives	1-6
Learning Electronics	1-8
Book Organization	1-14
Math Hacks: Units Prefixes, Orders of Magnitude, Checks, etc.	1-16
Review Exercises	1-19

Electronics is for fun and pleasure, work and creativity, and for profit too. It is everywhere...

This book provides multiple approaches to help with learning electronics, including visualization aids and shortcuts. The student is encouraged to learn electronics as a language.

At the end of the course, the student is expected to be:

> 1) Able to visually **_identify_** common audio components and connectors and obtain their values
>
> 2) Able to **_solder_** small to medium projects
>
> 3) Able to **_troubleshoot_** such a project using the theories learned
>
> 4) Able to **_communicate_** with (read and write) colleagues using the language of audio electronics (schematics, equations, graphs, terms, methodology)

For the latter, the student is also expected to also be:

> a) Able to articulate audio electronics **_terms and concepts_**, including time and frequency domains, decibel (dB), and analog and digital domains
>
> b) Able to perform **_DC Analysis_** on most circuits
>
> c) Able to perform **_AC Analysis_** on common audio circuits
>
> d) Able to understand **_amplifiers_** and **_op-amps_** and how they function
>
> e) Able to understand various sources of **_noise_** and how to mitigate

At this point in the semester, the student is advised to set his house in order by getting familiar with essential math skills.

1.1 WHAT IS ELECTRONICS?

Electronics is everywhere in our modern society…

Can we go a day without electronics?

If we go without electronics, that means we forego listening to music from iPad or Google Home or Amazon Echo. That means no more texting or calls on cell phones, and no more apps. No GPS to give directions. No television or video games for entertainment. No pleasure or work on computers. No modern automobiles, filled with modern electronics (which makes most of the profits for car manufacturers, by the way), including fuel injection, smart braking, automatic windshield wipers, and so on. No travel by planes, which are again filled with electronics for navigation, entertainment, and often fly-by-wire control of the airplane. No modern communications with email, or cell phone, or texting. Essentially for many, that means "no life" …

One may think, at least I can make music using an "old-school" guitar or piano. Yes, but wait until you need a metronome. Modern metronomes have multiple functions, are lightweight and easy to carry and control, because they are electronic. And they are cheaper too, which explains why it is difficult to find old-school mechanical metronomes these days. Likewise, when we need a guitar tuner, more than likely it is an electronic tuner, as it is cheaper and with better accuracy and easier to see, compared to old-school pitch pipes or other tuning devices. And if you sing, the microphones and amplified-speakers are all electronics. Most of the lighting in concert halls are also electronically controlled, as are recording studios.

Can you name three things you *do* or *touch* today that do not involve electricity (or charge carriers, which we shall study in chapter 3**) in any way?** Remember, most things were *produced using electricity* these days. And *our body uses electricity internally* for signaling (communication) and even in thinking…

Work or Leisure?

Most workers in an office use computers these days – that is electronics (despite the 2017 Apple ad disavowing the term computer, implying iPad is all a child needs to know). And most industrial operations use computers for control, including robots, Artificial Intelligence (AI), and Internet of Things (IoT).

Video games, cell phones, MP3 players, entertainment centers, karaoke machines, keyboards – they are all electronics. Even cameras, video cameras, and drones – they have significant amounts of electronics. And if you play chess or GO (an ancient Chinese board game considered significantly more difficult to crack than chess), the world champions are now all computers…

Were you woken up by an alarm clock or radio in the morning? Or was it a cell phone, or Alexa or Siri or the Google assistant? They are most likely all electronics…

A Brief History of the World of Electronics

One way to understand the history of electronics is to explore the evolution of professional organizations related to electronics…

In 1816, the first working electric telegraph system became available. In 1871, the Society of Telegraph Engineers was formed in England, arguably the first professional society in an area related to electronics, which at the time relates to what we now call circuit theory, and known as "electrotechnics" (or electrotechnology) in those days across the pond.

In 1880, its name was changed to the Society of Telegraph Engineers and Electricians, reflecting the addition of electric power to its focus. Electric poles then often carry *both* power and telegraph wires.

In 1889, its name was changed again to the (now) more familiar name of Institution of Electrical Engineers (IEE), the largest professional engineering society in Europe at the time.

In 2006, it became known as Institution of Engineering and Technology (IET) after a merger.

Across the pond a similar track occurred in America as well. In 1884, the American Institute of Electrical Engineers (AIEE) was formed by Nikola Tesla, Thomas Edison, Elihu Thomson, Edwin Houston, and Edward Weston (more in a few paragraphs), among others.

In 1912, the Institute of Radio Engineers (IRE) was formed in America, because its members felt the AIEE had too much focus on electric power. It eventually grew bigger than its older counterpart, because radio communications experienced great expansion.

In intervening years, telephony surpassed telegraphy in importance. Many present-day pro-audio standards were formed in the telephony years, and would be easier understood with such history in mind.

In 1946, the AIEE formed its Subcommittee on Large-Scale Computing, correctly recognizing the importance of computer engineering in the future.

In 1963, AIEE and IRE merged to form the Institute of Electrical and Electronics Engineers (IEEE), soon becoming the world's largest technical society. Electric power and radio communications were still the major players at that time, although they correctly understood the future dominance of electronics, which is reflected in its new name.

Electronics Continues to Evolve Rapidly

In 1958 and 1959, the integrated circuit (IC) was developed – "integrated" referring to the integration of more and more circuits in the same piece of semiconductor or package. From then on, the degree of integration has been increasing steadily, following Moore's law which is the empirical (the term explained in chapter 3) observation that the number of transistors in a dense integrated circuit doubles approximately every two years since 1975. Microelectronics is now a distinct subdivision of electronics.

In 1968, the microprocessor was first developed. The term Personal Computer (PC) was coined in 1975, with the IBM PC launched in 1981. The Compact Disk (CD) was developed in the following year, 1982, ushering in the age of digital audio. All these were possible due to the widespread and inexpensive availability of the microprocessor and IC in the first place.

Many people started replacing their cell phones every two years around the early 2010's. That is a testament to the rapid evolution of electronics (or at least its marketing).

Interdisciplinary Nature

Electronics had been influenced heavily by polymaths and people interested in interdisciplinary applications in its early days, as we can see from the roster of founders of AIEE in America.

Edward Weston is a good example of one who crosses many disciplines. He received his medical diploma in 1870, when he was twenty years old. He was noted as a chemist for advancing the work of electroplating, but also developed measurement instruments for electric current. He invented the Cadmium cell, which became the standard for measuring EMF (defined in chapter 3) in 1911. And no one will deny him as electrical engineer, as he co-founded and was President of the AIEE. His son, Edward Faraday Weston, is considered a photography pioneer with a special aesthetic, yet he also patented and manufactured several exposure meters for photography – in fact introduced the earliest use of electronics in photography. He also established the Weston film speeds, which is later the basis for international standardization.

Many of the people who dominated audio engineering in the early days, and which helped defined its terms and its historical course, were *both* musicians and engineers. Their legacy explains why we have a lot of complicated math and technology in musical applications today.

We'll cover some more history of the giants in electronics in the rest of the book.

Creatives and Electronics

It should be mentioned that electronics is a serious tool for electroacoustic composition, musical installations, spatial musical presentations (that are immersive with music appearing to come from multiple directions and distances), musical synthesis, and much more…

Another example of an interdisciplinary influence is Sir George Martin, occasionally called the "Fifth Beatle," who was record producer, arranger, composer, conductor, audio engineer, and musician. He arranged many of Beatles' songs, composing some of its most enigmatic chord arrangements. He performed the keyboard on stage (but not in the limelight). He experimented with state-of-the-art audio engineering and recording and composition techniques, pushing the studio equipment available then to their limits.

There are *two* ways a creative can handle electronics. This is best illustrated with an example musician and how she handles music and electronics. A non-disciplinary musician treats music as music, and electronics just like she would in her music world – a second-class citizen. An interdisciplinary musician, on the other hand, treats music as music and electronics as electronics, all as first-class citizens, and study and utilize both in creative and scientific ways. It is my hope that the reader will follow the latter example.

Making Electronics Easier to Study

While electronics seem so enticing, fun, growing, and creative, we must mention that some may find it difficult to study. It is nearly impossible to keep an electronics textbook up to date. Or for that matter to find an electronics textbook that covers one specific area of interest, like audio applications, fully. Because electronics is so interdisciplinary and covers so much ground, we must be careful to limit the material in this book to a reasonable level, and not try to include everything.

To keep a narrower field of study, electronics by convention is subdivided into advanced areas that are excluded in a beginning course: examples are radio frequency (RF), microwave, computer, control, communications, microelectronics, photonics, electromagnetics, etc. If one uses the definition of anything that uses or produces electrons as electronics, the field is simply too wide and too difficult for study. We emphasize audio frequency applications, which is usually excluded in a beginning course of electronics.

What are *not* electronics?

Not all things that use or produce electrons are electronics. For example, humans do, but we do not consider ourselves electronics.

Likewise, now we know *all* chemical reactions involve electrons. But by convention we call that chemistry and *not* electronics.

Electric motors and generators are the first devices engineers use that involve electricity. They are, in a sense, "electrics" but *not* electronics. Likewise, electricity transmission and distribution involve electricity, but are *not* considered electronics.

Static electricity started the revolution of electronics, but it and what is known as electrostatics are usually *not* considered part of electronics, although we do use and teach some portions of electrostatics just for continuity.

Switches, resistors, capacitors, and inductors are commonly studied in many disciplines, including electronics, but they are *not* the unique domain of electronics.

Electromagnetic solenoids and loudspeakers are commonly *used* in audio applications, and so we cover their schematic symbols. However, their electrodynamics aspects are also *outside* of electronics studies. On the other hand, the crossover filter or powered amplifier inside a powered speaker, and the pre-amplifier inside a microphone, would be considered electronics. Likewise, the box and the venting of loudspeakers, as well as the acoustics of the performance venue, are all acoustics related and outside of electronics, although they are part of the specific parts of audio engineering and live sound.

Supplement G summarizes some more relationships of electronics to commercial music curriculum.

Hopefully one can see from this discussion that **convention** dictates what is considered electronics. Usually one can tell that anything discovered before the age of vacuum tube or transistors are not considered exclusive domain of electronics, while anything that are associated with vacuum tubes or transistors would be. In a way, this is *not* a definition, or at best a circular one, associating electronics more with active circuits (which we shall define later.) which is its exclusive domain, with passive circuits (again to be defined later) a shared domain with other sciences.

1.2 Instructional Objectives

At the end of the course, the student is expected to be:

> 1) Able to visually **_identify_** common audio components and connectors and obtain their values
>
> 2) Able to **_solder_** small to medium projects
>
> 3) Able to **_troubleshoot_** such a project using the theories learned
>
> 4) Able to **_communicate_** with (read and write) colleagues using the language of audio electronics (schematics, equations, graphs, terms, methodology)

One may see that the above objectives are extremely practical and pragmatic; but one may wonder why so much "theory" is also required. For one thing, that has always been required in traditional curriculum. For another, how can one troubleshoot properly *without* understanding theory in depth? And how can one communicate properly if one doesn't learn to read and write the language of audio electronics? By the way, that necessarily means that we need to know enough theory - our coverage of theory is intended to be *just enough*, but *sufficient* to help with troubleshooting - one may call it "practical theory". That explains why we will spend more time with DC Analysis (which is also the case for traditional curriculum) - because in troubleshooting we often use DC Analysis more than AC Analysis!

Thus, the student is also expected to also be:

> a) Able to articulate audio electronics **_terms and concepts_**, including time and frequency domains, decibel (dB), and analog and digital domains
>
> b) Able to perform **_DC Analysis_** on most circuits
>
> c) Able to perform **_AC Analysis_** on common audio circuits
>
> d) Able to understand **_amplifiers_** and **_op-amps_** and how they function
>
> e) Able to understand various sources of **_noise_** and how to mitigate

Noise is handled throughout the chapters as they become consequential. The other topics have their own chapters. DC Analysis is probably covered more fully than traditional textbooks because it is the basis for much troubleshooting.

To synchronize expectations, here are examples of questions that students are expected to be able to handle – of course they are *not* expected to know at the beginning of the course, but if they *appear* intimating now, it *probably also will be later* – unless they treat electronics as *not* an easy course from the beginning...

Sample Questions

1. Read this schematic and explain its function. Draw its characteristic curve. For what type of audio circuit can this be a base?

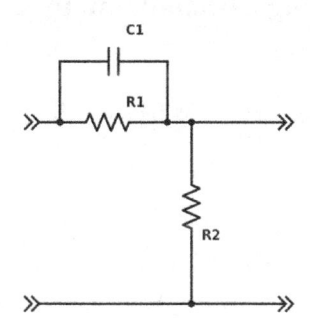

You of course need to know how to read the schematic symbols of resistors and capacitors. You'll need to know the DC and high frequency response of a capacitor (open and short circuits). You'll need to be able to do simple DC and AC analysis (DC: a voltage divider; AC: pass through). Armed with all those, and the usage of Bode Plot, you can then draw the characteristic curve, from which you can determine this is a shelving filter, which is the basis for an audio tone control circuit. And by the way, all these "analysis" can be performed in our head without a calculator. These are the types of questions one is expected to answer in a midterm exam.

2. Read this schematic and explain its function. Explain functions of essential components.

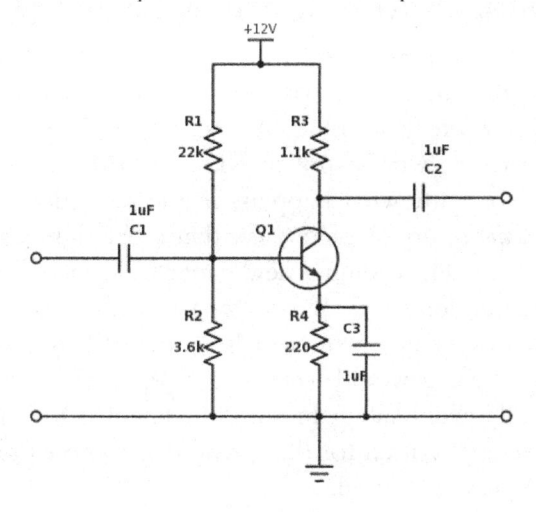

This is an active circuit; the active device is an NPN transistor. C1 and C2 are input and output coupling capacitors. R1 and R2 sets the DC operating bias for the base of the transistor. R3 is the load, and the voltage gain is roughly - R4/R3 = -5, this being an NPN common emitter amplifier circuit with inverting output (indicated by the minus sign). C3 is the emitter bypass capacitor, increasing the gain at high frequency to compensate and obtain a flatter frequency response. These are the types of questions one is expected to answer in a final exam.

1.3 Learning Electronics

Heard: "If you want to memorize names, become a biologist.
If you want to memorize exceptions, become a chemist.
If you want to memorize concepts, become a physicist.
If you don't want to memorize anything, become a liberal arts major."

We may add: "**If you don't want to memorize anything or handle math, avoid electronics at all cost.**" You have been warned. :)

Let us reverse the questioning: "We are going to study electronics. What are the best ways to do it?"
Here are some ideas…

Treat Electronics as a Language

Electronics behave like a living language. Americans call that thing a capacitor, while those across the pond calls it a condenser, yet when incorporated into a microphone, Americans would call it a condenser microphone and not a capacitor microphone. Likewise, the symbols for resistors and capacitors are different across the pond. And when everyone agrees to use the *same* term "amplitude", it could mean *different* things dependent on the context. For example, different textbooks will give different formulas for calculating the impedance of a capacitor. (Pardon the vocabulary we use; they will be explained later in the book.) These are all aspects of electronics as a language of communications, with many dialects and local adaptations. And thus, it is advantageous to learn it as a language.

Supplement A, "Electronics is a Language," explains how one person who failed to learn a subject (using her traditional way of learning) can learn it and master it after treating it as a language.

Take Advantage of Our Brain's Natural Order of Learning Language

Imagine you were a baby learning your mother tongue. Did you specify a learning order, and insist on tackling verbs *first* before nouns? Or even attempt to classify parts of speech? Or *demand* to have part of speech explained first? Or did you **tune out** any words you didn't understand?

A baby's brain appears to naturally **tune in** to *any* words spoken. That was the secret to the success of the baby in learning a language fast and easy. Imagine what happens if the baby decides to **tune out** any words not comprehensible, like most students *want* to do, or even *believe* that is the right way to go?

Now imagine an engineer at the peak of her career, reading a new paper published in a scientific journal. Will she tune out any words or ideas she doesn't understand? If she knew *everything*, there is no need to read the paper! My experience is that the typical paper contains many words or ideas I had not seen before – **that is normal**, and if I tune out on encountering them, I would have to give up reading the paper altogether! But those who persevere are the ones who would eventually gain the knowledge. One may even then come to the "strange" conclusion that scientific papers are written for them who don't know the subject matter, yet writing assuming the person already knew the words is normal…

Babies and experts alike do not tune out on encountering difficult words and ideas. But somewhere in between the two age groups, there are students who started to adopt the mentality that it is best to learn in a "systematic" manner, that it is best to tune out things one does not "immediately" comprehend. Is that the best way to learn a new language?

As an experiment, please read the above few paragraphs again, with the idea that **it is normal to encounter words we don't yet know**, and you may realize that the overall concept is still understandable!

Deep Dive in the Web of Language

There is a basic assumption many makes, that things *can* be explained linearly, proceeding with the easiest first, and then growing naturally more and more complicated, and learning *should* proceed that way. If language is linear, then that would be the ideal situation, and the best way to learn. But language is *not* linear! In fact, *computer* language research in the last few decades show us that we now usually define a new artificial computer language *recursively*. Recursive roughly means circularly, that there is no beginning and no end, that it cannot be defined linearly (or at least we don't know yet how to express it so, or at least *efficiently*). If that is the case for artificial synthetic language, how much more it is would it be for a *natural* language?

But *how* do people teach or learn a language linearly then? By untangling the web, breaking the circle into a line, *artificially* making it linear. That can only be done by introducing an "error" at the beginning of the line. The idea is that then learning can proceed systematically, and once the whole line is learned, then the "error" can be corrected and the circle is formed. Unfortunately, many students never knew that the "error" was there in the first place, and so never got to correcting the error and learning the language correctly...

I had a professor who was director of a major computer science lab and a member of congressional science panel. He related to us a work he personally did on neural networks. He simulated a brain of a fixed size using neural networks, and discovered that the brain was *never* fully utilized. Others had posited the same with real brains, but his work was with a simulated brain. He found the maximum usage of the simulated brain depended on what he called "initialization". Good stimulus at the beginning of learning initialize the brain well, leading to higher usage numbers up to 60% or so. But poor stimulus at the beginning initialize the brain to a bad start, and maximum usage would be 5% or less.

If that research is applicable to the learning of electronics, then I would suggest the student to simply **dive in**, and **deeply**, like a child, into the world of electronics, and *not* use their linear way of learning which may have significant limitations and may slow down or even limit learning. The breaking up of circles into lines *could* especially be harmful to learning electronics because of the poor initial stimulus hypothesis.

One may argue that "throwing away the baby with the dirty bath water" is a bad idea. Or that "if the baby can't eat steak, at least ground up the steak and give her minced steak." Thus, we still try to use the traditional order of learning electronics like everyone else – DC first, then AC, then amplifiers and op-amps, and likewise passive before active. (By now one should know there is no need to "know" what AC, DC, passive, and active means and still can get sufficient amount of information.)

Understand According to the Hierarchy of Language

One may legitimately ask what is the point of this chapter, or the purpose of the tradition of presenting a syllabus at the first class of a semester, if we *really* learn linearly. The answer is, we learn *hierarchically*. In the first day of class, we are *not* supposed to know all the terms used in the syllabus. But the words AC, DC, active, and passive (and impedance and capacitor by the way) are all there anyway. The reader of the syllabus was expected to understand only a *rough idea* of what the course is all about. They can refer to the syllabus as they move on in the semester, and then the concepts become clearer and clearer.

Understanding that we learn and communicate hierarchically is a key aspect of learning a language, and electronics is no exception.

There is another aspect of hierarchy that is particularly important to electronics. Electronics is too complex to be grasped linearly as a big "blob", and so it is usually communicated in small chunks. But small chunks do not communicate the big picture, and so the small chunks are organized hierarchically so that the big picture can also be understood easily. Notice that we had not even explained the word "hierarchy" - deliberately – to illustrate the concept we are trying to convey – one needs to get used to the web of language. A better way is to illustrate with examples. (However, again bear in mind that there are words and ideas in the example that may not make sense at this first day of the semester. The reader is *not* expected to understand everything, rather the key is to understand the main idea, which is how hierarchy works...)

A simple example is the word RADAR. We all know a *rough idea* of what it is – and in some instances, that is all we need to know to avoid a traffic ticket. RADAR is actually an acronym; some say it represents RAdio Direction/Detection And Ranging. When we look at the words corresponding to the acronym we sort of zoomed in to more detail at the next hierarchy level – and that is the way hierarchy works.

Examples of Hierarchical Decomposition

3. Read this schematic and explain its function. Explain functions of essential components.

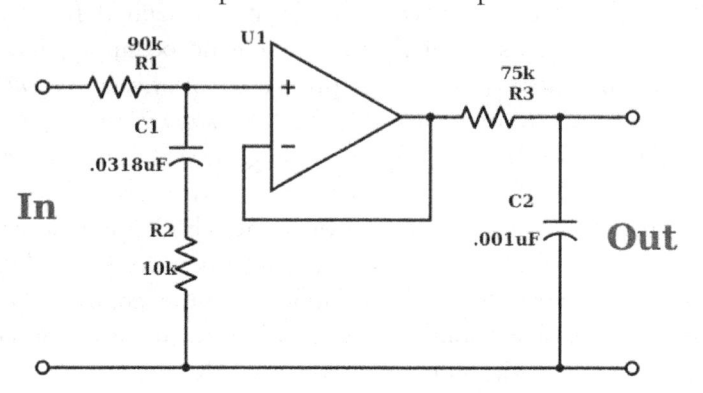

This is a good example of why we need to learn the skills of **hierarchical decomposition**. It looks complicated at first, until we realize we can decompose it into its three component parts: a shelving filter on the left (like example 1), an even simpler low pass filter on the right, with a time constant of 75k*0.001μ = 75 μs, plus a non-inverting voltage follower in between which acts as a buffer in between the two passive filters. By the way, U1 is an **integrated circuit (IC) op-amp**, with example 2 as prototype of an amplifier. The three component parts together forms a filter circuit once common in the front-end of home amplifiers.

4. Read this schematic and explain its function. Explain functions of essential components.

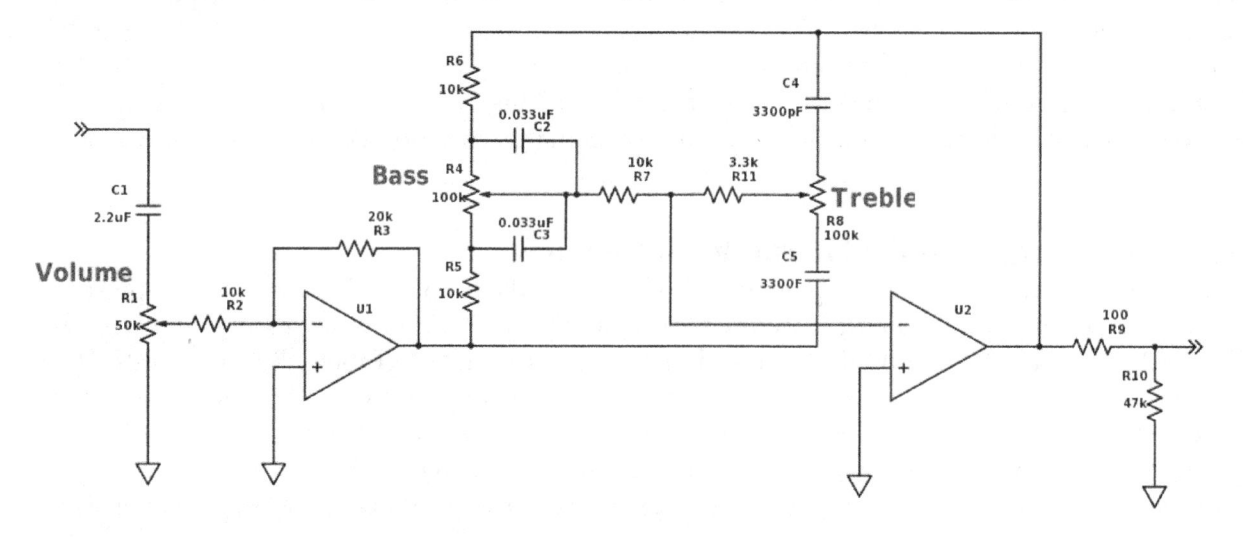

This is another good example of why we need to learn the skills of hierarchical decomposition. Had we not known how to decompose the circuit, it would have been too complex to analyze using common circuit theorems. The leftmost side is an input coupling capacitor connected to a voltage divider, which acts as an input gain level or volume control. It then feeds into an inverting amplifier with voltage gain of - 20k/10k = -2, the minus indicating an inverting amplifier with a gain of 2 which doubles the input voltage. To the extreme right is similarly a voltage divider which lowers the voltage by 47k/(47k+100). That circuit was also there to maintain a nominal output impedance. U2 acts as an inverting amplifier, with the James network in its feedback path. The James network is an excellent adjustable shelving filter (the basis of which is example 1), with the left-side handling bass tone control, and the right-side handling treble tone control. When we add parametric equalizers to this circuit, and a mixer bus explained later in this book, it becomes the prototype of a modern audio mixer board...

Chapter 1 – Getting Ready

Understand Different Approaches to Being "Systematic"

When a physicist offers advice on dairy farming, "first, we *assume* a *spherical* cow…" When an engineer offers advice, "first, we *create* an *equivalent* cow…" Most people won't think *either* way. But to communicate using this language of electronics, one needs to understand this: **engineers systematically solve problems through equivalence**. It is important to understand this language of **equivalent circuits**, and how to create them. Why? The concept of **equivalent circuit** is the key to easier DC and AC analysis. We teach *both* the analytical method, which is more cumbersome but does not require understanding of equivalent circuit, *and* the equivalent circuit method, which is easier to apply, but does require an extra understanding. It is up to you which approach to take…

Take Advantage of the Language of Shortcuts

When phrased that way, everyone would agree to take shortcuts. Who doesn't like shortcuts? Yet amazingly, those who said yes would at the same time *refuse* to learn a new law of electronics or to use a new domain of analysis or a new unit of dB, thinking that it is "more work" to learn such. (Again, please pardon the use of all these new words; the reason will be explained shortly.) What the reader needs to understand is that **the "extra" law or domain or unit are all invented as shortcuts to solving electronics problems**! The reason why we need to learn them is that they simplify our lives and makes the work of electronics easier! Plus, they are so useful that they become the language of electronics, so it would be foolhardy not to learn them. Trust the wisdom of the engineer crowd in seeing that what we learn are mostly shortcuts and most *essential*… :)

Take an extreme example. Some cultures created a simple counting system by using one stick for one object. That is really *all* we need to represent *any* arbitrary number. It is the easiest to learn, but is it practical? What if we need to represent ten, or hundred, or a thousand? That is where shortcuts come into play. Think about it: ten and hundred and thousand are *shortcuts* that make our life easier and simpler and more enjoyable! But if we think it is too complex, we'll never utilize the shortcut nor be able to communicate with those who decided to learn the language fully.

Another way of saying this is that engineers make Zen-like tradeoffs that most likely will confound most entry-level Jedi. They simplify their work by going through a more complex path! By learning a new trick, it simplifies life.

Take Advantage of Visualization & Alternative Approaches

Other than the concept of equivalent circuit and hierarchy, another common shortcut used by engineers are visualization methods. Bode Plot is one such example being taught in chapter 5. One needs to invest extra time to learn it, but in return pays big dividends in easier and shorter analysis. Again, the analytic method is *also* taught, but those who *decides* to *not* learn visualization because of the extra work would be condemned to the hell of the analytical method, and when one decides not to do that either because of its tediousness, then no tools are left to handle the subject matter.

Another alternative approach we teach is the time constant method, again widely used in the industry. We also teach the traditional analytical method. But again, those who decide not to learn an *extra* approach is condemned to the hell of the tedious method…

In addition, we often provide *additional* formulas for those who do *not* want to convert formulas into their equivalent forms. It is better than other textbooks who do not give the **option**. Often students get the impression that we are requiring them to remember *extra* formulas. It is really for their **convenience**. One can choose that option if they decide not to bother with using the superpower of math…

Take Advantage of the Superpower of Mathematics

Mathematics should be viewed as serving the language of electronics in dealing with relationships - **the language of relationship is mathematics**. (See supplement B, "Language of Relationship" where formula, table, graph, and text can all be used to represent a relationship.) And the shortcut of studying relationship is through formulas and graphs. If one refuses to do so because they are math, then again, we are condemning ourselves without the language of shortcuts.

1-11

You May Have Used That Method Before!

How did you learn the use of emoji or text message abbreviations? Did you learn "systematically" or "intuitively" with delayed resolution, just like what was described above? And how did you learn to use software so much faster than the generation of your parents and grandparents? If you LOL or SMH now, just realize that it is better to use the *natural* way to learn a new language and give similar flexibility and approach to learning electronics.

Imagine why parents and grandparents were having troubles learning these new technologies? Most likely it is *not* because of intelligence or old age, but rather because they refused to approach them as a language...

Immersion in the Language

To learn a language, the best way is to immerse ourselves in the culture. That is where it helps to interact with practitioners of electronics. Projects also help contextualize the study. If occasional lab sessions are scheduled for class, please treat it as time for immersion in the culture, and *not* simply as just a time to do their projects. In fact, recently one student asked, "Can you do your lecture first so I can skip lab and leave early? I can be more efficient doing my project at home." That misses the main point of the lab, which is interaction within the language culture. One learns by seeing how other students tackle their project. Your instructor will most likely suggest you do things differently or learn from some other fellow students' approaches. More importantly, you may be saving a lot of time by troubleshooting in the presence of your instructor. We say troubleshooting is mostly learned through osmosis. We are probably one of the few who systematically introduce troubleshooting concepts, like separating DC and AC analysis. Still, there is a lot to be learned via the interaction of the lab hour. Please do not waste that time by *not* doing the preparation necessary. For example, if you do soldering during lab hour for the final project, you are not taking advantage of the interaction time - solder at home so that you can do troubleshooting at the lab hour... (In contrast, for mid-term projects you may want to do soldering during the lab session, so the instructor can help critique your soldering.)

Practice Makes Perfect

Students seldom like or appreciate this tried-and-true advice, that practice makes language learning perfect; instead most seem to prefer just taking risk that they'll get it right the first time, or even without any practice at all. It is important for the student to come up with practice questions on their own (or look them up in books or on the internet), and keep on practicing until they are comfortable with the idea. The exercises provided at the end of each major chapter are mere subsets of what may be needed to perfect the skills.

Critical Mass

Language students often discover that they are exposed to "a lot of stuff" at the beginning. It is because one needs *enough* command of various aspects of the language before things start to make sense; the earlier that "critical mass" is reached, the better off it is. "Front-loading" is the secret to language learning.

Commit Essentials to Memory

Every language student will realize that the essentials of the language needs to be memorized, or else there would be little improvements. This is a corollary to the "practice makes perfect." Supplement F contains top 26 items to commit to memory. Please do not delay, as one idea builds on another, and anything not committed to memory will make future ideas less difficult to comprehend.

Troubleshooting and Planning is Key

Students often prefer to just execute a project in "steps A to Z" as provided by someone. They prefer not to think or plan or solve problems. But the fact of the matter is that the key to electronics lies in troubleshooting and planning. Students need to learn to plan their projects, figuring out always how best to tackle and which steps to be taken first. (For example, once you finish soldering connectors on both ends of a cable, it is too late to put on the heat shrink tube.) One also needs to learn to keep planning for testing at intermediate stages, making sure that no errors are accumulated along the way, and to troubleshoot early on to avoid cascaded failures. The earlier a failure is detected, the easier it is to fix the problem! Those who insists on "pick-and-choose" will most likely pick and choose a "linear" method of ignoring troubleshooting altogether, thus likely dooming the project.

If you are required to do a final project, bear in mind that more than likely the project is designed so that you will be required to troubleshoot as it is unlikely you'll make perfect soldering or assembling all along. Plan accordingly. Also note that you may be required to maintain a project notebook and write a final report. (See supplement T "Project notebook and report".) Also, an enclosure is usually required. That is the traditional part often missed, as components may not fit if planning for the enclosure was not done from the very beginning.

If you order a kit, make sure that you order early and that international shipping will not delay your schedules.

Metacognition

If you are scared by such "big words", you need to learn to calm such fears. I was introduced to this term by my professor in Honors English I. But it turns out metacognition is useful for every subject, including electronics; and a 2017 paper considered it useful for taking some of the most feared standardized exams like the General Chemistry and Organic Chemistry exams administered by ACS; by analogy it can be successful for other exams like SAT and GRE administered by the College Board.

Metacognition is often defined as "cognition about cognition", "thinking about thinking", "knowing about knowing", becoming "aware of one's awareness" and higher-order thinking skills in general. What is interesting is that metacognition has been applied to the study of language and linguistics, exactly where we say electronics is also characterized with.

Practically speaking, the idea is that during an exam, I don't just simply answer the questions. I think about what the persons setting the questions were thinking. When I continue to think that way for every question, I come up with an overall plan from the point of view of the examiner. When I don't understand certain aspects of the questions, I can use this overall canvas of the examiner's point of view to figure out what the question may be asking. Surprisingly, **I learn something during the exam**! And in doing so, it gave me a better strategy to answer the questions...

Many students think like primary school students - they assume everything had been taught in class. That is the case probably up to high school level. By college level, it simply won't work that way. This textbook is unique in that it covers far more subject areas in general than other textbooks. But still, there are not enough time during class to cover all the subjects in this book. The student should take the initiative to read the whole book on their own.

Answering Essay-style Questions as Language Communications

Many students are so accustomed to multiple choice questions that they may not understand how to tackle essay-style questions. Typically, such exams are short-answer and fill-in-the-blank, not the "multiple guess" type of exams. You'll not only have to remember basic concepts and facts but you'll also need to be able to express these concepts and facts in three or four sentences along with diagrams when relevant. The spirit of the short-answer exam is to give you practice articulating the concepts to someone (such as a future employer). This future person may oversee the budget and purchases for a company you work for, or they may be a consultant you've hired to address issues in your recording studio, etc.

For calculation-based essay-style questions, see supplement E, "Individualism or teamwork," for why the individualistic approach of "this is my final answer" may not be beneficial. One should quote a theorem (e.g. KVL), show the equation, show the relevant steps, and then repeat (Ohm's law, etc.) until the answer is reached. One should *not* assume that only *one* formula is needed; in real life (IRL) usually multiple laws and formulas are needed to solve electronics problems.

1.4 BOOK ORGANIZATION

Do you like to be weighed down while running an obstacle course? Electronics is an obstacle course or a perseverance race, *not* a sprint. This book is organized to help you systematically *carry less unnecessary weights* while running. If you don't do so, you may find out eventually you can't move even though you exerted yourself, because you may be thoroughly weighed down! As shown in the figure below, as you accomplish said goals, you'll win merit badges…

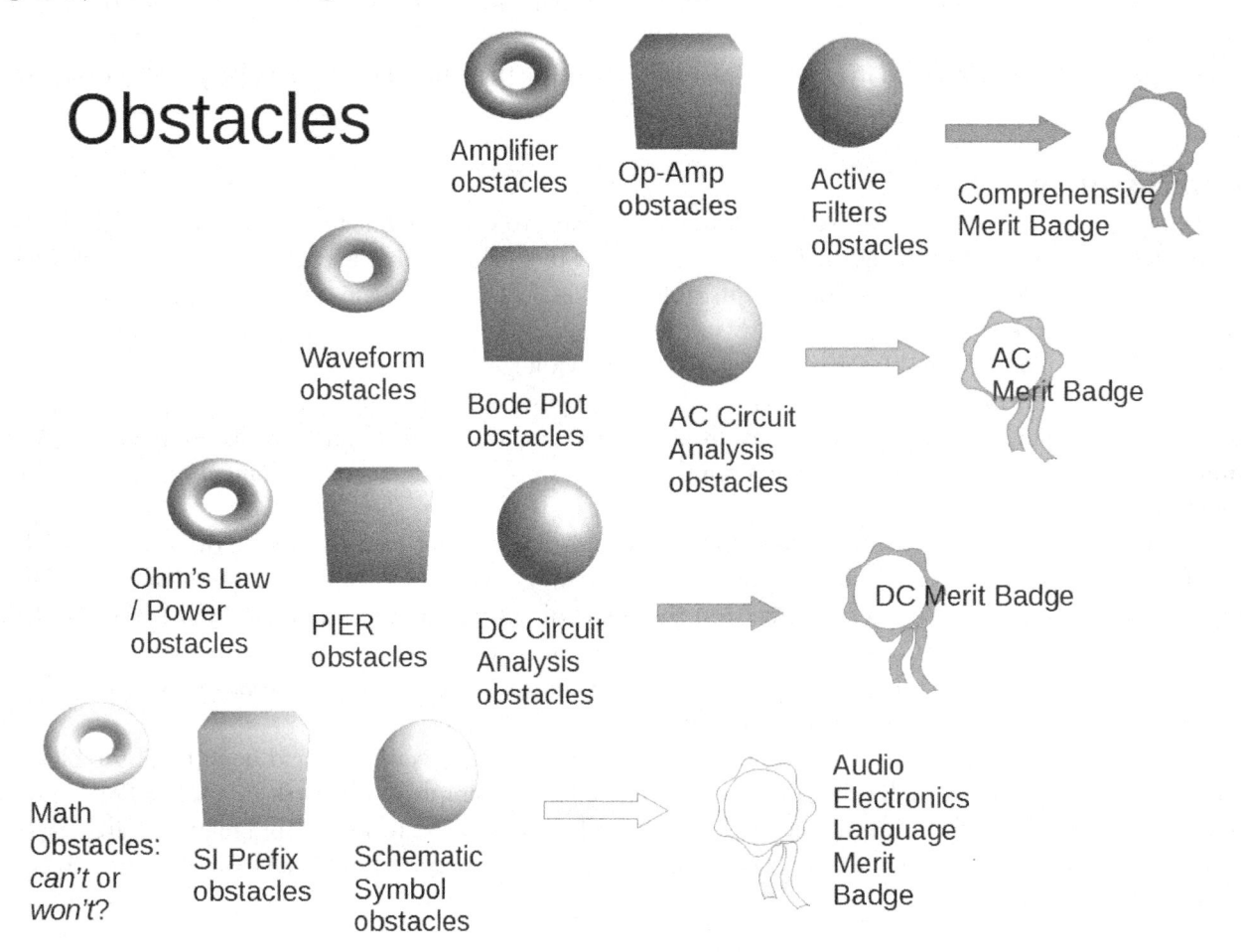

Obstacles

- Amplifier obstacles
- Op-Amp obstacles
- Active Filters obstacles
- → Comprehensive Merit Badge

- Waveform obstacles
- Bode Plot obstacles
- AC Circuit Analysis obstacles
- → AC Merit Badge

- Ohm's Law / Power obstacles
- PIER obstacles
- DC Circuit Analysis obstacles
- → DC Merit Badge

- Math Obstacles: *can't* or *won't*?
- SI Prefix obstacles
- Schematic Symbol obstacles
- → Audio Electronics Language Merit Badge

On the foundation (bottom) row of the above diagram, you'll see you need to overcome the fear of math, actively counter the idea of "I can't" when in reality it is "I won't". If you need help, talk to your professor or to go Math Lab, if that exist in your college, to help you. Then you need to overcome the fear of SI prefixes; and then the biggest task of the rush week, to commit to visual memory the set of schematic symbols we gave you. You may use flashcards and online quizzes to help you. Once you accomplish those, you win your first merit badge - audio electronics as a language.

Your first week is *rush week*, unfortunately (as if you just joined an audio electronics fraternity). But your seniors are lenient. You may have just a simple rush-week quiz, and we'll be lenient in grading. This is to *build your confidence* and get you ready for your future mid-term exams and finals.

On the next row, you learn Ohm's Law and power equation, and then double up and learn PIER and double PIER, all the while making use of math and SI prefixes. Finally, you'll be asked to tackle general DC circuit analysis, and that's where you'll have to have a good grasp of schematic symbols and how to read schematic diagrams as well. With that accomplishment, you earn the DC merit badge - that will be your midterm I exam.

Then on the row above, you'll learn waveforms. This demands math and SI prefixes again; students should pay more attention to this obstacle. Next is how to read frequency response or Bode Plots. This is basically visual recognition again, but again a challenge to some - that's why we separate out this portion to show the emphasis. And once those are accomplished, we can then take advantage of the DC merit badge to perform AC analysis. Yes - you need to do well in DC to handle AC. With that accomplished, you earn the DC merit badge - that will be your midterm II exam.

Finally, with AC and DC merit badges, you are *finally free* to learn amplifiers, op-amps, and active filters on the upper row. Remember you can't deal with those until you have a good grasp of AC/DC (not the band). If you get weighed down with obstacles from earlier, you'll have trouble at this point. But if you persist, you'll win the final comprehensive merit badge of this course - at the final exam.

So, I just explained roughly the overall organization of the major chapter divisions in this book - chapters 1 and 2 for the first badge, chapter 3 for the second badge, chapters 4 and 5 for the third badge, and the remaining chapters for the final badge.

> **DC** = Direct Current = electric current that is relatively steady, flowing in one direction = unidirectional flow or movement of electric charge carriers
> **AC** = Alternating Current = an electric current that changes direction regularly
> **Passive** = functions without extra added external power
> **Active** = functions only with extra added external power
>
> **Major milestones of audio electronics course**:
>   Components, Symbols, Circuits, SI prefixes (1-2 weeks)
>   DC Passive Circuit Analysis (1 month)
>   AC Passive Circuit Analysis (1 month)
>   Active Circuit (1 month)

Notice in the box above, we finally define the terms DC, AC, passive, and active. This is an example of **delayed resolution**, and we occasionally do this to help students learn to tackle such common issue.

There are also supplementary chapters in this book, labeled chapters A through Z. The students should read them on their own even if not covered by the instructor during class. There is simply not enough time during class lecture to cover every aspect of electronics or all material in the book.

We use a **units-first** methodology in this book. In other words, for any new topic, we consistently start with the unit. That gives us a lot of insight and hint into the subject. In addition, for any calculations, you should consistently perform a check on the units. This helps us make sure we apply the right formula and get the right result. The student is reminded to **always provide a unit for each calculation** in your exams.

Each major chapter (not including supplements) begins with a front page which includes a table of contents, a summary of the chapter, and an expected learning outcome. The chapter ends with a review guide and review exercises.

A heading with this symbol ♪ indicates a practical topic relating to live sound, audio engineering, or sound synthesis, basically anything with audio emphasis.

How would you use this book? I suggest that you read the material *before* class. How much? Roughly 25 pages will be covered for each two-hour class sessions. If nothing else, at least read over the chapter front page to get an idea of what the learning outcome is expected to be. Then pay attention during class. After class, as soon as possible re-read the material in the book *again*. Then do the exercises at the end of the chapter - only the portion covered during class (it is most likely already subdivided into class-sized chunks). If there are issues with the exercises, then review the material one more time...

1.5 MATH HACKS: UNITS PREFIXES, ORDERS OF MAGNITUDE, CHECKS, ETC.

We use a units-first methodology in this book. In other words, for any new topic, we consistently start with the unit. That gives us a lot of insight and hint into the subject. In addition, for any calculations, we consistently perform a check on the units. This helps us make sure we apply the right formula and get the right result. The student is reminded to always provide a unit for each calculation in your exams. (See supplement C, "SI Units," for SI base and derived units.)

Unit Prefixes

Please remember the unit prefixes listed below, especially for the range from 10^{12} to 10^{-12}. Please learn to **read** *all* the prefixes in the language, but **write only using prefixes that are multiples of three**, with the exception of dB, which is still allowed (and preferred) in electronics.

Factor	Name	Symbol
10^{24}	yotta	Y
10^{21}	zetta	Z
10^{18}	exa	E
10^{15}	peta	P
10^{12}	tera	T
10^{9}	giga	G
10^{6}	mega	M
10^{3}	kilo	k
10^{2}	hecto	h
10^{1}	deka	da

Factor	Name	Symbol
10^{-1}	deci	d
10^{-2}	centi	c
10^{-3}	milli	m
10^{-6}	micro	µ
10^{-9}	nano	n
10^{-12}	pico	p
10^{-15}	femto	f
10^{-18}	atto	a
10^{-21}	zepto	z
10^{-24}	yocto	y

Taking Advantage of Prefixes in Calculations

Most students will sigh at this point and complain that this is "*one more thing to remember*". (What is worse is some students will simply "pick and choose" and decide to *not* do anything about this.) I encourage you to think of this as an opportunity to simplify work and improve communications skill in the new language of electronics. If your instructor allows you to use prefixes in your calculation, here is how…

Chapter 1 – Getting Ready

Examples
1) Calculate 6 mV / 2 mA
   We can eliminate the two "m" right away. Answer = 6 m / 2 m = 3 Ω. The industry convention is that the unit must be attached at the final answer, but can be skipped during intermediate steps. Note that the units V/A = Ω, which you will learn in chapter 3.
   The traditional approach is to first convert to engineering representation, which takes up far more space and thus error-prone: $6\,mV / 2\,mA = 6 \times 10^{-3} / 2 \times 10^{-3} = 6/2 \times 10^{-3-(-3)} = 3 \times 10^{0} = 3\,\Omega$.
2) Calculate 6 mV / 2 A
   Answer = 6 m / 2 = 3 mΩ.
   The traditional approach is to first convert to engineering representation: $6\,mV / 2\,A = 6 \times 10^{-3} / 2 = 3 \times 10^{-3} = 3\,m\Omega$.
3) Calculate 6 V / 2 mA
   Here the best way is to learn that **1/m = k**. Thus, answer = 6 / 2 m = 3 kΩ.
   The traditional approach is to first convert to engineering representation, which produces a double negative and thus error-prone: $6\,V / 2\,mA = 6 \times 10^{0} / 2 \times 10^{-3} = 6/2 \times 10^{0-(-3)} = 3 \times 10^{3} = 3\,k\Omega$.

Convince yourself that the following are true, and then memorize them:

1/k = m;	1/m = k
1/M = μ;	1/μ = M
1/G = n;	1/n = G
1/T = p;	1/p = T
k/M = m;	M/k = k

Elementary Math Necessary

Some students have developed various syndromes through "pick-and-choose" mentality. (See supplement D "Counterproductive Math Syndromes".) As adults, it is *not difficult* to learn (or relearn) these basic skills. Failure to face this reality typically means rising further in electronics become very difficult.

You need to be able to:
1) **Calculate reciprocals well (e.g., 1/8)**
2) *Not* be bothered by unfamiliar terms (e.g. RMS) or long or complex sentences - it is common in learning a new language that we face a lot of new and unfamiliar vocabulary; no need to be scared and become defensive and then decide to avoid anything unfamiliar or give up too easily
3) Calculate logarithmic arithmetic (associated with dB)
4) Willing to do "reverse calculation"
5) Willing to substitute values into a formula

As hinted earlier, we will *not* be requiring the use of differential equations, calculus, or even advanced algebra in this course. Specifically, college algebra is *not* a prerequisite or a co-requisite. However, you need to have mastery of sufficient **high-school algebra** and especially **grade school arithmetic** to do well. Essentially whatever is required for college entrance you should have mastered by now. The major stumbling point for students are handling of **ratios** in equations with three or four variables, and **logarithms** together with its inverse, the **exponents**.

For your convenience, we shall supply you with alternatives that allow you to handle the required ratio equations *without* using algebra. On your part, you need to **practice** the memory aids sufficiently to avoid clerical errors, if you so choose to use them instead of standard algebra. For that matter, the same is true if you choose to plain algebra: you need to **practice** their use to make sure you do not make clerical errors in your exam. The reason why we highlight **practice** is because we found students making clerical mistakes all the time. For example, in one exam, two students got 8 or 8000 respectively for 1000/8 (the correct answer is 125). If you make such clerical mistakes, then you need to **practice** (there again) the use of calculator as a checking aid. Sadly, the major obstacle is really students often don't allocate sufficient time to practice; hope you will not be one of them…

Remember Formulas and Definitions

Many students think they can get away with *not* memorizing definitions (who *needs* it?) or formulas (who *wants* it?), but being unclear about definitions of current or power or voltage is the key reason many can't grasp concepts in electronics, or why they don't feel confident about calculations (for one thing you won't be able to check answers if you don't have a good grasp of definitions and units). Many are slow in calculations because they did not have formulas in their head, and often unsure even as to their application - if one has not committed to memory, one will most likely be unsure! But we want to limit the amount of material we have to remember; supplement F "Essentials to Memorize" gives us the top 26 items for a semester… :)

Another aid we can utilize to reduce memorization by *half* is to invest in the understanding of the duality principle in Supplement W. It's like buy-one-get-one-free, but you need to show the "coupon of duality understanding" to reap such benefits.

How to Double Check

We should learn to double check our calculations. By double checking, I don't mean just repeating the calculation. For an accountant, summing for a table means that each row is totaled and given a row subtotal, and then the row subtotals totaled to give the final total. The double checking consists of summing each column to give a column subtotal, and then the column subtotals totaled to give a second opinion on the final total. The two final totals should match.

Likewise, we should attempt always to double check our calculations using multiple means. That is one reason why we present more than one way to solve a problem in this book.

One of the most obvious math problems in my classes are handling simple reciprocal or reverse problems. Checking reverse problems are easy; we simply reverse the reverse problems, yielding simple problems that are easy to double check.

1.6 Review Exercises

Chapter 1 Review Guide

Theory

- **concepts** of SI units, orders of magnitude, DC, AC, passive, active
- **concept** of hierarchical decomposition
- **concepts** of how to do basic algebra and logarithm needed for audio electronics
- **units**, **symbols**, **abbreviations** of basic **quantities** in electronics
- concepts of **functions** and **graphs**, including labeling and converting axis
- **avoidance of wrong concepts**

Practice

- ability to **understand** and **convert** SI units and prefixes, specifically, read *all* SI prefixes, but *write* all except d, c, da, and h, other than the exception of dB
- ability to **calculate** using basic algebra (or alternative) for audio electronics
- ability to **convert** a linear graph into a logarithmic graph or vice versa

Quick Diagnostic - Caught any of the syndromes?

1. If a metronome set to *allegro* is actually beating at 120 bpm, how many beats does it make per second?

2. peak = $\sqrt{2}$ * RMS and peak_to_peak = 2 * peak. If RMS = 1, what is peak_to_peak?

3. peak = $\sqrt{2}$ * RMS and peak_to_peak = 2 * peak. If peak_to_peak = 2, what is RMS?

4. E = I * R. If E = 5, I = 0.1, what is R?

5. dB_{field} = 20 * $\log_{10}(V_1/V_0)$. If V_1 = 10, V_0 = 5, find dB_{field}.

6. dB_{field} = 20 * $\log_{10}(V_1/V_0)$. If dB_{field} = 10, V_0 = 5, find V_1.

7. The f_Z of a shelving low-pass filter is f_Z = 1/ (2π * $C_2 R_2$). If R_2 = 12.1 k and C_2 = 18 n, what is f_Z? (reminder: k = 1000, n = 10^{-9})

(Trivia: What pitch is f_Z? Answer: it is a slightly flat $F^\#5$ or $G_b 5$.)

Exercise

(Note: you gain *nothing* by copying answers. Instead, try as best as you can; if still you can't get it, then read the text and try to remember it; then attempt the exercise again. But *don't copy*!)

1. Has the U.S. gone metric? What system of units shall we use in audio electronics? Explain your reasoning.

2. List four SI base units that you may use during this semester.

3. List seven SI derived units.

4. List the SI prefixes from 10^3 to 10^{12}, and from 10^{-3} to 10^{-12}.

5. List four ways relationships between quantities can be expressed. Give the order of priority we use in audio electronics between them.

6. Define DC, AC, passive, active.

7. Explain hierarchical decomposition with an example or illustration.

8. Draw the sine wave and the cosine wave, labeling the axis and making sure their values are correct at zero.

Chapter 1 – Getting Ready

[The following represents the calculations expected in midterm and final exams. None of them are difficult or tedious. The exams cover theory and the steps in solving problems as well, so the above should take *only* a small portion of the time allocated for the exams. For a two-hours exam, you should not spend more than 45 minutes on these calculations, preferably half as much. Please work on improving your speed as well as accuracy.]

(Optional) Extended Diagnostic/Exercise

(Please start your timer at this point, and record duration at the end of this exercise here: _____ min.)

Make sure you can calculate the following quickly and accurately, with the result expressed in SI prefixes that are related to ten to powers of integers of 3 (in other words, you are expected to *read* all SI prefixes, but *write* all except d, c, da, and h, except for dB):

1 k / 4 m

3 M * 2 n

2π * 1 k * 0.22 μ

1 / (2π * 10 k * 470 p)

6 m / 3 cm

Given the equation to the left, make sure you can solve the problem quickly and accurately.
f = 1 / T. T = 0.5 n, calculate f.

f = 1 / T. f = 2 M, calculate T.

f = c / λ. c = 300 M, λ = 500 n, calculate f.

f = c / λ. c = 340, f = 1 k, calculate λ.

X = ωL and ω = 2π × f. f = 1 k, L = 0.01 m, calculate X.

X = ωL and ω = 2π × f. X = 1, L = 0.1 m, calculate f.

1-21

$X = -1/\omega C$ and $\omega = 2\pi \times f$. $X = -1$, $C = 0.1\,\mu$, calculate f.

$X = -1/\omega C$ and $\omega = 2\pi \times f$. $f = 1\,k$, $C = 220\,n$, calculate X.

$R = E/I$. If $E = 5$, $R = 1\,M$, calculate I.

$R = E/I$. If $E = 5$, $I = 0.1\,m$, calculate R.

$R = E/I$. If $R = 100\,k$, $I = 10\,n$, calculate E.

RMS = peak / $\sqrt{2}$. If peak is 0.707, what is RMS?

RMS = peak / $\sqrt{2}$. If RMS is 1, what is peak?

V1/VT = R1/RT and RT = R1 + R2. If VT = 3, R1 = 100 G, R2 = 200 G, calculate V1.

V1/VT = R1/RT and RT = R1 + R2. If V1 = 1, R1 = 100 m, R2 = 200 m, calculate VT.

V1/VT = R1/RT and RT = R1 + R2. If V1 = 1, VT = 3, R1 = 10 k, calculate R2.

$dB_{field} = 20 * \log_{10}(V_1/V_0)$. If $V_1 = 2$, $V_0 = 1$, find dB_{field}.

$dB_{field} = 20 * \log_{10}(V_1/V_0)$. If $dB_{field} = -12$, $V_0 = 2$, find V_1.

SUPPLEMENT A - ELECTRONICS AS LANGUAGE - LEARN DIALECTS TO COMMUNICATE!

Barbara Oakley, a professor of engineering who researched neuroscience and cognitive psychology, cracked the science of learning, declaring that "**the biggest limitation is often just your own thoughts that you can't do something.**" I may add it is the attitude of pick-and-choose and then giving up too easily.

She suggested that maybe the best way is to learn math and science is **as if learning a new language**, "as it calls on the same underlying neural approach, the best approaches to learning."

In a nutshell, this book addresses **learning audio electronics as a new language**. You'll need to visually identify components and symbols, draw schematics and interpret their meaning, and use equations and graphs with understanding, all part of a new language of audio electronics. This chapter helps the student overcome the limitation of one's thoughts that you can't do it.

We'll use metaphor and analogy to convey difficult ideas - not to dumb things down - but "mathematical equations themselves are simply metaphors, and that Neural Reuse Theory shows us that when you understand something using a metaphor, you're actually activating the same neural circuitry you need to understand the in-depth concept itself." The key is we do not dumb down the metaphors, which would otherwise defeat the whole purpose.

The **equivalent circuit** mentioned earlier is another metaphor, a trick that engineers used in the language of audio electronics. Please treat it as your interpreter friend and *not* an additional burden. In fact, there are cultural and technical foundations and infrastructures that you need to learn when you learn a new language; the same is true for audio electronics. This book is written to clue you in on those matters. Do not skip them, or else you would have difficulty with the language.

If you can master English and Music, then you already have *two languages* under your belt of learning. Use the same language skills to learn audio electronics as well.

Regardless whether you use the physicist' or engineer's approach, you'll find the "learning a new language" approach useful.

But this is an *exotic* language; as an analogy: from the perspective of someone speaking English, it is less like French or Spanish or even Greek; rather it is like more Hebrew or Chinese, where one must learn more than grammar, but also the culture and implications before we can fully understand the language. And that is essentially the challenge - one needs to be **fully immersed in the culture** - of audio electronics, and not attempt to pick-and-choose what to learn. That is why we start with the "pick apart" project in the next chapter.

It may come as a surprise to many, to find out in chapter 5, that the language of audio electronics is *not* precise. Just like there are Southern and Northern accents, there are various interpretations of the words *reactances* and *impedances*. Or to find out that Europeans draw schematic symbols differently than Americans in chapter 2. Or to even call the same device by different names, *condenser* versus *capacitor*!

Barbara, by the way, flunked math and science for a decade, until after she learned Russian, and then used the same language skills to become an engineering professor.

The student should realize that on the one hand **college algebra is *not* a prerequisite in this course**, but on the other hand that **the foundation of audio electronics is math**. Probably the only practical way to amass a good foundation and understanding of theory is through math. Yet National Public Radio in July 2017 reported that at America's community colleges, while 60% of those enrolled are required to take at least one math course (at college algebra level or above), yet nearly 80% of which never completed that requirement. So how can we accomplish our task with only minimal level of math prerequisite? The trick is we use advanced math in preparing for this textbook, but we do *not* show them in the explanation! In other words, we digested the material for you. There are two main higher level math operations in physics: we use

Fourier transform to come up with the example material in chapter 4, and we use **Laplace transform** to come up with our formulas in chapter 5, **so you do *not* need to know either math operations**. The Fourier transform of a function is a complex function of a real variable (frequency); the Laplace transform of a function is a complex function of a complex variable, it takes a function of a real variable t (often time) to a function of a complex variable s (frequency), and is used to solve linear ordinary differential equations such as those arising in the analysis of electronic circuits. That way, you do not need to know the math, but you *do* need to know the language. Reasonable?

What is even more interesting is that when the student later decides to take further engineering courses or higher math, all the derivations begin to make sense, as what we did is completely compatible with standard engineering practice - it is *not* a water-downed approach.

Just to be clear, we do *not* require *intermediate* algebra, which is the level of college algebra. But **we do require *beginning* algebra**. We'll soon explain what specifically are needed. And bear in mind it has mostly to do with language and the necessity to communicate…

Specifically, **a significant part of the language of audio electronics is visual**. In chapter 2 you will be required to visually identify components. In chapter 3 you will be required to visually read schematic diagrams and do simple math of DC analysis, based on visual identification of the schematic topology (i.e. visual shape). In chapter 4 you will be required to visually read graphical plots in time and frequency domains. In chapter 5 you'll use all the prior skills learned to do AC analysis, through visual manipulations…

A Physicist, An Engineer, A Musician Each Wrote a Book…

A physicist offers advice on dairy farming, "*First, we assume a spherical cow…*"

That may sound stupid, but one needs to recognize that the physicist's way to tackle a complication problem is to start small and simple. By assuming a spherical cow, there is a higher chance of success in solving the problem. Then the assumption of a spherical cow will be changed to something closer to reality, and the problem is solved again with progressively better assumptions. That iterative approach is tedious and may turn off *some* people, if not most.

An audio engineer offers advice on dairy farming, "*First, we create the equivalent cow…*"

That may sound even more stupid. Why not *directly* tackle the issue of dairy farming, but instead delve into creating an equivalent cow? Isn't that extra effort spent? Well, by creating the equivalent cow, even though there is some effort spent there, and an *apparent* diversion, it actually saves *overall* time in solving the problem, as the equivalent cow is a *simplified* abstraction from which easy table-lookup solutions can be obtained. The characteristics of engineering is that a lot of canned solutions are already available, so once we turn the problem into the form of a canned problem, then that canned problem is easily solved (or I should say, had been solved by someone else in the past, and we don't need to repeat all those extra work). In other words, the equivalent cow concept is one that expends extra effort to find a shortcut and thus finishes quickly and easily.

Notice that I did *not* say, "*First, we choose the equivalent cow…*" People in this culture tended to have a "pick-and-choose" mentality. But *not* for the engineer. The equivalent cow is *created*, with clear derivations and steps. The student is advised to follow all the steps, and clearly argue the case and show all the steps in your exam.

Throughout this book, we'll follow the engineer's concept of an **equivalent circuit**. We'll discuss that further in chapter 2. Do not be afraid of the *extra* work involved in learning and using an equivalent circuit, as in the end it'll simplify your *total* effort. In fact, without a good understanding of equivalent circuits, we can't do much in a one-semester course. It is because of that concept that you can handle relatively complex real-world circuits and analyze its DC performance by the end of chapter 3, for example.

Why did the engineer write the theory section like a physicist? Maybe it is because he or she was taught by a physicist, or his or her professor was taught by a physicist. But more likely it is because of a traditional one-size-fit-all instructional strategy: that engineers may do practical work (which does *not* require a lot of the detailed theories) or research and development (R&D) work (which *does* require the detailed theories). This one-size-fit-all approach of instruction is probably the reason for the continued approach. But the work for an engineer has changed over the years: in the past, the engineer may switch between practical and research several times throughout a career; now work in R&D is scarce because of lack of funding from government and corporations.

More importantly, the main point is that in practical work, all the detailed derivations and theories are *not* necessary. Engineers has been using a *simplified* workflow through equivalent circuits and other techniques for more than a century. And that simplified workflow is what we will teach in this book.

This book is designed to be useable as a textbook or course manual for use in a college environment. It is designed to be the first book in a series for sound and music engineering, with various students electing to emphasize live sound, studio recording, music synthesis, or MIDI programming. Few prerequisites or co-requisites are required.

The Takeaway

Engineers work differently than physicists. People often lump them together and miss the point altogether. Engineers cared about getting done the quickest way possible, using shortcuts and even empirical formula where it makes sense; physicists typically shun away from those options.

If you think you can create a better shortcut than the *engineers*, think again - most likely your shortcut will *not* yield correct results, or will *not* be simple or fast.

Once I was doing a music synthesis project; I was about 90% done, and I decided to switch to another tool and start all over again. Why? It seems ridiculous, but the main reason is that with my engineering training, I realize that there is still 90% effort left in the 90% done project, and it is far better off to start from scratch again; once I choose the right tool, as I'll get done sooner. It is sort of like a photographer would rather shoot a new photo in a better light than to spend more time post-processing, or a musician starting off a new composition rather than keep hammering at the same problematic passage hoping that it'll work soon. And if a musician wants to learn audio electronics, what kind of a book would one be looking for?

And if a musician wants to learn audio electronics, what kind of a book would one be looking for?

	Audio Electronics Book		
	by Physicist	**by Engineer**	**by Engineer-Musician**
Theory	- Long, detailed, - systematic, - a lot of derivations	- similar to the physicist: - Long, detailed, - systematic, - maybe less emphasis on derivations	- short summary of theory, - no derivations, - a different kind of systematic (handles all components at once, handles all related theories at once) - modern, up-to-date

Practice	- lots of calculations (from principle, requires higher math), - may not relate much to audio electronics - does *not* emphasize reading of schematics - does *not* emphasize visual recognition of components - does *not* emphasize soldering and troubleshooting	- lots of calculations (practical), - but too deep into audio electronics that musicians may not need to know - emphasize reading of schematics eventually - does *not* emphasize visual recognition of components - does *not* emphasize soldering and troubleshooting	- still some calculations (but practical, and only requiring high school algebra and logarithms) - contains *more* practical areas of audio electronics than either two, but not too deep, just the right level for musicians - emphasize reading of schematics from the beginning - emphasize visual recognition of components - emphasize soldering and troubleshooting

B. THE LANGUAGE OF RELATIONSHIP: GRAPHS & MATH

In our experience, some students have an aversion towards graphs and math, avoiding them like a plague. What one needs to understand is that they are **concise forms of expressing relationship**: nothing more, nothing less. So it is essential to deal with that aversion at the very start. It language terms, it is just a metaphor, explaining why some people don't get it.

It is very true what Math professors tell us: you can express relationships using an equation (function), a graph, a table, or verbally. For the sake of clarity, conciseness, and avoidance of miscommunication, that order represents the preferred order in audio electronics. That explains why we usually express relationships in the form of equations. However, when the relationship becomes too complex to be expressed in simple equations, then we resort to graphs. Tables are only used when we have limited or approximate data, insufficient to even produce a graph. And verbal or textual would be the last option, only when we cannot express effectively in the other forms.

You do not need to know differential equations or calculus, or even college level algebra or trigonometry. But you do need to know high school algebra and logarithms and exponents and how to graph sine and cosine waves. You also need to know how to label the x- and y- axis of a graph, and how the graph will change when the axis are changed, for example, from linear to logarithmic or vice versa.

Example 1

The relationship between current I and voltage V for a resistor R can be represented four ways:
1) Equation: $E = I R$
2) Table:

Voltage E	0	1	2	3
Current I	0	0.2	0.4	0.6

3) Graph:

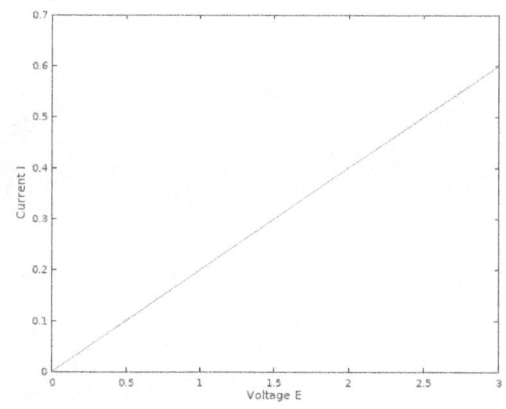

4) Verbally: For ohmic components: voltage drives current; resistance impedes current flow; the current through a conductor between two points is directly proportional to the voltage across the two points, and inversely proportional to the resistance; the potential drop across a resistor is produced by the current flow; and resistance is the ratio of voltage and current, which is a constant independent of voltage and current.

As you can tell, the most concise representation is the equation. The table is only a snapshot of only a few points. The graph covers only the region it covers (the domain and range covered). The text is verbose and is difficult to remember. Hopefully, now give all the options, you'll see why the equation is the best option.

Example 2

There is no simple relationship between current I and voltage V for a diode and a battery. Thus we do not use an equation. Graphs would be the next best option:

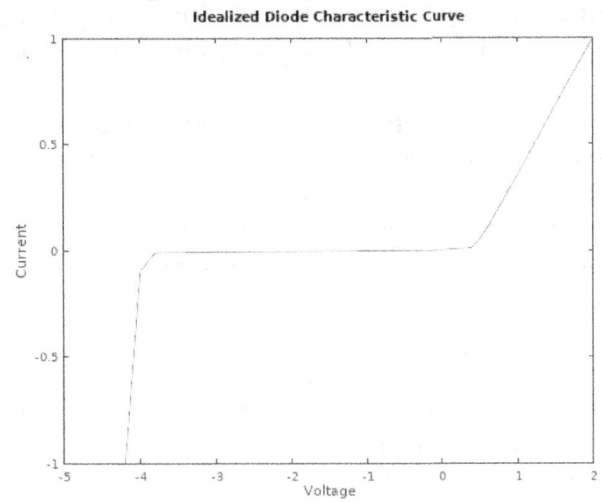

Example 3

A graph written with a linear axis can be converted to a logarithmic axis (or vice versa) as follows:

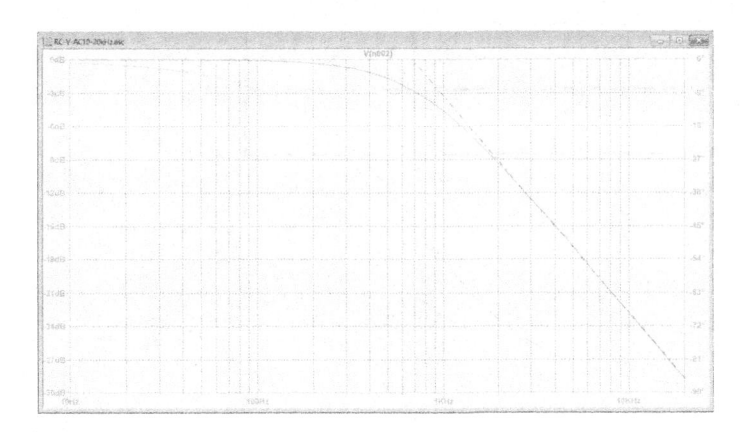

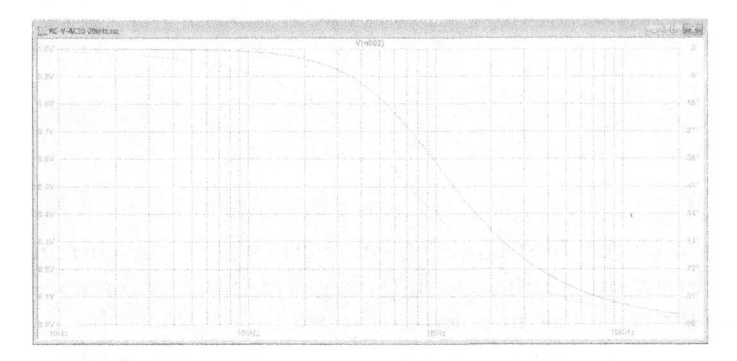

Notice they represent the *same* data, yet the first one can be approximated by two straight lines while the second is an S-curve! That explains why *every* graph needs to be labeled and with *sufficient* indication of its exact scale.

Supplement C - SI Units & Prefixes

We use a units-first methodology in this book. In other words, for any new topic, we consistently start with the unit. That gives us a lot of insight and hint into the subject. In addition, for any calculations, we consistently perform a check on the units. This helps us make sure we apply the right formula and get the right result. The student is reminded to always provide a unit for each calculation in your exams.

We shall use **SI units** (from the French *Le Système International d'Unités*) in this book. In the U.S., President Gerald Ford signed into law the Metric Conversion Act on December 23, 1975, declaring the Metric system "the preferred system of weights and measures for United States trade and commerce", but permitted the use of United States customary units in non-business activities. The Metric Program within the National Institute of Standards and Technology (NIST) "helps implement the national policy to establish the SI (International System of Units, commonly known as the metric system) as the preferred system of weights and measures for U.S. trade and commerce." NIST, which kept the standards for the United States, described the SI as the "modern metric system of measurement" and "long the language universally used in science, the SI has become the dominant language of international commerce and trade." It is especially true for audio electronics.

SI Base Units

There are only seven base units, the first four being more commonly used in this book (Source: NIST):

Base Quantity	Name	Symbol
length	meter	m
mass	kilogram	kg
time	second	s
electric current	ampere	A
thermodynamic temperature	kelvin	K
amount of substance	mole	mol
luminous intensity	candela	cd

SI Derived Units

For ease of understanding (and convenience), there are 22 SI derived units; these are the ones we may use in this book (Source: NIST):

Derived quantity	Name	Symbol	Expression in terms of other SI units	Expression in terms of SI base units
plane angle	radian	rad	-	$m \cdot m^{-1} = 1$
frequency	hertz	Hz	-	s^{-1}
force	newton	N	-	$m \cdot kg \cdot s^{-2}$
pressure, stress	pascal	Pa	N/m^2	$m^{-1} \cdot kg \cdot s^{-2}$
energy, work, quantity of heat	joule	J	$N \cdot m$	$m^2 \cdot kg \cdot s^{-2}$
power, radiant flux	watt	W	J/s	$m^2 \cdot kg \cdot s^{-3}$
electric charge, quantity of electricity	coulomb	C	-	$s \cdot A$
electric potential difference, electromotive force	volt	V	W/A	$m^2 \cdot kg \cdot s^{-3} \cdot A^{-1}$
capacitance	farad	F	C/V	$m^{-2} \cdot kg^{-1} \cdot s^4 \cdot A^2$
electric resistance	ohm	Ω	V/A	$m^2 \cdot kg \cdot s^{-3} \cdot A^{-2}$
electric conductance	siemens	S	A/V	$m^{-2} \cdot kg^{-1} \cdot s^3 \cdot A^2$
magnetic flux	weber	Wb	$V \cdot s$	$m^2 \cdot kg \cdot s^{-2} \cdot A^{-1}$
magnetic flux density	tesla	T	Wb/m^2	$kg \cdot s^{-2} \cdot A^{-1}$
inductance	henry	H	Wb/A	$m^2 \cdot kg \cdot s^{-2} \cdot A^{-2}$
Celsius temperature	degree Celsius	°C	-	K

SI Prefixes

It is important to use SI prefixes in our work. Note that since the **kilogram** is already **kg**, for SI prefix, we assume that the base is **gram**, with symbol of **g**. Notice also that starting with 10^6 capital letters are used; this has to do with the historic order the prefixes were developed.

Factor	Name	Symbol
10^{24}	yotta	Y
10^{21}	zetta	Z
10^{18}	exa	E
10^{15}	peta	P
10^{12}	tera	T
10^{9}	giga	G
10^{6}	mega	M
10^{3}	kilo	k
10^{2}	hecto	h
10^{1}	deka	da

Factor	Name	Symbol
10^{-1}	deci	d
10^{-2}	centi	c
10^{-3}	milli	m
10^{-6}	micro	μ
10^{-9}	nano	n
10^{-12}	pico	p
10^{-15}	femto	f
10^{-18}	atto	a
10^{-21}	zepto	z
10^{-24}	yocto	y

Concept of Deprecation

You are supposed to be able to *read* all the above prefixes properly and understand what they mean. But does that mean you are free to *write* with all the above prefixes? The answer is ***no***!

An analogy: you probably can *understand* many profanities, but are you supposed to freely *use* them in polite company? Of course not! One needs to follow the conventions of our culture.

Similarly, there are conventions in each of the specific industries concerning which prefixes are recommended. In most circumstances, the current preference is to **use only prefixes that have multiples of three in the order of magnitude**.

In other words, we are to avoid using **da, h, d** and **c**. The latter (in the form of cm) may be a surprise to many. For commonplace measurements, we'll of course be able to use it, but in scientific discourse, that is discouraged. The only exception in audio electronics is the use of dB, which we will cover in the next section.

The concept of **deprecation** turns out to be *very* common in engineering and science. You need to turn your ears towards **convention** used within your organization and your culture.

You will notice a somewhat similar situation to be presented in the next chapter: you are supposed to be able to *read* schematic symbols based on either IEEE (American) or IEC (International), but you need to be able to *write* only one convention - based on what your boss (or professor) tells you to.

Units Outside the SI
They are also accepted for use. Those we may use in this book are (Source: NIST):

Name	Symbol	Value in SI units
degree (angle)	°	1° = (π/180) rad
minute (angle)	'	1' = (1/60)° = (π/10 800) rad
second (angle)	"	1" = (1/60)' = (π/648 000) rad
neper	Np	1 Np = 1
bel	B	1 B = (1/2) ln 10 Np
electronvolt	eV	1 eV = 1.602 18 × 10^{-19} J, approximately

As mentioned earlier, other than the exception in the case of **decibel**, which is 1/10 of a **bel**, or B, we shall avoid prefixes that do *not* have multiples of three in the order of magnitude. The use of **dB** in audio electronics is so widespread that there has not been any talk of a deprecation.

IEC Prefixes for Binary Multiples
The SI Units only use prefixes based on ten, as indicated earlier. However, computers often use a base of two. A megabyte according to SI prefix would be 1 000 000 bytes, while according to computer convention, may be 1 024 000 bytes or even 1 048 576 bytes. That created a lot of confusion. As a result, the International Electrotechnical Commission (IEC) approved the following prefixes in 1998 (Source: NIST):

Factor	Name	Symbol	Origin	Derivation
2^{10}	kibi	Ki	kilobinary: $(2^{10})^1$	kilo: $(10^3)^1$
2^{20}	mebi	Mi	megabinary: $(2^{10})^2$	mega: $(10^3)^2$
2^{30}	gibi	Gi	gigabinary: $(2^{10})^3$	giga: $(10^3)^3$
2^{40}	tebi	Ti	terabinary: $(2^{10})^4$	tera: $(10^3)^4$
2^{50}	pebi	Pi	petabinary: $(2^{10})^5$	peta: $(10^3)^5$
2^{60}	exbi	Ei	exabinary: $(2^{10})^6$	exa: $(10^3)^6$

Note that the IEC is consistent: every prefix starts with a capital letter (unlike the SI unit, with kilo starting in lower case).
Thus, 1 kbit is 1000 bits while 1 Kibit is 1024 bits (notice the capitalization).
1 MiB = 2^{20} B = 1 048 576 B while 1 MB = 10^6 B = 1 000 000 B. (Notice there is no option for 1,024,000 bytes.) And 1 GiB = 2^{30} B = 1 073 741 824 B.

Supplement D - Counterproductive Math Syndromes to Avoid

Let us start with a simple division using SI prefixes (consider a "pre" example using milliseconds and music):

"I'm a hammer, everything is a nail" syndrome
"1) *Divide 1 second by 8.*"
Many students like to pick and choose in daily life; they often use the same approach in exams. Many prefers to multiply than to divide. Division does not fit in their regular tools belt. So, when they see a question requiring division, they unconsciously substitute multiplication.

Answer i) If you don't like division, how about just express your answer as a fraction? Ask your professor if that is acceptable. Some may accept it, and if so, the answer is 1/8 **s**.
Answer ii) If you divide 1 s by 8, getting **0.125 s**.
Answer iii) If you convert 1 second into 1000 ms first, then dividing it by 8, yielding **125 ms**. In this question, you have a choice, but in other cases, the question may specify your answer has to be in milliseconds. Just remember the materials in the section above (SI Units & Prefixes) and you'll be fine. You may also want to remember the word millipede which doesn't really have as many as a thousand feet (the record is 750), but that should help convert milli to a thousandth.
Answer iv) you convert the 0.125 s into **125 ms**.
All four answers are acceptable to me (your instructor may provide further input though). However, 8 ms, 8 s, 800 ms (common answers found in exams) are not correct.

"I won't touch anything unfamiliar" syndrome
"2) *RMS = peak / $\sqrt{2}$. If peak is 10, what is RMS?*"
Many students like to pick and choose in daily life; they often use the same approach in exams. Many prefers to skip anything they don't understand or zone out with anything unfamiliar. But in the industry, people use unfamiliar terms a lot. Your future boss or colleague probably will do the same. So now may be the best time to get used to it. It's not as bad as it seems...

Just looks at the question again. Does it **require** you to know what is RMS? Or what is peak? No! To verify, just substitute RMS with A, and peak with B. This is what we get:
A = B / $\sqrt{2}$. If B is 10, what is A?

Answer i): Just substituting as above, A = 10 / $\sqrt{2}$ = 10 / 1.414 = 7.071 and then substitute back to get RMS = 10.
Answer ii): Alternatively, you'll see through that there is really no mystery to the variables RMS and peak and just do the calculation directly without even messing with substitution: RMS = 10 / $\sqrt{2}$ = 10 / 1.414 = 7.071

"I won't reverse course" syndrome
"3) *RMS = peak / $\sqrt{2}$. If RMS is 7, what is peak?*"
Many students like to pick and choose in daily life; they often use the same approach in exams. Many prefers the question to be easy, that variables will all line up in the right side of the equation like in 2) above. But that happens only about half the time. So now may be the best time to get used to it. It's not as bad as it seems...

Substituting with A and B as above:
A = B / $\sqrt{2}$. If A is 7, what is B? Here what is given is on the left-hand side; what we need to solve is on the right-hand side; it is in reverse. So, we just need to reverse course. How? Multiply both sides by $\sqrt{2}$ and we get $\sqrt{2}$ A = B. Reverse left and right (since it is an equality), and we get B = $\sqrt{2}$ A.

Answer: just substituting, B = √2 * 7 = 1.414 * 7 = 10

"It's too complex" syndrome
"4) $|Z| = |\sqrt{(R^2 + X^2)}|$. If R is 3 and X is 4, what is $|Z|$?"
Many students give up when they see square roots and squares, or when they see an unfamiliar symbol like $|Z|$. What is shown here is simply the Pythagoras theorem. $|Z|$ simply means dropping the sign for Z. It's not as bad as it seems...

Substituting with R and X, we get $|Z| = |\sqrt{(3^2 + 4^2)}| = |\sqrt{(9+16)}| = |\sqrt{25}| = 5$

Remember a symbol is a symbol - that's all. You don't need to know what is $|Z|$ to solve this problem! Just like the term RMS earlier. In fact, you may want to know you just performed something like an RMS calculation. Did you see that you did a root after you do an addition, after you did the two squares? That is root-mean-square (without the full mean which would have another division)!

A more complicated example
"5) *R = E/I. If E = 5, R = 1 k, what is I?*"
*Many students are scared when they see an unfamiliar equation like R = E/I... But think through the above syndromes again. Why should we? It is a **given**! Just treat it as A, B and C - they are variables after all. Do not be scared...*

Answer i) you convert the algebraic equation into I = E/R, then substitute to get I = 5 / 1 k = 5 m (If you are not familiar with the SI prefixes, I = 5 / 1000 = 0.005 = 5 m)
Answer ii) you recognise that this is a ratio of three variables, and thus can fit into the Ohm's Law memory aid, so you draw E on top, and R and I on bottom, cover the I, and then you arrive at the answer as I = E/R (same as algebraic approach, obviously), and then substitute as before to get the answer. You still need to get the SI prefixes right! You are just saved from the use of algebraic approach...
Note: most students stumbled through the SI prefix conversions. You are advised to get used to their use.

"TL;DR syndrome"
"6) *V1/VT = R1/RT. If VT = 5, R1 = 1 k, and RT = R1 + R2, where R2 = 4 k, find V1.*"
*Many students are scared when they see something too long and so they don't read; this is even codified into the shorthand notation of TL;DR... But think through the above syndromes again. Do not be scared; you will see this type of problems a lot in the industry - it is nothing more than a **serial** application of what is already known...*
V1/VT = R1/RT. If VT = 5, R1 = 1 k, and RT = R1 + R2, where R2 = 4 k, find V1.
Answer i) you convert the algebraic equation by multiplying both sides by VT, thus obtaining V1 = R1/RT * VT, then substitute to get RT = 1 k + 4 k = 5 k, and V1 = 1 k / 5 k * 5 = 1
Answer ii) you recognize that this is a ratio of four variables, and thus fit into the four-variable ratio memory aid, so you get V1 = R1/RT * VT (again obviously same as algebraic approach), and then substitute as above to get the answer.
Note: some students find the presence of *two* equations too threatening. You just have to get used to it...

"Don't dB me syndrome"
"7) *dBpower = 10 * $\log_{10}$(P1/P0). If P1 = 2, P0 = 1, find dBpower.*"
Many students are scared when they see anything with dB or logarithmic or exponential... But we'll be frank: dB is the secret handshake; if you don't know dB, you can't be initiated into the audio engineering community.
dBpower = 10 * $\log_{10}$(P1/P0). If P1 = 2, P0 = 1, find dBpower.
Answer i) just substitute and use your calculator, dBpower = 10 * $\log_{10}$(2 / 1) = 10 * $\log_{10}$(2) = 3.01 (or 3 to one decimal).
Answer ii) just substitute and remember that $\log_{10}$(2) = 0.301 (or just 0.3 to one figure). Engineers remember this number well, and usually by convention remembers it to only one figure for simplicity.

Supplement D – Counterproductive Math Syndromes

"Don't reverse dB me syndrome"

"8) *dBpower = 10 * log₁₀(P1/P0). If P0 = 1, and dBpower = -3, find P1.*"

Many students are scared when they see anything with dB or logarithm that requires a reverse course, meaning requiring the use of exponential... But we'll be frank: dB is the secret handshake; if you don't know dB, you can't be initiated into the audio engineering community.

dBpower = 10 * log₁₀(P1/P0). If P0 = 1, and dBpower = -3, find P1.

Answer i) convert the logarithm to an exponential - that is the fundamental theorem of logarithm: thus P1/P0 = 10 ** (dBpower/10); substituting, P1 / 1 = P1 = 10 ** (-3/10) = 0.501 by using the calculator. Or, by convention, we use only one figure, yielding simply 0.5 as the answer.

The key, of course, is you need to **practice** using the calculator to get this right.

Answer ii) you can use the answer you already got from 6), thus: since 3 = 10 * log₁₀(2), and if you know the property of logarithms, then -3 = 10 * log₁₀(1/2), thus getting the answer P1 = ½ or 0.5 In other words, it is worthwhile getting back into logarithms and exponents to get yourself familiar. It is *not* required for the exam (as you can simply use the calculator), but you'll discover most audio engineers do this conversion in their head, using the method of answer ii). It may be worthwhile to imitate their behavior so you can talk to one another in the studio, when they say let's reduce the power by half and you know to move 3 dB down...

"Won't substitute syndrome"

"9) **The frequency output of the 555 multivibrator is f = 1.49 / (RA + 2*RB) * C, if RA and RB = 10 kΩ, and C = 0.1 µF, find f.**"

*Many students are scared when they see long equation requiring substitution. But we'll be frank: that's mostly what we do in audio electronics. Plus, it is **not** difficult...*

The frequency output of the 555 multivibrator is f = 1.49 / (RA + 2*RB) * C, if RA and RB = 10 kΩ, and C = 0.1 µF, find f.

Answer: just substitute, f = 1.49 / (10 k + 2 * 10 k) * 0.1 µ = 1.49 / (30 k * 0.1 µ) = 1.49 / (3 m) = 2.51 k

Note: again notice you need to be very familiar with SI prefixes and realize that the m in the denominator becomes a k as numerator.

For music application, the values for 10 kΩ and 0.1 µ F would create roughly 2.51 kHz wave.

These are the calculations that stumbled most students. Please make sure you can handle these (plus the SI prefixes) with ease, and can use calculators when needed. You'll notice that quite often, students are **scared** and then decided not to tackle the problems. I hope once you see that all these are within your reach, you'll execute them like a pro. The main take-away: **Do not be scared and then act irrationally**...

You do not need a calculator for midterm I, although you can use one if it is a calculator approved for use in exams by the college. Your cell phone is *not* an approved calculator. You will be required to use an approved calculator for midterm II and for final exams. Check with your college library (ahead of time of course) to see if you can borrow an approved calculator for use during the exams.

(page deliberately left blank)

Supplement E - Individualism vs. Teamwork in Electronics Industry

Most of us had been taught to value individualism since grade school. The most dramatic moment in TV game shows often comes when the contestant says, "*This is my final answer!*" and "*locked in*" the answer.

This kind of individualism often gets reflected in how students tackle examinations. First, students often answer a complex essay-style question with a single number as the "*final answer*". Second, when one sees that there is not much chance of getting the final answer correct, they often won't even waste time starting to answer the question at all. This is often the case *even despite* the instructions clearly saying, "partial answer is acceptable; the final answer is worth only very little points; most of the points lie in the steps shown in how to solve the problem."

Industry Practice

Let us look at this from the point of common practices in the industry. When you work in the future, will it be individualism or teamwork that is treasured?

The answer obviously depends on the nature of the work. But in general, in past centuries, individualism was highly treasured. That yielded most of the work ethics in America. That idea remains in many areas of work, especially in research and advanced development, and is expected for some managers.

However, in the more common lines of work in audio electronics, it is worthwhile taking note that most companies are now organized *more* on **teamwork** and *less* on individualism.

Through the last century, managers have discovered that individuals are less and less likely to be able to carry a project from beginning to finish all by themselves. The complexities of today's technology mean that we need to collaborate with a large array of specialists and utilize wide varieties of information. So as a result, employees are often put within a team, with team members checking on one another. Even without the team structure, managers had discovered, and used, **project reviews** at key milestones to help with projects. Other staff essentially will ask questions, make recommendations, and like it or not, the result is to change our "individual project" to go a *different* direction or conclusion.

Wisdom of the Crowd

So how do we work within such environment, when we are expected to change directions and collaborate with others we don't even know? The simple key idea is to contribute what we know, documenting every step of how we come up with our answer, making it easy for others to collaborate and critique and make use of our work - even if our answer is not chosen, most of the way we come up with the answer *may* be correct, and that would be our main contribution, keeping us employed. That is how we can make our work relevant, even if the final answer is different from what we propose. That is also the idea in writing scientific papers/reports (see supplement T, "Project Notebook & Reports"). Just like the 2017 TV show, companies often believe more people looking at the problem will arrive at a better answer...

How Then Shall We Act in Exams?

This is also the key we want to emphasize in our essay-style exams. The student needs to **show all the steps**. The steps carry the most points, the final answer just a little. That is how they are evaluated in college, just like how they will most likely be evaluated in their future employment - not just by their individual "final answers", but also by how they articulated clearly their steps in coming to their conclusion, so that others can join in and participate and change to an even better answer. One needs to thoroughly realize this fundamental value of their contribution in clearly documenting their steps to solving a problem, instead of focusing just on "this is my final answer!"

Resistor Value example

Many students tend to give up when given a picture of a resistor and asked to give a value. Why? They quickly judged that the chance of giving a correct final answer is low, very low. The colors are difficult to discern, with red looking like brown, brown looking like orange, and yellow looking like gold, violet looking like brown, and silver looking like white...

But one should notice that the situation is common in the industry as well. That is a common problem for audio electronics.

So, what is the response I expected of my students? Simply to show their best problem-solving skills, and document the steps in such a way that a manager can evaluate and agree or change the assumptions and come up with a different assumption.

In other words, the key is to document the assumptions. Say we look at the first color band, and we state our assumption that the color is red. Then one can proceed to go through the number coding and arrive at the result, step by step.

One should also use various tests to ascertain the assumptions. For example, if the coding for the color turns out to be 29, more than likely it would be wrong, because 29 is not a common value within the E6 or E12 scheme (see supplement J, "determining component values"). In that case, we should change the assumption and try again...

I give full credit to students who clearly document their assumptions of the colors and show clear steps on how they arrive at the answer, even if the final answer is incorrect.

DC Analysis example

When asked what is the resistor value for a circuit with a battery and a diode, many students simply give the answer as 100 ohms. How do I know if it is a simple guess, or it was the result of a careful calculation?

The proper approach is to document all the steps. First the student should give credit to Kirchhoff's Voltage Law, and state KVL as first step. Then the voltage is calculated. Then the student should state Ohm's Law, and then the current is calculated. The final answer is then shown with the proper unit given.

SUPPLEMENT F - ESSENTIALS TO MEMORIZE

What if there are flash quizzes (which average one minute each) that could verify that students have memorized essentials that need be memorized in this course? Those that if not memorized will impede the student's progress in continuing with the course. Here are the top 26 that may be considered **'essentials'**...

A	Show the SI Unit prefixes used for: 10^{12}, 10^9, 10^6, 10^3, 10^{-3}, 10^{-6}, 10^{-9}, 10^{-12}
B	1) Draw schematic symbols for a) NPN and b) PNP transistors and 2) label their terminals
C	Show the color-to-number coding chart for resistors and capacitors
D	Show the equations to calculate: 1) current (given voltage, resistance) 2) resistance (given current, voltage) 3) voltage (given current, resistance)
E	Show the equations to calculate: 1) current (given voltage, power) 2) resistance (given current, power) 3) voltage (given power, resistance)
F	Calculate the voltages across each of the resistors in the voltage divider shown, and the voltage at point A
G	Calculate the voltages across each of the resistors in the circuit shown, and the voltages at points B and C
H	Show how to calculate the voltage corresponding to 0 dBm for pro-audio
I	State Kirchhoff's two Laws in a way a peer can follow
J	Properly *define*: 1) Current, 2) Voltage, 3) Power ('alternative facts' *not* allowed)
K	1) Show the formulas for calculating a) wavelength, b) period (given frequency and speed in media); 2) annotate your symbols used in the formulas
L	Show the impedance formula for a) a capacitor of value C, and b) an inductor of value L, and c) a resistor of value R
M	Calculate time constants for circuits with these components only: 1) R = 2k, C = 0.1µF 2) R = 2k, L = 0.4µH 3) C = 0.1µF, L = 0.4µH
N	Calculate the corner frequencies of the circuits given the component values or just the time constants

O	Explain how 1) a capacitor 2) an inductor behaves at a) DC b) high frequency
P	Show how to estimate frequency response of the RC circuit shown. What kind of filter is that?
Q	Properly *define*: 1) Capacitor, 2) Inductor ('alternative facts' *not* allowed)
R	Draw the schematics for common-emitter, common-base, and common-collector amplifiers using NPN transistor
S	Draw the schematic for common-source amplifiers using n-channel JFET
T	Draw the schematic for a common op-amp amplifier
U	Draw the schematics for a 3-channel audio mixer using one op-amp
V	Calculate voltage gain of the amplifier circuit shown
W	Calculate voltage gain and bandwidth of the op-amp circuit shown
X	State the three kinds of coupling that an amplifier can use, and their advantages/disadvantages
Y	Explain Class A, B, and C amplifiers and their advantages/disadvantages
Z	Explain impedance a) bridging and b) matching *properly* and examples of when they should/could be used

SUPPLEMENT G - How this Book Relates to Other Audio Courses

How does this book relate to Synthesis, MIDI and Audio Engineering courses? Let's look at some specific areas this book covers and its overlap or differences in emphasis…

- **Waves**: Synthesis talk about differences in sine, square, triangle and sawtooth waves, and how to use them to synthesize more complex sounds. MIDI uses the complex sounds in musical applications. This book **lays the foundation** of time and frequency domain and how to move between the two domains, and how the harmonics and overtones of these waves relate in the two domains. That in turn lays the foundation for AC analysis.
- **Filters**: Audio Engineering and Synthesis use filters to achieve their goals. Given a schematics, this book explains how to analyze it to **determine its filter characteristics**, or alternatively, given a set of requirements, **how to create active and passive filters** using physical electronics.
- **Studio Equipment**: Audio Engineering, MIDI and Synthesis all use studio equipment, and maybe explain the practical need for "matching" impedances. This book explains the **fundamental concepts of impedance matching and bridging**, and how they affect sound output and quality. It explains the **levels** used in studio and consumer equipment, and **how they were defined and calculated via application of Ohm's Law**. It explains **how noise may be generated**, and **what fundamental principles are used to mitigate against it**, while Audio Engineering and Live Sound may just give the approaches used (e.g., "balanced cable") without explaining the principles clearly. It explains how electronic components like op-amps, filters and voltage dividers and are used to **implement the controls and buses** used in studio equipment, while other courses explain *how* they are used. Audio Engineering are more concerned with how the studio is set up, while this book explains the **theory of operations** of many of the circuits used in such equipment.
- **Practical labs and projects**: Soldering and Troubleshooting are emphasised in this book. While troubleshooting is also essential in the other courses, they are usually not given as much structured emphasis. The labs are constructed to build confidence and ability to solder cables and electronic projects, plus how to troubleshoot simple audio problems.

In other words, this book lays the foundation for the other audio courses by giving solid theory of operations.

One may say, I just need to pay money and buy the stuff; that's why I don't need to **study** audio electronics. To a certain extent it is true, just like one can just pay money and buy the services of an audio engineer to do audio engineering, or an electronic musician to do synthesis; one with money doesn't need to pay attention to anything… But of course, how do you **know** that you are buying the right stuff? This book aims to **develop educated people that can discern truth from lies**. Is that carbon microphone with slick sound samples in a website a good reason to fork over $500 for a special promotion? When we understand fundamental principles it is easier to discern what is good and what is bad. That is why we write this book.

The secrets of the success of electric vehicle manufacturer Tesla's design language, according to its chief artistic designer, was that they decided to think from *first principles* instead of through *analogies*, which is how artists usually think. He believes that only when an artist think *first principles* can breakthrough design language happen.

(page deliberately left blank)

Supplement H - "Take Apart" Project

This book follows a hands-on curriculum. In this project, you are asked to find an old piece of junk electronic component, and take it apart at home or in class so that everyone will benefit from learning a wider example of what circuits and components may be present in typical audio electronics. Your mission is to learn visual identification of the components, without a detailed understanding of what the components do (instead they will be presented throughout the semester).

Optionally, another purpose of this project is that you will also learn proper soldering technique. You and your instructor will together select a couple of components to de-solder in one session, and then solder them back in another session.

Not every audio electronics circuit is suitable for the take-apart project. First, we don't want students to spend a lot of money. Try to ask your parents or uncles or friends to get a used or junk circuit board. If that does not yield anything, consider getting it from an estate or garage sale or a used product store instead of buying it new. Second, there is a reason why we suggest old circuit boards. They are easier to work with. Do not attempt to use a cell phone or laptop or similar types, as more than likely they will be implemented using surface mount technology and not through-hole technology, making it too difficult for first-timers to work with such de-soldering or soldering.

In **through-hole** technology, leads are put through holes in the board (thus the name) and then soldered onto pads located on the other side of the board. In **surface mount** technology, the component itself is mounted directly onto the board and then soldering on the underside via pads, with no leads going through holes. The dimensions for surface mount are much smaller than through-hole; our standard soldering iron tips are too big for through-hole applications.

There are two versions of this project: A) Just take apart electronics to get a feel of what's inside; B) Take apart some old electronics, then de-solder a component, and then re-solder it back on - this in addition provides a feel to the difficulty of DIY soldering and repair, especially of old electronics.

Just Take Apart

Bring any used electronics that you don't mind losing or being damaged to class. Optionally students may be asked to prepare and take it apart at home. The idea is to let the class see the wide diversity of components and organization of circuit boards in different electronic gadgets.

Take Apart and De-solder/Solder

This is a multi-week project.

1) On the day previously announced, you are to bring an old, used electronics that you don't mind losing or being damaged - it should have some electronic components (not mechanical only), and enough through-hole components (not all surface-mount) - typical examples are clock radios, recorders. Newer electronics (like cell phones, iPod, Walkman) are nearly always surface-mount). You will take it apart and show the class what components are in it.

2) On another day, you'll learn how to clean a soldering iron, check the temperature, tin the tip, tin the wire to be soldered, and orient the tip for maximum contact. You should practice those skills further on your own...

3) On yet another day you'll, with agreement from instructor, select one component to be de-soldered from that electronics, de-soldered it, show the instructor, then solder it back, and test to see if everything is still working (if the electronics was previously in working condition). You'll be graded primarily on participation - whether you destroyed anything or not, hope you learn through this experiment :) That explains why we specify using electronics you don't mind losing/destroying, and preferably cost like nothing to you...

Chapter 2 HELLO, WORLD!

> This is a "hello world" equivalent, where one is *exposed* to the world of electronics *quickly*, without necessarily explaining *fully* every aspect all at once. The reason why this is done is because we have a chicken-and-egg problem - we can't learn electronics until we understand its components, and we can't understand components until we understand electronics. We have to start somewhere...
>
> *Expected Learning Outcome*: The student will be able to visually identify electronics components, their values, and be able to read and write schematics and understand how they are connected, and whether circuits are open, closed, short, and why we use equivalent and lumped circuits... The student will also be able to classify components intelligently. However, the student is *not* expected to know how the components work or their purpose at this point.

CONTENTS

Hello Electronic Components & Schematics	2-3
Components and Symbols - Visual Identification	2-5
Classification: Passive, Active, Electromechanical, Solid-State	2-26
Circuits: Open, Closed, Short, Equivalent	2-27
Review	2-28

This is an introduction to electronic components and how to read their symbols and schematics.

It's easy to feel overwhelmed, but just pay attention and try to get as much as you can "through osmosis."

2.1 Hello Electronic Components & Schematics

For communication purposes in electronics, the most important piece is the schematic, which uses symbols of electronic components connected together in certain ways. Here is an example of a schematic.

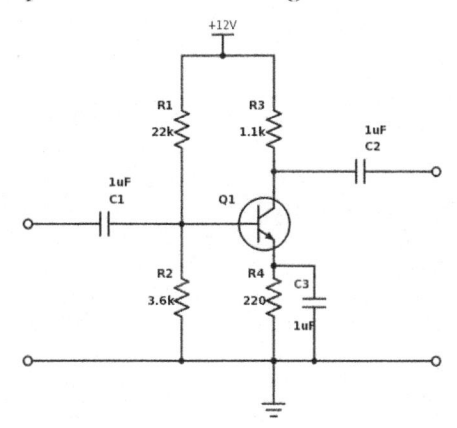

In the diagram, each component is represented by a **reference designator** (e.g., C1), a **schematic symbol** (for example, the double plates), and **components value** (e.g. 1 µF). Occasionally other data may also be listed, or omitted when not required.

Here is another example, the most conspicuous item in the schematic is Q1 - that's it. The reference designator starts with a Q, hinting that it is most likely a transistor. But the schematic symbol contains much information: one needs to learn to read that as indicating an NPN junction transistor. Often the exact component reference may also be stated.

The key to note is that this is a **visual identification** process. One needs to learn how to visually identify components, and then **visualize** the connections.

Block Diagrams
An example of a block diagram:

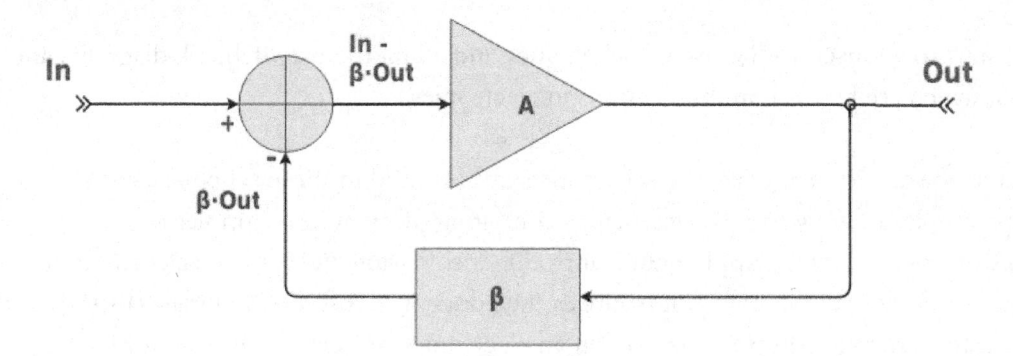

Note that there are three "blocks" in this diagram: the circle with a + sign is the summing junction, the block with a triangle shape pointing towards the right is the amplifier block, and the rectangular block is that of a "generic" network.

Hierarchical Decomposition

The amplifier block can be decomposed further into a schematic like that in the previous page. The generic network could be represented by a schematic like the following:

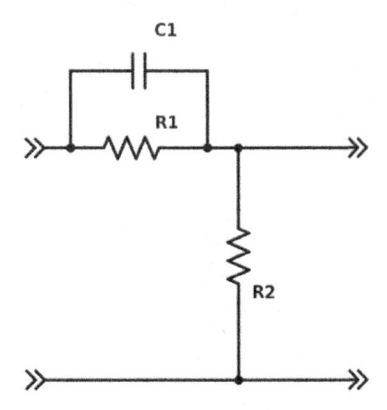

The summing junction can be decomposed further into a schematic like the following:

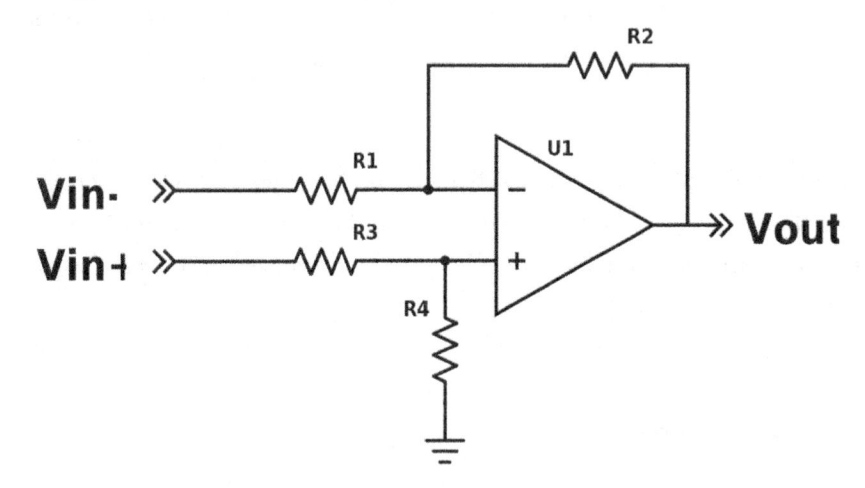

Again, you are not expected to understand everything about these schematics, but rather *only* the hierarchical decomposition aspect. The filter network shown above as "generic" will be covered in chapter 5, the amplifier example will be covered in chapter 6, and the summing junction example in chapter 7.

The point is, one needs to start to get used to the use of schematics, including the use of block diagrams and hierarchical decomposition, which are key to communications in electronics.

And to begin with, one needs to start to memorize the schematic symbols used in these schematics. Supplement I give further examples of how to read schematics consisting of switches. Supplement J explains how to read values of components. Supplement L introduces soldering and how to select soldering irons, needed for electronics projects. Supplement L introduces the concept of Bill Of Materials (BOM), a common communications vehicle for the components used in an electronics project. Supplement M introduces the pricing aspect of electronics. Often students thought because they have put in effort to make a DIY project, and so they convinced themselves that it would be cheaper that way than buying directly. After doing the BOM analysis and reading supplement M, hopefully the student will realize it is *not* the case.

2.2 COMPONENTS AND SYMBOLS - VISUAL IDENTIFICATION

Your Mission
Your mission is to learn to **visually identify components and schematic symbols commonly found in audio electronics, and to relate them together**. They are part of the *language* of audio electronics.

Can I *select* what symbols to learn?
The most common question reveals a *pick-and-choose* mentality. It is especially true when the student realizes that there are **variants** to some symbols, especially variants across the pond. Isn't it just simple *linear* logic to think that we can just *choose* American and ignore European or International conventions?

The simple answer: we shall learn to be *consistent* and ***write* in only one variant**. But we should learn to **read all common variants** we shall see - it is part of the language environment. Here is why…

1) Think of the final project: from where would you be ordering your project kit? Most of the project kits that our students ordered in the past came from Asian and European manufacturers. Which means most of the time their schematic (if included) will be using international symbols.
2) Think of the mixer consoles that one may be using or encountering. From where would they be manufactured? I typically use large British consoles and small American audio interfaces, the former because of their distinctive British sound, and the latter for easy computer integration. Their schematics tell me I should learn to read both conventions.
3) Some American audio companies also choose to publish schematics in international convention. Why? So that they can reach out to clients worldwide. (And so that their manufacturer in Asia can understand them. Remember even American companies only *design* them in America, but manufacture them elsewhere.)
4) Even the IEEE, which currently maintains the American standard, has indicated a desire to merge in the international standard managed by IEC, in the future. Their rationale is the same as what we stated above. Internationalization is now a fact of life. There are no TV manufacturers in the United States, so all TV schematics are International. Ditto for consumer audio electronics.
5) Because audio electronics is a language, we can only constrain what we *write* - we can't constrain what we *read* unless we say we will not read anyone who does not use our dialect. There are variants even within the American standard (as we shall soon learn below)! We should do what language learners do - learn to *hear* most of the dialects likely to be encountered, but *speak* only one targeted dialect…
6) In summary, learn to read all the components and symbols you'll encounter during your Pick-Apart Project and during Final Project, at a minimum.

If so, one may legitimately ask why textbooks like this one are still written using the American convention? Simple: because the schematic symbols are easier to write on the blackboard (or whiteboard) and on paper, which is important because time during exams is of the essence, and ease of writing is a virtue. The Euro symbols for capacitors are especially scary if you had not encountered them before, and consumes a lot of time to write…

Resistor

	Symbols - American	Symbols - Euro
	R1, R2, R3, R4, R5, R6	R1, R2, R3, R4, R5, R6
	For the symbols, please remember the reference designator format also. For resistors, the common format is Rx where x is an integer. They can be drawn in vertical (R1) or horizontal (R2) format. There is also the variable resistor (or potentiometer, often shortened to pot) with *a center* terminal, and can be drawn in four common directions (R3-R6). A diagonal arrow drawn through the resistor is another form when only two terminals are available. The reference designator for variable resistors can also be of the form VRx. For other components, we shall *not* show all the orientations - you just recognize them on your own. And don't worry about excessive memorization - Americans and Europeans disagree on only a *few* symbols…	

The American symbol for resistor resembles an ideal **wire-wound** resistor or a printed circuit resistor. When the circuit zigs and zags, the effective length increases within a given volume, thus allowing for higher resistance values. Note that a practical wire-wound resistor would be wound first in one direction and then in the opposite direction, so that the electromagnetic field generated by the former will cancel out the latter.

Another historic form of resistor is carbon composition, where fine carbon particles are mixed with a binder material like clay and then baked and coated to make the resistor. One can visualize the Euro symbol as indicating a carbon composition resistor.

Nowadays it is usual to use carbon or metal deposition to make resistors. The thickness of deposition can be controlled to achieve the target resistance.

The wire-wound resistor remains the favorite for achieving best sound in the audio paths of a circuit. It is also the typical form for precision resistors.

The **size** of a resistor usually does *not* correlate with its resistance value, but rather with its **power dissipation** – 1/2 watt resistors are larger than 1/8 watt resistors.

(Generic or Non-polarized) Capacitor

	Symbols - American	Symbols - Euro
	C1, C2, C4 (variable)	C1, C2, C3, C4, C5, C6 (variable)
	For capacitors, the common reference designator format is Cx where x is an integer. C1 is the common format, with C2 (or C3) variants. C4 is variable capacitor, with again variants in C5 and C6. They can be drawn in various orientations (not shown). Like variable resistors, variable capacitors sometimes are notated VCx.	

The shape for a capacitor reflects its construction of two parallel metal plates in between a dielectric material. Of course, to increase capacitance, the surface area of the metal plates need to be large, and it is usually rolled and folded to achieve such high density.

One of the earliest audio tone capacitors is constructed **paper-in-oil** (PIO). Vintage "tone capacitors" like that have names like bumblebee, black beauty, orange drop, tropical fish, and so forth, based on how the component looked. The latter is constructed from **metallized PET film**, with the metal being typically aluminum, and the dielectric PET being the plastic "polyethylene terephthalate." Orange drops were usually made with **metallized polyester**. The more vintage ones perform less ideally than modern capacitors, yet were optimized for their specific tone circuit, with a fatter sound when compared with modern capacitors, and so highly sought after. In other words, if one is trying to replace tone caps from a by-gone error designed for a non-ideal capacitor, it is best to replace it with another non-ideal capacitor instead of a "better" capacitor for which the circuit wasn't designed for.

Other common capacitors are made with **ceramic**, **mica**, and other metallized plastic films.

Capacitor is often abbreviated as cap.

The **size** of a ceramic capacitor usually correlates with its capacitance value, because larger surface area means larger capacitance. Likewise, larger film capacitors of the same family usually have larger capacitance.

Polarized Capacitor

	Symbols - American	Symbols - Euro
	C7 +⊥ ⌣	C7 ⊥+ C8 ⊥+
	For capacitors, the common reference designator format is Cx where x is an integer. C7 is the common format, with C8 variant. Notice the addition of "+" sign compared with the non-polarized capacitor.	

The polarized capacitor differs from non-polarized version by just a + sign. In the case of Euro C8, the two plates also have different colors to emphasize the different polarity.

Electrolytic capacitor is the most common implementation, using liquid electrolytes between the two metallized plates. As such, the component is *not* a solid-state component, with a shorter operating life than most other common components. When one sees electrolytes leaking out of the polarized capacitor, or even a bulging body, the cap should be replaced. Likewise, when we hear extra noise from an old amplifier, at least one of the electrolytic capacitor needs to be replaced. If not, the cap will fail shortly. Electrolytic caps have the highest failure rate of most common passive components, and so is the first component to consider replacement.

When we see a large electrolytic capacitor, it usually means high **working voltage** and large **capacitance value**. For power amplifiers, the capacitors for power supply filtering is usually large because power amplifiers operate at high DC voltages, and thus requiring high working voltage capacitors, plus they need high capacitance to smooth out the output voltage.

Inductor

	Symbols
	L1 L2 L3 L4 L5 L6 L7
	For inductors, the common reference designator format is Lx where x is an integer. L1 is the common air-core format, with L2 variant. L3 is the common iron-core format, with L4 variant. L5-7 are variable inductors.

The inductor can be visualized as just a coil of wire. The core is also represented, as two solid lines as iron-core, nothing as air-core, and double dashed lines as ferrite-core.

Transformer

	Symbols
	T1　　　T2 T3　　　T4 L1　　　L2
	T1 emphasizes step-down, while T2 step-up transformer. T3 emphasizes single primary and dual secondary, T4 has center-tapped secondary. L1 is generic (does not show if step-up or -down) but emphasizes its mutual inductance, hence labelled L1, while L2 is air-core. Essentially examine the symbol carefully and deduce its corresponding function.

Diode

	Symbols
	D1, D2 — **Diode** DIAC1, DIAC2 — **Diac** T1, T2 — **Triac** D3, D4 — **Tunnel Diode** D5, D6 — **Schottky Diode** D7, D8 — **Zener/TVS Diode**
	For diodes, the common reference designator format is Dx where x is an integer. D1 is the common general diode format, with D2 variant. Diac's have their own reference designator, with DIAC1 is the common format and DIAC2 variant. Triac's have *three* terminals, T1 common and T2 variant. D3 and D4 are tunnel diodes; D5 and D6 are Schottky diodes; D7 is zener or unidirectional TVS, D8 is bidirectional TVS diode. The key is to remember the general and zener diode symbosl, and know that there are many other forms of diodes, each with unique symbols, *some* of which can be seen here; in particular make sure you draw the zener diode unlike that of Schottky or tunnel diode - that's why it is presented last... :)

Visualize the symbol for the diode as the "cat's whisker," the earliest form of diode used in the earliest "crystal" radios, where a thin wire lightly touches a crystal of semiconductor mineral like galena. The direction of the arrow indicates the direction as if P material proceeding to N material.

If the arrow points to a non-flat line, then the shape of that line reflects the rough characteristic curve of the specific diode.

Optoelectronics

	Symbols
	L1 is the common LED, with L2 its variant. Often, we use LEDx as reference designator. D1 is the symbol for photodiode; its counterpart without the circle seldom used. L3 is a photoresistor; Q1 a phototransistor. Notice the direction of the light tells one if it is light-emitting or photosensitive.

Semiconductor diodes, when exposed (not enclosed by encapsulation material), are transducers between light and electricity. Visualize the double arrows, which represents light, as either going out or coming in.

Bipolar Junction Transistor

	Symbols
	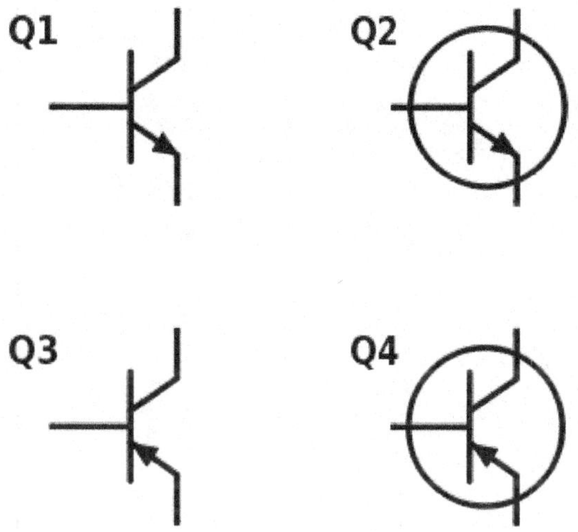
	Qx is the reference designator for transistors. Q1 is NPN, with Q2 variant. Q3 is PNP, with Q4 variant. Essentially some prefers the circle, others skipping it. To decide which is NPN or PNP, look at the junction created by the arrow and write down P and N (arrow from P to N), then add the remaining P or N to the remaining connector to arrive at the answer.

Visualize the BJT symbol as three layers of semiconductor material, with two of the bottom layers (in the diagram above) as a diode, thus having the standard diode arrow of P and N. More visualization will be shown in chapter 6.

Field-Effect Transistor

	Symbols
	Q1, Q2, Q3, Q4, Q5, Q6
	Qx is the reference designator for transistors. The major division are junction and MOS transistors. Q1 and Q2 are P-channel JFET, and Q2 and Q4 N-channel. Q5 is P-channel MOSFET, and Q6 is N-channel MOSFET. There are more symbols for specific kinds of MOSFETs (not shown).

More visualization will be shown in chapter 6.

Vacuum Tube

	Symbols
	V1　　V2
	Vx is the reference designator for Vacuum tubes, or Valves. The filament at the bottom is often omitted. V1 is the diode; the middle screen is often omitted. V2 is the triode. There are more symbols for tetrodes, pentodes, dual triodes and so forth (not shown). Basically, visually observe the symbol construction and deduce the corresponding vacuum tube type.

More visualization will be shown in chapter 6.

Amplifier

	Symbols
	U1 (triangle symbol) U2, U3 (with V+/V-), U4 (with V+/GND)
	Ux is the reference designator for generic blocks, including amplifiers and ICs. U1 is the general amplifier with single-ended output. U2 is the single-ended op-amp with differential input, power supply not shown. U3 shows support of a dual power supply, while U4 shows a single power supply. The symbol for the fully differentiated amplifier can be found in section 8.7.

The symbol for the amplifier came from its early days used in analog computing. See chapter 7.

Integrated Circuit

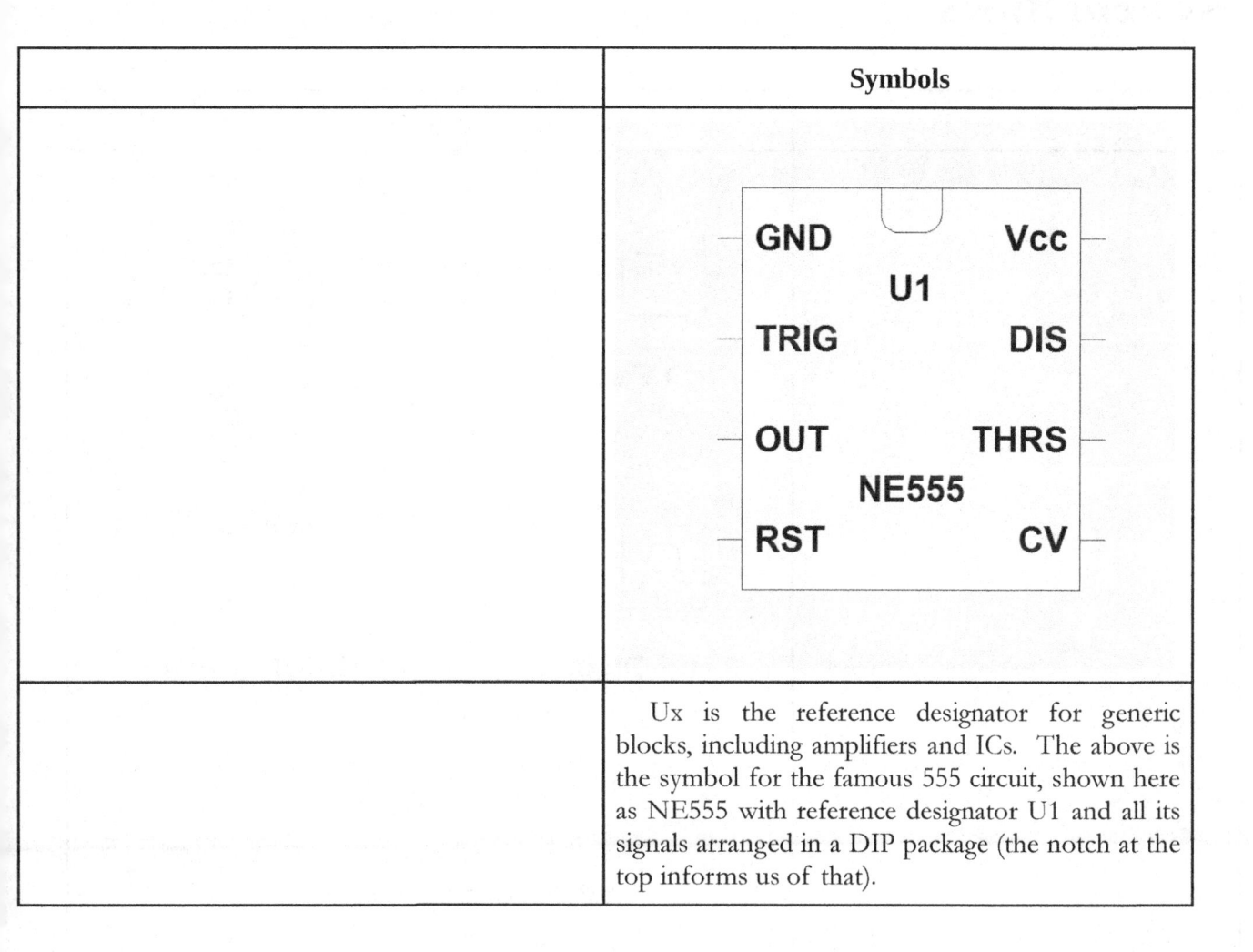

	Symbols
	GND — U1 — Vcc TRIG — — DIS OUT — — THRS NE555 RST — — CV
	Ux is the reference designator for generic blocks, including amplifiers and ICs. The above is the symbol for the famous 555 circuit, shown here as NE555 with reference designator U1 and all its signals arranged in a DIP package (the notch at the top informs us of that).

System Block

	Symbols
	F1 High Pass **F2** Low Pass **F3** Band Pass **F4** Notch Filter Summing Junction
	Highpass, lowpass, bandpass, and notch filters are shown, as well as the summing junction system block.

    Block diagrams often uses just a rectangle with words or figures in it to represent what the block is about. The summing junction, amplifier, and converter (shown next) are exceptions where specific symbols are used.

Converter

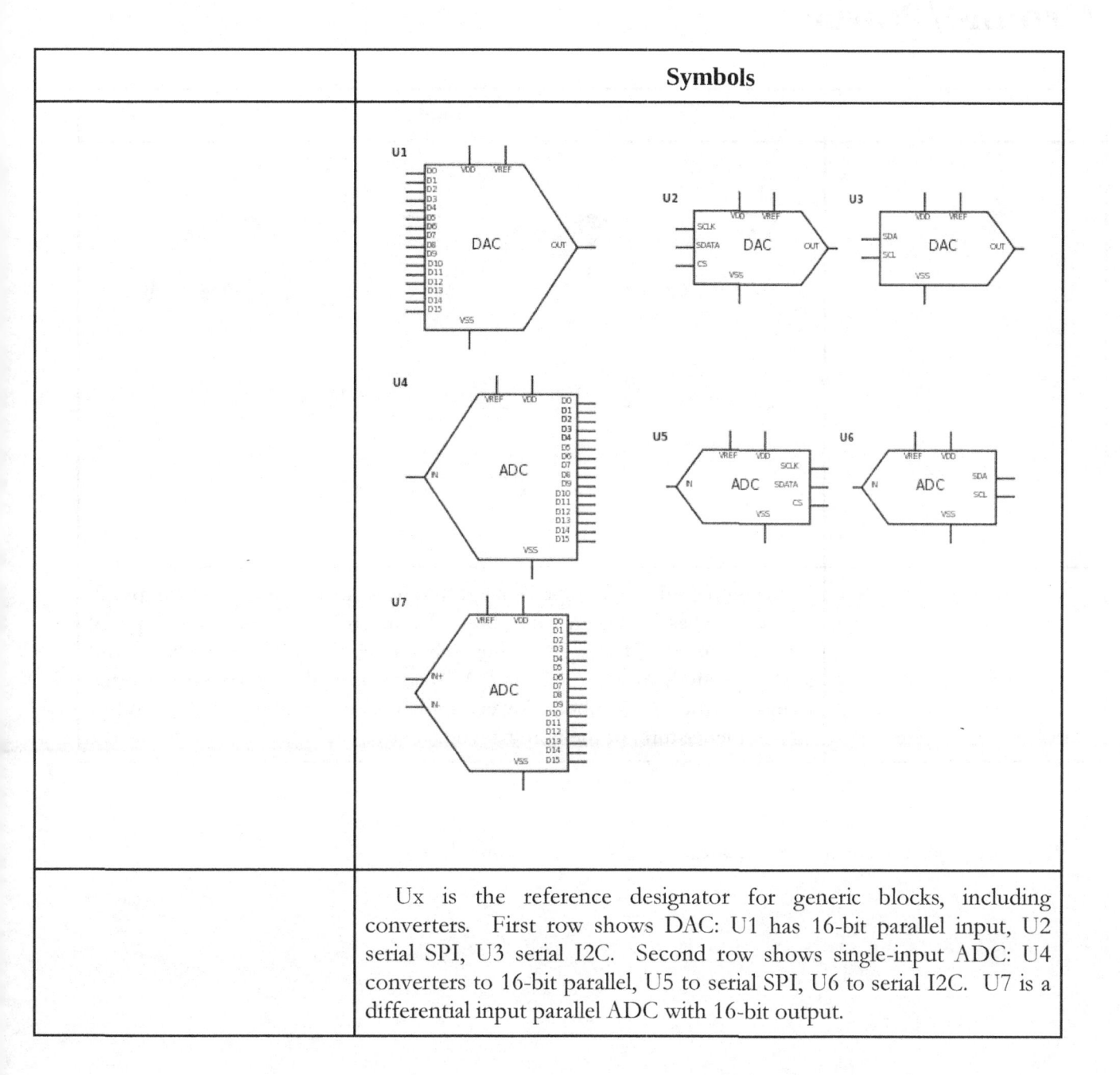

	Symbols
	[Symbols: U1 DAC with 16-bit parallel input (D0–D15), VDD, VREF, VSS, OUT. U2 DAC with SCLK, SDATA, CS, VDD, VREF, VSS, OUT. U3 DAC with SDA, SCL, VDD, VREF, VSS, OUT. U4 ADC with IN input, VREF, VDD, VSS, 16-bit parallel output D0–D15. U5 ADC with IN, VREF, VDD, VSS, SCLK, SDATA, CS. U6 ADC with IN, VREF, VDD, VSS, SDA, SCL. U7 differential ADC with IN+, IN−, VREF, VDD, VSS, 16-bit parallel output D0–D15.]
	Ux is the reference designator for generic blocks, including converters. First row shows DAC: U1 has 16-bit parallel input, U2 serial SPI, U3 serial I2C. Second row shows single-input ADC: U4 converters to 16-bit parallel, U5 to serial SPI, U6 to serial I2C. U7 is a differential input parallel ADC with 16-bit output.

Ground/Power

	Symbols
	Ground **Earth** **Chassis** Vdd +12V **+12V** **Power Supply**
	Ground and earth represents different vocabulary across the pond. However, chassis ground may be different from ground, and a special symbol is used. If the grounding scheme separates between digital and analog ground, then an "A" or "D" would be added inside the inverted triangle in the first symbol. Power supply is either indicated with a Vdd or Vcc nomenclature, or the nominal voltage (e.g., +12V).

Visualize a ground rod on the ground or earth, and you can visualize the symbols.

Switch

	Symbols
	S1, S2, S3 (switch symbols) S4, S5 (pushbutton symbols) S6, S7, S8 (DIP switch symbols)
	Sx or SWx are the reference designator for switches. S1 is SPST; S2 is SPDT; S3 is DPDT. S4 is normally open pushbutton; S5 is normally closed. S6 is single DIP switch; S7 is double; S8 is slide DIP switch, double.

Visualize the actual switch connections represented as a line drawing, and that is the schematic symbols for various forms of switches. It is a good idea to learn reading schematics beginning with switches, using supplement I.

Battery

	Symbols
	(cell symbol) (battery symbol)
	A cell, or single unit of power source, is indicated on the left. A battery, or multi units of power source, is indicated on the right. Occasionally people use the latter even though only a cell was used.

Visualize a series of anodes and cathodes as electrodes in a bath of electrolytes, and you get the symbol for cell and battery.

Connector

	Symbols
	Phono Jack Phono Plug 1/4" Phone 2 conductor 1/4" Phone 3 conductor Stereo 2.5 mm Stereo 2.5 mm tip switch 5-pin MIDI Jack 5-pin MIDI Connector 3-pin XLR female 3-pin XLR male 1-pin screw terminal 2-pin screw terminal
	Jacks (female) are shown on the left, and Plugs (male) are shown on the right. Notice *phono* and *phone* are *different*.

Historically, we started with **binding terminals**. Then we have **banana jacks** and **plugs**, **RCA connectors,** the **TR** and **TRS** series - **1/4 "**, 1/8 " (**3.5 mm**), and **2.5 mm**. Over in Germany, we have the **DIN** series - which was used in **MIDI** and **XLR/XLP**. And more recently, we have the **SpeakOn** and the **USB** series.

More...

Pictures	Symbols
	SPK1, MIK1, Microphone S1, S2, S3 X1, OSC1, OSC2 LS1, LS2, RY1
	The first row and last item are electromechanical: speaker SPK1, Microphone MIK1, or its alternate symbol, and relay RY1 with SPDT switch. The second row are fuses S1 and S2, with S3 showing the fuse holder. The third row shows crystal X1, and oscillators OSC1 and OSC2. The fourth row shows incandescent lamps LS1 & LS2.

Simply visualize the symbols...

Test Equipment

	Symbols
	(voltmeter symbols V vertical and horizontal; ammeter symbols A vertical and horizontal; DMM; Oscilloscope; Spectrum Analyzer)
	Used to be that a meter is simply shown as a meter face with a needle. Now mostly we want more preciseness: in the upper left quadrant, first row shows a voltmeter in vertical or horizontal orientation, second row shows similarly an ammeter, and on the right a modern Digital Multi-Meter (DMM), followed by oscilloscope and spectrum analyzer. To the casual observer, the latter two looks alike: in fact, modern test equipment often performs *both* functions. In this book, most of the **DC voltage measurements come from a voltmeter, AC time domain measurements from an oscilloscope, and frequency domain Bode Plots from a spectrum analyzer**.

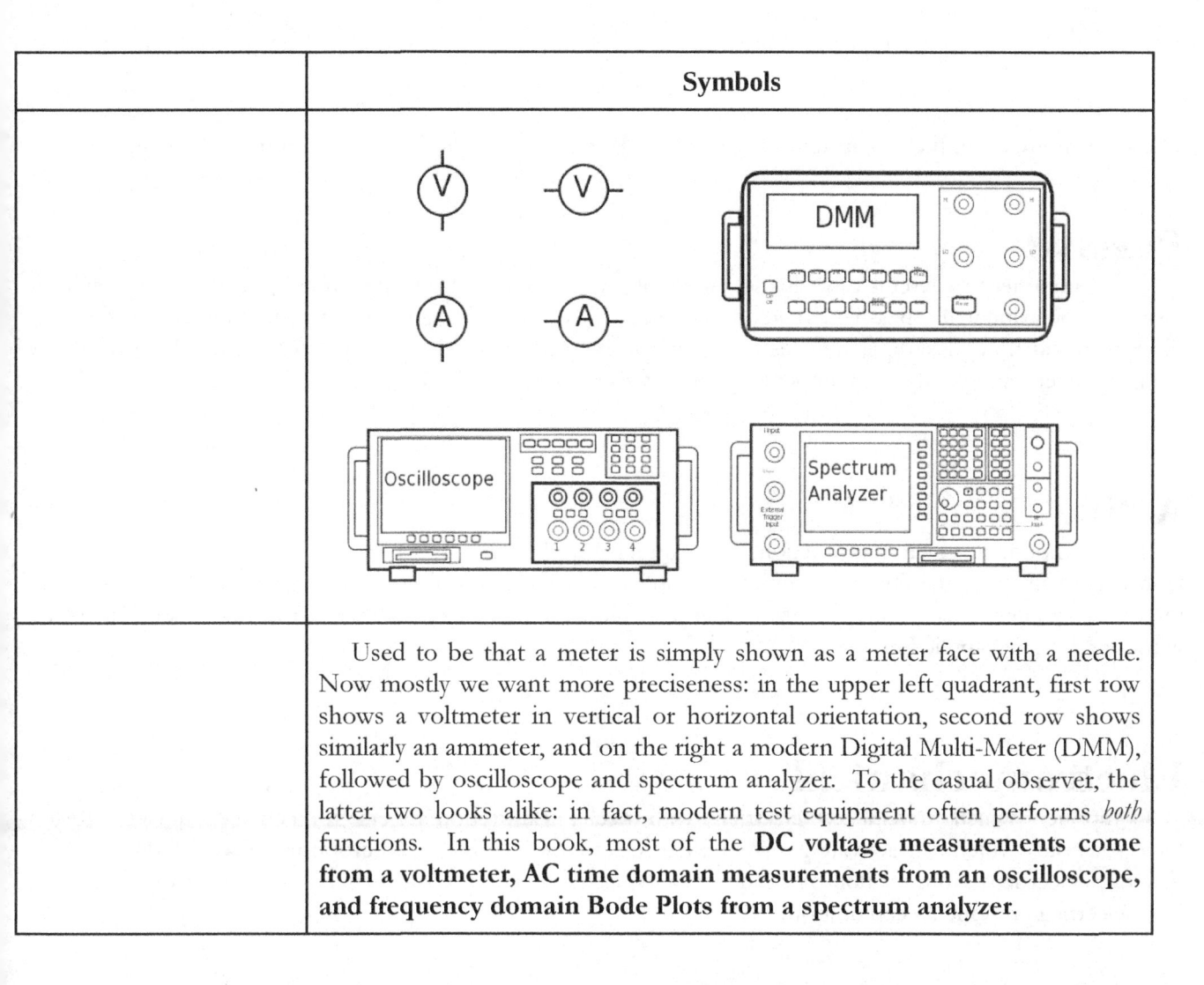

Simply visualize the symbols...

2.3 Classification: Passive, Active, Electromechanical, Solid-State

These terms are often thrown around. That's why it is important to understand them from the beginning...

Passive

The component or circuit can function without any "applied power" (whether applied from outside or inside). Consequentially, passive circuits cannot magnify or amplify signals. It can only impede or resist or degrade signal flow. Examples are resistors, capacitors, inductors, and passive filters - they don't need any applied power - and yet they can impede signal flow in ways according to the design.

Active

The component or circuit functions only under "applied power." This is the opposite of passive. Consequentially, it is possible to magnify or amplify signals. Examples are transistors, vacuum tubes, and integrated circuits usually. Diodes are usually considered active components even though they can rectify without additional applied power.

Electromechanical

Component which "transduces" electricity with mechanical motion are electromechanical devices. Relays have solenoids which can activate a switch when electricity is applied. Microphone is a transducer which generates electricity from motion of sound. Loudspeaker is the opposite, generating sound from electricity. Piezoelectric components can do both.

Solid-State

For a solid-state component, its main function occurs in a single solid state. For a solid-state circuit, its major active component is a solid state component. Example of solid state components are transistors.

Examples of non-solid-state components are vacuum tubes and electromechanical components. During the reign of vacuum tube, it suffers from short life, interference from even mechanical noise, and many other issues attributed to its non-solid-state. Plus, it is also expensive. When transistors came out, their slogan was solid state was better, more reliable, and have longer lives.

Other examples of non-solid-state components: the electrolytic capacitor and alkaline and current lithium batteries. They all wear out, may heat up, or even explode when charged improperly. Solid-state battery is now a high priority for future electric vehicle development.

2.4 CIRCUITS: OPEN, CLOSED, SHORT, EQUIVALENT

Again, we need to learn the language of audio electronics for communications, which includes terms like open, closed, short, or equivalent circuits.

Circuit

Ideally, a **circuit** can be envisioned as allowing current to flow around, eventually back to the same place. When the circuit is **opened**, say by action of an opened switch or because of a breakage of a wire, current no longer flows, and it becomes an **open circuit**. When said situation is **closed**, we say it is now a **closed circuit**.

A more important usage of the terms come when we say a capacitor becomes "like open circuit" - that means it acts as if it is **open**, or in other words, there is no connection. It appears as if we erase the capacitor from the schematic under the open circuit condition (which occurs under DC, by the way). Likewise, the capacitor becomes "like short circuit" at high frequencies, acting as if it is a **short**, or that it can act as a wire. We can simply visualize the capacitor as a wire under such condition.

Equivalent Circuit

As hinted in chapter 1, engineers use the concept of equivalence to solve problems. It is important to under such concept - especially if one does not use equivalence concept at all - as it may be difficult to grasp. (One may keep on asking "why?" - but the answer always is engineers use complexity to simplify…)

For example, below are schematics for four equivalent circuits that we'll study in chapter 3. We will transform from one complicated circuit to one simple equivalent circuit of one resistor of value 33.3 ohms.

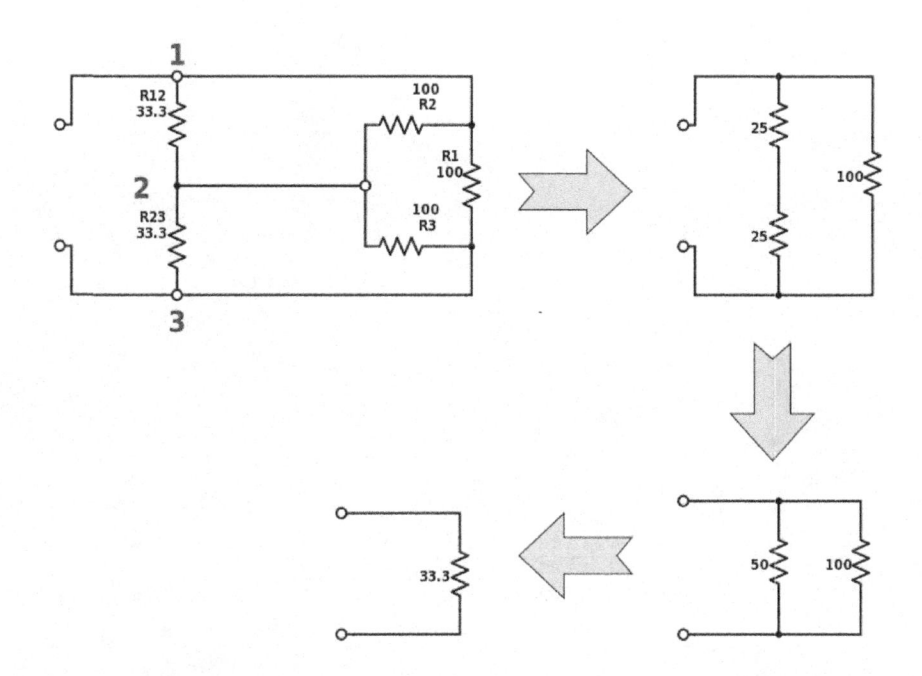

2.6 Review

Chapter 2 Review Guide

Theory
- **concepts** of component, reference designator, component values, schematic, block diagram
- **concepts** of active, passive, electromechanical, solid-state
- **concepts** of open, closed, and equivalent circuits
- **concepts** of hierarchy and hierarchical decomposition
- **avoidance of wrong concepts**

Practice
- ability to **read** most simple audio electronics **schematics**

SUPPLEMENT I - THREE WAY, TWO THROW, ONE POLE?

One of the best ways to start learning to read schematics and troubleshooting skills is to start with switches. That is *not* how most electronics textbooks start. Let us start with the most common switch in typical homes, where one switch controls one lamp…

One Two-Way, SPST Switch

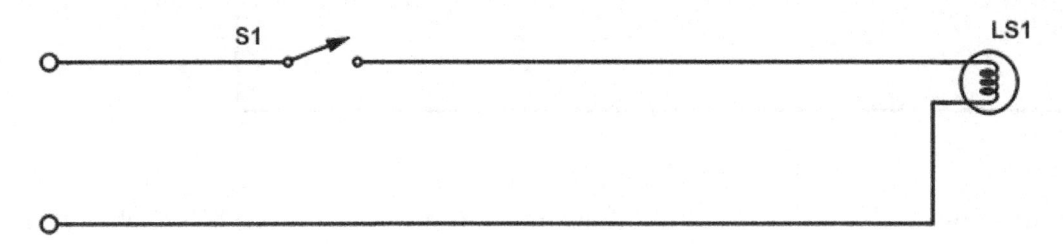

The mains is connected to the left of the schematics. The "hot" wire is connected to the switch (designated as S1), which we categorize as single pole single throw (SPST) because there is one moving pole that can be thrown to one position, and it is also called a two-way switch because it is used with two connections (in this case on the left and the right of the symbol). Then the hot wire continues to the lamp (designated as LS1), and then returns through the "neutral" wire back to the other mains connection. When the switch S1 is in the indicated position, which is known as "open", the lamp is off, because a circuit is not completed.

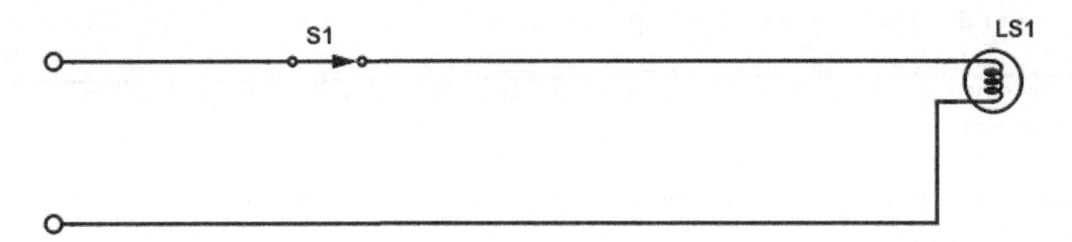

When the switch is in the new "closed" position shown, the circuit is completed, current flows, and the lamp is lit.

What if we want to have *two* switches to be able to control *one and the same* light bulb?

Two Three-Way, SPDT Switches

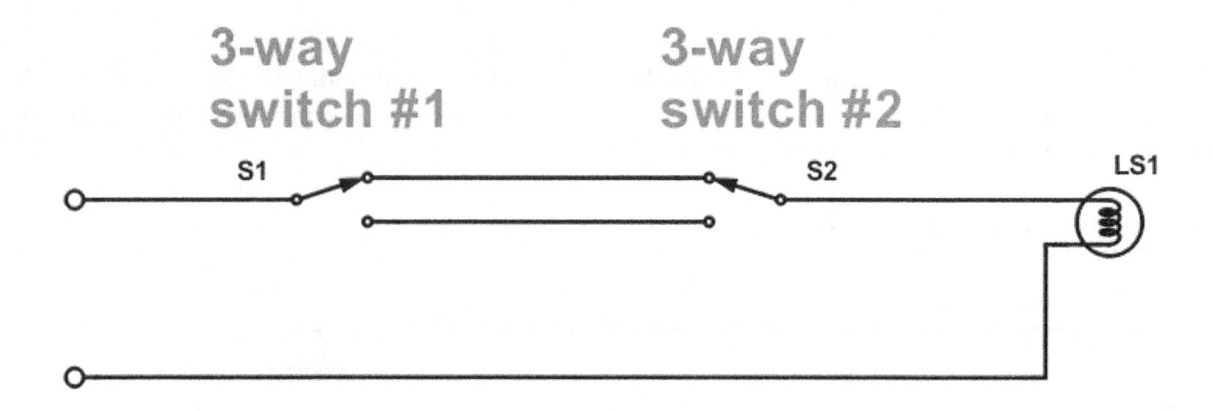

The most common way to have *two* switches to be able to control *one and the same* light bulb is to use the circuit shown above. The mains have hot and neutral wire coming into switch designated as S1, which is a single pole double throw (SPDT) switch because it has one moveable pole and two positions it can be thrown to, and it is also known as a three-way switch because in practice it has three terminals. After the run to S1, the cable uses three wires, with the neutral wire continuing to switch S2, and likewise the hot wire, but an additional wire similar to the hot wire also goes all the way to S2. That extra wire connection is the trick for two switches to control one lamp. After S2, wiring resume as previously, using hot and neutral wires going to the lamp LS1.

There are no open or closed positions for a SPDT switch. When S1 and S2 are in their indicated positions above, the circuit completes and so the lamp would be lit when connected to the mains. If either S1 or S2 is switched to the other position, the lamp is turned off. However, if both S1 and S2 are in the other position, the lamp is also turned on as a complete circuit also results, but this time using the alternate hot wire. Thus *either* switch can be used to turn on or off the lamp, but there is no designated "on" or "off" positions on either switch.

What if we want to have *three* switches to be able to control *one and the same* light bulb?

Add a Four-Way, DPDT Switch

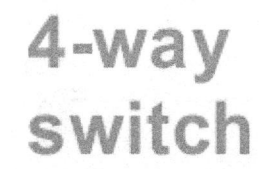

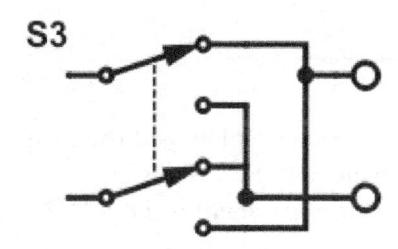

We need to first create a four-way switch, as shown above, with four connecting terminals. It consists of a Dual Pole Dual Throw (DPDT) switch connected as shown to reduce its normal six terminals to only four. When in the position shown, the wire from the let connected to the top also connects to the top wire to the right, and ditto the bottom wire. When the switch is changed, the top left wire connects to the bottom right, and the bottom left connects to the top right.

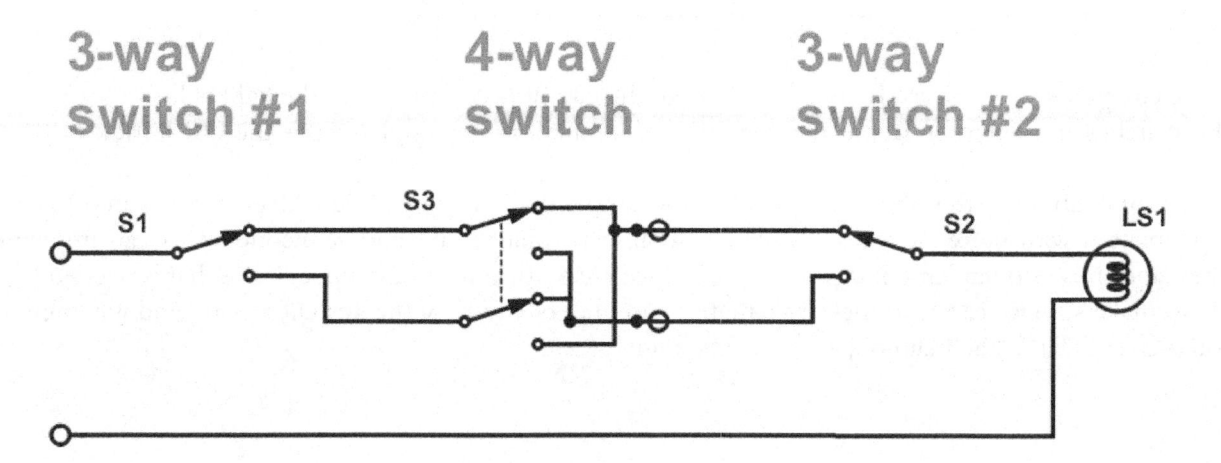

With the four-way switch put in between two three-way switches, as shown above, we can now have three switches control the lamp. *Any* switch can change the state of the lamp (from on to off or from off to on).

What if we want to have *four* switches to be able to control *one and the same* light bulb? We simply add one more four-way switch in the middle! In fact, we can keep adding as many four-way switches as we desire…

Two-way in Series = AND Logic

What if instead of *either* of the two switches controlling one lamp, we want the two switches *together* control the lamp?

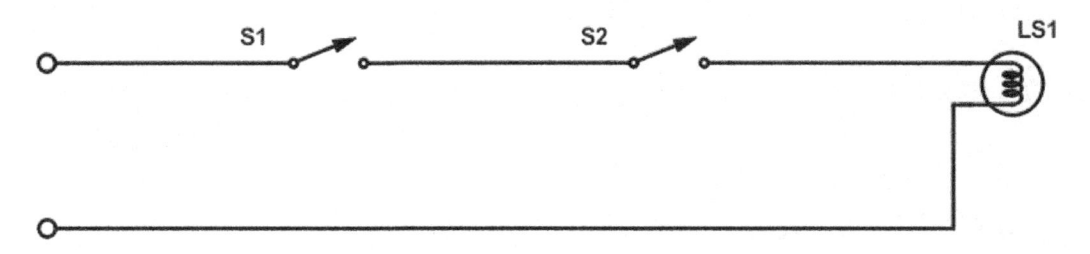

In the circuit above, unless *both* S1 and S2 are closed, the circuit is not complete and the lamp would not be lit. However, when both S1 and S2 are closed, the circuit is complete and the lamp is lit.

The two switches are said to be **in series**, and the logic function they implement is the **AND logic**.

Two-way in Parallel = OR Logic

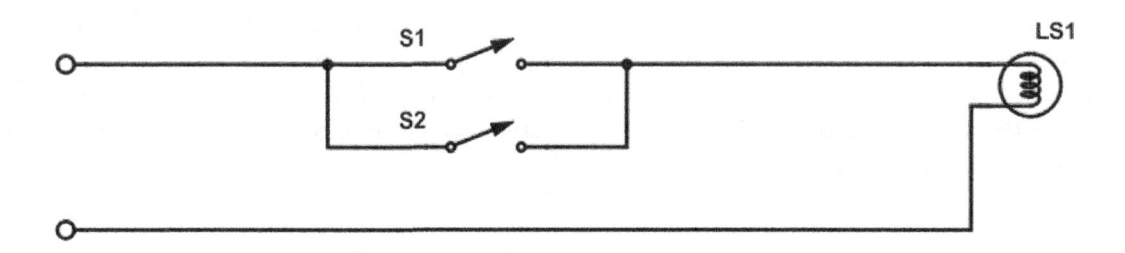

In the circuit above, if *either* S1 and S2 are closed, the circuit is complete and the light is lit.

The switches are said to be **in parallel**, and the logic function they implement is the **OR logic**.

If you think about it, the physical switches above implement **digital logic**. More complicated logic can be implemented with more switches. Later on, we'll learn that diodes and semiconductors can implement switches, and thus also implement digital logic. But we are getting ahead of ourselves. What is needed at this time is to make sure we can read these schematics and understand how the switches work, and when circuits are completed. That is the main objective at this point…

SUPPLEMENT J - DETERMINING COMPONENT VALUES

Components are marked primarily for readability. However, bear in mind that in the past, there are limitations of what is writeable as well, especially on cylindrical components. That explains why on cylindrical and curved components often color bands are used for marking component values, while on relatively flat components, coded values or direct component values are used, depending on the real estate available.

Decimal Point Replacement

One aspect for readability has to do with the decimal point, which is easily rubbed off. Thus the decimal point is seldom printed. Instead of the decimal point, a character is often used instead, based on the unit. For example, a 4.7 pF capacitor is often marked "4p7", while a 1.0 nF is marked "1n0".

Number Codes

Where numbers larger than 100 is used (and where decimal point is not needed), usually a number code is used to save space. Historically, the three-digit code is easy to understand: the third digit is the number of zeroes attached to the first two digits. For example: 221 means 220 (one zero), 220 means 22 (no zeroes), and 222 means 2200 (two zeroes).

Likewise, the four-digit code consists of three digits followed by the number of zeroes represented by the fourth digit.

For ceramic capacitors, the 3-digit code is often followed by a letter to represent its tolerance: for example, Z for +80%, -20%, J for ±5%, K for ±10%, M for ±20%. Working voltages are usually marked directly on capacitors as well.

Color Codes

Color dots or bands are simply used to replace the number codes; it is used when it is difficult to write actual numbers (for example on cylindrical surface).

Black	Brown	Red	Orange	Yellow	Green	Blue	Violet	Grey	White
0	1	2	3	4	5	6	7	8	9

Real-life color bands

Resistors (formally) have four to six bands, depending on its precision. However, the four band version can have the last band "*omitted*" - thus one sees only three bands, but it is still considered a four-band resistor! The last band in a 4-band or 5-band resistor is tolerance. For a 6-band resistor, after the 4-digit code, the 5th band is tolerance, and the 6th band is TCR (Temperature Coefficient of Resistance).

For real-life components, one needs to view the color bands from **both** directions and evaluate which one is appropriate. When one sees silver or gold bands for a resistor, one can determine that side is the end of the bands. Likewise, when one sees a double-width silver band, one can determine it is a mil-spec inductor and that side is the end of the bands.

Note also that even if one is not color-blind, it may be difficult to differentiate between brown and red, red and orange, yellow and gold, and violet with many other colors, especially when the resistor had been overheated previously. Use the EIA preferred coding to help eliminate unreasonable colors (see below).

For tropical fish capacitors, one needs to understand if there is a thick orange band, it actually means two consecutive orange bands! Plus the bands go all the way to the bottom of the tropical fish. The bands always start from the top to the bottom.

One must *not* assume a cylindrical color band would be for resistors. If it is cylindrical yet it has hollow center, it is most likely a capacitor. If it has more bands than common, and the last band is a thick silver band, more than likely it is an inductor. Google to find the specific color band coding for each of the specific components.

Preferred Numbers

Do we stock *any* values for components? No! There are definite preferred values. The most preferred are the E6 series: 10, 15, 22, 33, 47, 68 - these six numbers are most commonly used in the first two digits of the 3-digit number code and color code. These are the only six that needs to be stocked for 20% tolerance.

The next are the E12 series, which includes the values in the E6 series, plus: 12, 18, 27, 39, 56, 82. Those twelve numbers are the only ones possible for 10% tolerance.

10		15		22		33		47		68	
	12		18		27		39		56		82

When we are not sure if we have oriented the cylindrical bands properly, we can check the first two numbers against the preferred E-series values. If they are not, then reverse the color bands and if it fits, then the reversed order is the proper order...

Is "100" larger than "10"?

We have to first figure out the coding scheme. A "100" ceramic capacitor is most likely coded in 3-digit number code, which means it has a value of 10 pF (10 plus zero zeroes). A "10" ceramic capacitor is most likely uncoded (remember a code has three digits), which means it is a 10 pF capacitor as advertised. Surprise! That means "100" has the same value as "10"! A single manufacturer obviously would not be inconsistent, but different manufacturers for ceramic capacitors are known to use either scheme. Do *not* be fooled!

(this page deliberately left blank)

Supplement K - Soldering

Soldering electronic components typically involves three materials:

1. a copper pad or a metal contact
2. a component lead or a wire
3. solder

During soldering, only the solder melts. This is different from other forms of soldering where all three materials melt.

For the traditional 60/40 solder, it is 60% tin and 40% lead, with a melting point of 188 °C (370 °F). Lead-free solders which meets European RoHS requirements have melting points about 5 to 40 degrees °C higher. The most common is Tin-Silver-Copper or SAC (based on chemical symbols Sn-Ag-Cu), with melting point of 217 °C.

Note that tin-based solders readily dissolve gold, so when soldering gold-plated contacts, try to do it quickly and right the first time.

Electrical solder usually comes with an integrated rosin core - the rosin flux is a reducing agent which helps return oxidized metals to their metallic states. The dull layers developed on top of solder on electronics or solder tips are due to oxidized metals.

Steps in Soldering

1. **Prepare the soldering iron**: Obviously one needs to first heat the soldering iron; it should not be left on for too long, as doing so will overheat the tip, oxidize it, and even pit the surface, rendering it useless. To test if the proper temperature is reached, dampen a sponge and let the soldering iron tip touch the dampened sponge - if it sizzles with the right sound, then the temperature is right. If it doesn't pass the test, wait a little longer, or increase the temperature if you have waited long enough, or if by visual inspection the solder iron tip is oxidized, then you may have to scratch the dampened sponge or a copper wire ball to clean the tip. Only when the tip is clean will it reach a good temperature. If it is still not clean, you may use rosin flux to clean it.
2. **Tin the tip**: Once the temperature is reached, and the tip already clean enough, then it is time to tin the tip. Place a tiny bit of solder very quickly over the solder tip, and the solder should immediately melt and flow freely over the tip. This is the ultimate test for temperature and cleanliness of the tip: if it does not flow and thinly coat the tip with a thin coat, it means either the temperature is not right or the tip is not clean. In that case repeat the preparation step to make it ready. Worse case scenario you may have to replace the tip. (This explains why again one needs to take care of the solder tip - power down the soldering iron when you do not expect to use it in the next 5 minutes.)

3. **Tin the wire/lead and pad/contact**: Once we have tinned the tip, then we should tin the two other materials that are to be soldered. Again the tin should flow thinly and freely. If not, then we should clean the areas to be tinned and repeat until the tin flows.
4. **Heat the wire/lead and pad/contact simultaneously, and apply solder**: Next we place the soldering iron tip in a way that provides maximum contact with the *two* materials to be soldered together, to provide maximum power and temperature gradient. When temperature is reached, put solder *not on soldering iron tip* but on either the wire/lead or pad/contact or both where there is no soldering iron. Because of the maximum contact designed earlier, the solder will melt and flow around all areas previously cleaned and previously tinned. In fact, new solder may not be necessary, or only minimum amount is needed, since the previously-tinned tin will melt.
5. **Remove solder iron and hold everything still while cooled**: Once we obtain the proper free-flowing tin which is shiny and with it wicked in such a way that the tin is at an angle of 20 to 70 degrees to the wire/lead, making it look like a tent or a volcano, then it is time to remove solder iron and thus heat, and hold everything still while everything cools. Holding still for only a couple of seconds is essential to not making a cold solder joint.

Visual Inspection

It is easy to visually inspect the solder joint to see if it is properly soldered. Many students use too much solder; a few use too little. Many put solder directly on the solder iron tip, and so there are huge blobs of solder appearing like bubbles, but not solidly connected onto the surfaces. Other do not position the solder iron tip for maximum contact, and so the surface of the solder looks blotchy and not shiny, meaning correct temperature has not been reached. Still many others do not tin the tip or the contacts or the wires or leads, and it is easy to see that the surfaces had not been prepared properly or cleaned and thus there are not free flows of solder that are thinly coated.

Again on the main solder joint, it should appear as a tent or volcano, with shiny coat when just soldered, and angled at around 45 degrees or so. Then it should spread thin around the area all the way around where the pad or wire is - as if they are wicked. No solder should appear on the solder masks (where solder is not supposed to be), especially no bridge shorts. Not much solder flux should remain - whatever that remains should also spread thin and outline the solder.

Supplement K – Soldering

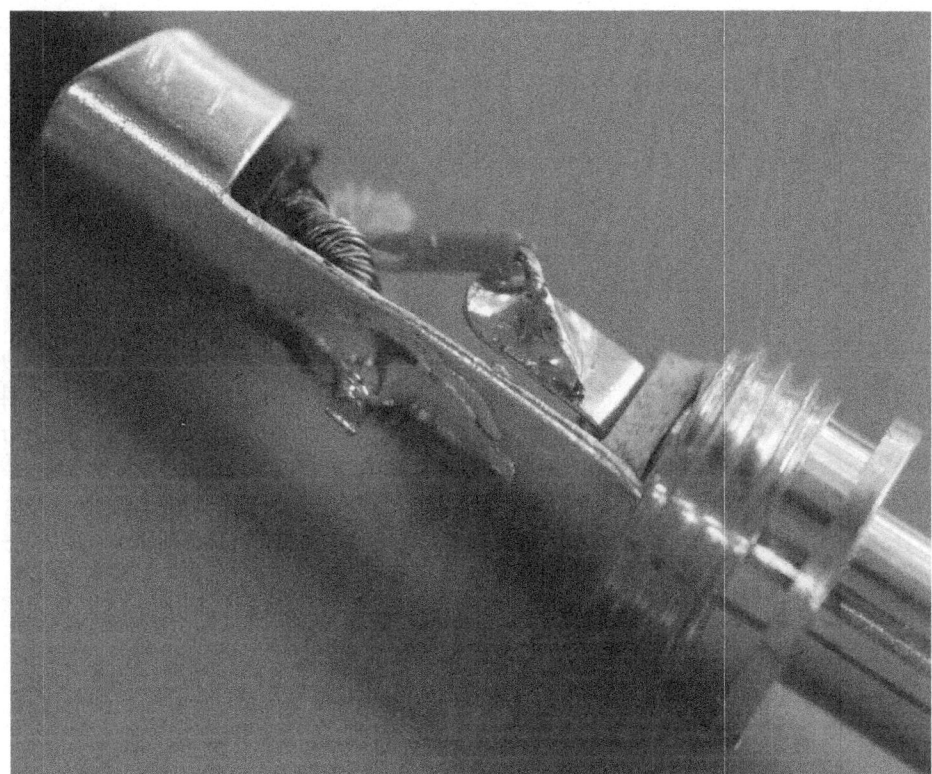

Relatively good soldering

Poor soldering

K-3

Tip

The most common mistake is overheating - overheating the soldering iron tip may damage it; overheating the copper pad may lift it (and then you may have to buy a new circuit board!) - so be careful *not* to overheat!

Soldering cables may require higher temperature and thicker flat screwdriver type tip for larger contact surface. The reason is that there is a temperature gradient between the soldering iron tip (which is the hottest) and the end of the cable (which is coolest) - the temperature at the solder is somewhere in between. So if the temperature at the coolest end is 200 degrees, the desired temperature of the solder is 400 degrees, the temperature indicated for the soldering iron tip may have to be 600 degrees to 800 degrees (depending on the way it is calibrated). Significantly lower temperatures are sufficient for soldering components on printed circuit boards, maybe with a small conical tip.

Choosing a Soldering Iron

A soldering iron may be selected using these parameters:
1. **Tip size and shape** - for soldering through-hole or large surface-mount components on circuit boards, a **small conical** tip is usually preferred for precise operation. For de-soldering of components or soldering of power supply boards, heat sinks, shielded or braided or thick cables, a **large flat blade** is desirable to maximize contact area to exert maximum power transfer. The consideration is a continuum from precision of contact to amount of power transfer, and multiple kinds of tips are available, ranging from butter-knife shaped to partially-flattened conical tip. Soldering irons can either be fixed with one kind of tip, or can support interchangeable tips. Tip material also varies, but affects usability, solderability, and working life.
2. **Maximum power available** (measured in **wattage**) - low-power ones go down to 15 watts or so, while high-power ones for electronics go up to 70 watts or beyond; 150 watts power can be purchased. Medium ones are typically around 30 watts. The power determines what *kind* of soldering can be performed with the soldering iron. Items that conduct heat away from the solder joint quickly (for example, large copper wires or heat sinks) will require higher power.
3. **Usage styles**:
    1. One piece, pencil style - this is often the cheapest, but operator fatigue may set in after a while as it is heavier, and it may be harder to manipulate as a result.
    2. Soldering station style - a lightweight soldering pencil tethered to a heavy main control unit.
    3. Soldering system style (typically for rework/repair) - for "production" use, it includes hot-air guns and de-soldering irons and essentially everything needed for production in a system.
    4. Portable styles - previously butane based, now usually powered by rechargeable batteries. Butane based soldering iron has been reported bursting into flame, so think twice before going that route.
    5. Soldering gun - previously popular because it heats up really fast and when not needed will immediately cool down when the finger is released; however, temperature control is not very precise.
4. **Temperature control** mechanisms:
    1. Fixed wattage - this is the original, but is usually also the most reliable because of its simplicity (no control circuitry!) - but it also means that multiple soldering irons are required, from low to medium to high wattages.
    2. Coarse temperature control - this is now typically the cheapest style on the market, and most popular with hobbyists as it gives the impression that one soldering iron can do all. It typically comes with a small variable temperature control knob, sometimes marked with 3 to 5 settings, but does not offer fine temperature control. More critically, reaction time is usually slow and so temperature cycles through large ranges, and students often didn't realize the thermostat was off when they attempt to solder, leading to inability to solder or cold solder joints.
    3. Fine power control - the power unit provides a fine and *constant* power level to the soldering tip. The power control dial is usually large and precise, leading to repeatable performance. Often preferred by soldering technicians, as they know from experience how to set the power just right, which is the minimum needed to get the soldering job done. But given such a wide range of power control, beginning students tend to abuse it and move to the highest power, resulting in overheating, lifted pads, and reduced longevity for the soldering tips.
    4. Fine temperature regulation - the power unit supplies power only when the sensor at the soldering tip tells it the temperature is lower than expected; once temperature is reached, the power is shut

off. Some units can control to within one or two degrees Centigrade, and temperature cycling is fast, within seconds, so it is beneficial to beginners and experienced technicians alike.

5. **Special requirements**:
    1. It should be noted that when lead-free solder is used, the soldering iron needs to reach higher temperatures than when the traditional leaded 60/40 solder is used. Such soldering iron is typically advertised with something like "support high thermal recovery needed for lead-free soldering".
    2. The soldering iron should also be ESD-safe to protect ESD-sensitive components during soldering.
6. Special **management features** - some soldering stations have automatic inactivity power-save feature, sensing no movement of the soldering tip, thus preventing overheating and promoting longevity, as well as saving power. Other soldering stations have lock keys or digital passwords which prevents students from using higher temperature/power, thus preventing abuse and reduce oxidation of tips, thus improving service lifetime.

At the hands of an expert, *any* soldering iron can be useful.

The usual recommendation for someone serious in soldering electronics, or for use in class labs, is to get at least a low-cost soldering station with fine temperature regulation, with ample available power around 50 watts, and interchangeable tips. Soldering guns or portable ones, and those with coarse temperature control are not recommended because students may not be able to control them well, or may even abuse them, leading to overheating and short operating life for the soldering iron, and potential damage to the circuit board. One student had shown lifted circuit pad due to overheating. Another student came back and show his soldering iron, used only three times, with tip pitted and the whole heated surface with a blue-grey sheen, indicative of overheating. One may think it good to buy an inexpensive soldering iron like that, but the cost of ownership is very high when it is easily abused; thus buying the more expensive soldering stations may be more cost-effective, and with less maintenance headaches, especially for educational institutions. Likewise at employment in electronics companies, most likely soldering stations at a minimum, or even complete soldering systems, will be used.

SUPPLEMENT L - BOM PROJECT

Bill of Materials (BOM)

BOM is used to communicate what materials are used by a design. Often it is used formally across organizational boundaries, e.g. to let a technician or assistant procure parts, to get purchasing or supply-chain specialists to evaluate manufacturing cost of a design, and is often used in design reviews, thus considered part of the design itself. A capable Compute-Aided-Design (CAD) software will automatically spit out a BOM from a schematic capture program upon request.

BOM columns:

- Quantity—The number of times a component is used in the design.
- Description—Includes its type and value (for example, "resistor, 6.8 kΩ" or "resistor, metal film, 6.8 kΩ, 5%").
- RefDes—The reference designator of each component in the design.
- Package/Type/Part Number/Notes—The component's other characteristics important at the BOM level.
- Unit Price
- Subtotal Price

Then the total price is shown on the last row.

Example

RefDes	Qty.	Value	Description	P/N	Est. price ($ea)
R1	1	10k	Resistor	xyz	0.02
U1	1	555	IC	NE555J	0.37
LED1	1	green	LED	Yx23	0.07
					0.46

Occasionally, there may be multiple unit price columns, indicating the effect if purchased by 1,000 units or by 100,000 units, for example. In larger projects, the BOM is made hierarchical, where each board or section is their own BOM's, and then the subtotals are further totaled together. But typically, all the components are also presented in one single BOM. Why? The hierarchical BOMs are more useful to engineers designing the system, while the linear BOM is more useful to buyers and supply chain managers.

BOM Project

The objectives:

- Learn where to find a *local* electronics store so that when you need a component when a project is due the next day, you know where to go
- Learn how to build a BOM
- Learn through first-hand experience the business side of electronics
- Discover if is it cheaper to buy individual parts? Kits? Whole electronics?

Your mission:

1) Find a *local* electronics parts store that has most of the parts indicated in a selected kit - use your detective skills to determine all 8 parts including the PCB, 555 IC, and battery clip. Your instructor may give you examples of local stores, or you can google, but please note that many listed are distributors or may not be general vendors for electronics parts.

2) Assume that you are going to find and buy the components yourself. Take snapshots of each part with its price tag (please do not disturb the sales associates unnecessarily and be polite when you ask for help).

3) Complete the BOM as a table with the estimated prices (explain how you come up with the estimated prices; substitutes are acceptable - just explain your reasoning). For PCB, find a blank board as a substitute.

4) Submit as ONE Word document (or a format accepted by Canvas or whatever IT software used in your college - the snapshots should NOT be separately loaded, but rather included in the one document). It should contain your name, the name and address of the electronics parts store, your completed BOM, the snapshots, and your conclusion whether it is cheaper to buy kits or to buy parts on your own for DIY projects.

TIP:

- Consider carefully your conclusion - make sure it is defensible with the observations you had made.
- Do *not* consider an online store a viable substitute for local store - again re-read the instructions - can your online store deliver before tomorrow's project deadline? You need to know where to find a *local* electronics store in case you need a component just one day before your project deadline...

Supplement M - Pricing Electronics

This is only a quick survey, intended to dispel obvious myths and post-truths that students *may* have collected over the years…

Supply and Demand

Supply and Demand affects prices. Competition in supply depresses prices. Competition in demand raises prices.

Middleman

It is *not* smooth sailing from manufacturer to consumer in traditional stores. Middlemen are involved, each demanding their cuts. Stores are where consumers buy and return goods; it is the same whether for online or brick and mortar stores. Distributors provide warehouses for goods near the stores, so goods are available whenever demanded; they also provide negotiation for pricing, and supply to stores. Brokers and representatives introduce manufacturers to distributors. Sales, marketing, advertising folks help market to consumers. Banks provide capital for transactions, warehousing, marketing, etc. Governments demand accounting and legal requirements. All these add to the cost.

Costs = warehousing + transportation + capital + logistics support + accounting/legal requirements + the material to be sold + margins and profits for the store + marketing

Rise of National Chains

How to provide cheaper pricing to consumers? By removing middleman. The first to be removed is the distributors. For a large national chain, they could be their own distributor - which is exactly what happened.

Rise of Contract Manufacturers

How to lower manufacturing cost? By vertically integrating the activities associated with warehousing, transportation, capital, logistics support, and accounting/legal requirements on the supply side before the manufacturers. Proper supply chain management is the key to lowering cost. Many people think that manufacturing had gone to China or Asia because of lower labor cost. That may be true, but not the only reason. If so, why would the largest contract manufacturers decide recently on investing in manufacturing in America? The real reason why manufacturing had gone to Asia was primarily because of poor supply chain management. Thus, with good supply chain management, manufacturing in America actually has more benefit than manufacturing in Asia! Why? Because of proximity to the final consumer market, the cost of transportation is reduced significantly, which also means the time spent in warehousing and cost of capital is also reduced. These lowered costs more than offset the slightly higher labor cost.

Rise of Design Services

With the rise of key contract manufacturers, it becomes beneficial to have design services tied to specific manufacturers, which can take advantage of specific processes they are more knowledgeable about compared to anyone else because they have done it more times than any other general customers. Because of such concentration of opportunities, it becomes cheaper for them to design the circuits than general designers who do that only occasionally.

Final Cost is Far from what we may Expect

When we buy a resistor from retail, it may cost us anywhere from about 50 cents to 10 cents, depending on the quantity we buy at a time. If we buy in quantity, it costs us less than a cent. It costs even less for a manufacturer who do this in huge volume with specially negotiated rates.

Stores manage pricing in strange ways. For example, they price entry-level computers competitively, which means quite often they squeeze the manufacturer's' margin to 3-5%, and they mark up at similar margin levels. That explains why most manufacturers loathe building entry-level machines.

In contrast, stores often sell hard-to-find but needed accessories at margins up to 20:1, meaning a special cable that cost them $1 they will sell for $20.

That explains why it is often cheaper to buy a whole machine and then dismantle it to obtain parts, then to buy the parts individually. Ditto for buying kits - it is often significantly cheaper than buying individual parts - even if we have to upgrade some components.

However, things change regularly. Recently, national chains had essential monopoly because competitors had all gone bankrupt - their only major competitor is now online stores. Many electronics products are now selling at margins of 3:1 or so again, as it was several decades ago. With such margins, even when they have a 50% sale, they are still making a profit...

Also, contract manufacturers do not get the same price as we do on Intel processors, or their advertised quantity prices. There are special deals and "rebates" - whatever is legal they shall use to enhance their competitiveness. That explains why a new iPhone costing us nearly $1000 may cost Apple only a little over $300 - when one buys from the carrier we should not expect the carrier to be paying as much either.

Chapter 3 DC & NETWORK ANALYSIS

> "DC Analysis" is the first foray into the world of analytical circuit analysis. Instead of derivations and endless theorems, we'll provide a proven workflow - you just need to follow "the method" to perform basic DC analysis. This chapter begins with Atomic and Quantum Theories.
>
> *Expected Learning Outcome*: The student will be able to handle these levels: 1) Ohm's Law and Power equation individually, 2) in combinations, 3) series-parallel and dividers, 4) series and parallel in combination, 5) general circuit simplification with single source, 6) with multiple sources, 7) perform DC analysis from typical schematics.

CONTENTS

Fundamental Definitions	3-3
Potential Difference - Cell & Battery in Series	3-4
Current, Conductivity, Resistivity, Resistor, Resistance, Conductance	3-6
Ohm's Law Explained & Clarified, Linearity, Noise of the 1st & 2nd Kind	3-13
Fundamental Network Theorems	3-17
Ohm's Law in Practice: An IER to an EIR	3-21
Power Equation in Practice: To PEI or not PIE	3-23
Double the Laws, Double the Fun: PEIR on a PIER	3-25
Series & Parallel Topologies: Be Simplified	3-30
Voltage & Current Divider Topologies: Be United	3-33
Combining Series-Parallel & Dividers: A Wall of PIER or PEIR	3-37
Star, Delta, and Bridge Topologies: Reaching for the Moon...	3-41
DC Analysis for General Audio Circuits	3-43
Multi-Sourced, Changing Load: More Network Theorems	3-46
Review Exercises	3-56

This chapter was initially titled "Basic DC Analysis." Students typically learn DC first, and then AC, and *then* they get confused as to *when* to apply DC or AC analysis. Because when they learn waveforms, they identified them as AC. Audio, for example, are shown as waveforms. And so they thought they *must* use AC analysis for audio. The fault to this over-simplified logic is that signals (for example, audio) consist of *both DC and AC components*, so *both* DC and AC analysis applies - we should do DC-style analysis for DC components, and AC-style analysis for AC components. Thus we introduce the concept of *impedance* in parallel with *resistance* in this chapter titled "DC and Network Analysis," earlier than most other textbooks, but help the student understand that the *style* of analysis is what matters.

Some may think that, since they have been acquainted with Ohm's Law in high school, they already *know* Ohm's Law. Here are levels of understanding of Ohm's Law and power equation for your considerations:

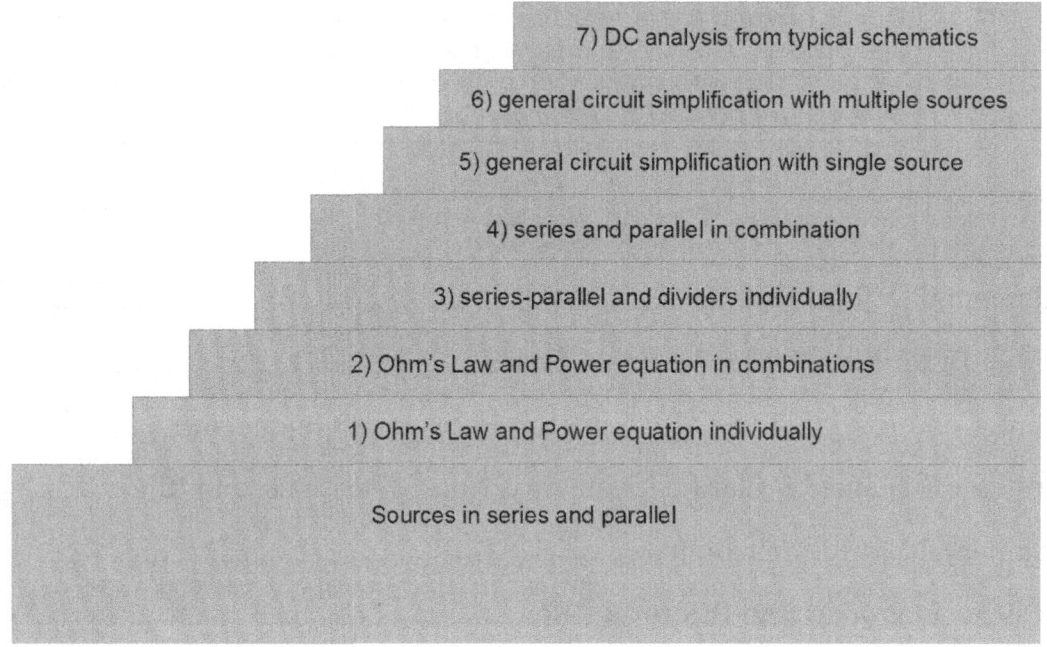

Levels in DC Analysis

7) DC analysis from typical schematics
6) general circuit simplification with multiple sources
5) general circuit simplification with single source
4) series and parallel in combination
3) series-parallel and dividers individually
2) Ohm's Law and Power equation in combinations
1) Ohm's Law and Power equation individually

Sources in series and parallel

This chapter progresses from the ground floor all the way up to level 7 in a systematic manner.

Can you perform DC analysis on circuits like these below? When you finish this chapter, you should be able to, plus handle more general circuits as well...

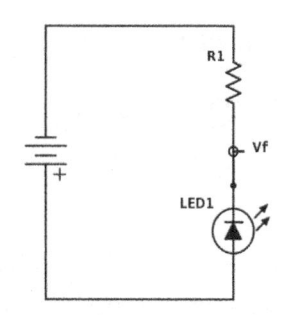

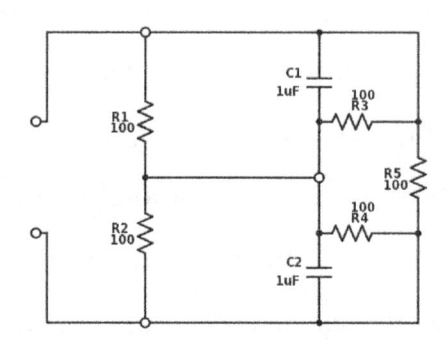

3.1 Fundamental Definitions

Supplement N introduces us from **matter** to **current**; supplement O introduces us from **energy** to **voltage**, and with it **power** as well. Here is a summary of how they are defined:

Current

Current is the rate at which charge moves past a point on a circuit. $I = Q/t$.

> Current is the rate at which charge moves past a point on a circuit.
> $I = Q / t$ (Definition of Current)

The dimension symbol for current is I; the unit for current is the Ampere, abbreviated A. The unit for charge is Coulomb, abbreviated C. The unit for time is second, abbreviated s. Thus A is equivalent to C/s.

Notice that the basis for SI units do *not* necessarily relate to how the *quantities* are *defined*. The ampere is a base SI unit. Thus as far as SI units are concerned, the Coulomb is a derived unit, expressed in terms of **s·A**, and derived from the second and the ampere using $Q = t \cdot I$

Potential Difference, Voltage

The electric potential difference across the two ends of a circuit is the **potential energy difference per charge** between those two points. We also use the term voltage to represent the result.

> Potential Difference = $\Delta PE / Q$ (Definition of Potential Difference)

Potential difference is expressed in **volts (V)**, which is an SI *derived* unit. The unit for potential energy is **Joule (J)**, and the unit for charge is **Coulomb (C)**. Thus volt can be derived from Joule per Coulomb, or J/C.

Power

Power is the rate of doing work, or the rate of generating or consuming energy, or the rate of producing or dissipating heat.

> Power = change in energy / time   (Definition of Power)

Power is expressed in **Watt (W)**, an SI *derived* unit. It is expressed in terms of other SI units as J/s. Potential difference is expressed in volt, another SI *derived* unit; and current in ampere, which is an SI *base* unit. Thus from the power equation (introduced later), watt can be derived and expressed as V·A. Alternatively, we can also derive volts as W/A from this equation.

3.2 POTENTIAL DIFFERENCE - CELL & BATTERY IN SERIES

We'll introduce an alkaline cell (the chemistry *only* to give the reader a feel for the concept of potential), then define potential difference, then explain cells and battery in series.

An Alkaline Cell

In a so-called alkaline "battery" that we can buy from a store, we actually have a single cell consisting of a zinc rod in the middle, manganese oxide on the perimeter, and potassium hydroxide (the namesake alkaline) in between. The zinc rod is connected to the negative terminal, and the manganese oxide is connected to the positive terminal.

The cell diagram is a shorthand notation describing the oxidation reaction in the left half-cell, and the reduction reaction on the right half-cell (the two half cells of course would together be one cell), with | | separating the two:

$$Zn(s), ZnO(s) \; || \; MnO_2(s), Mn_2O_3(s)$$

The anode is on the left hand side and cathode is on the right hand side.

The two half reactions can be represented thus:

$$Zn(s) + 2OH^-(aq) \rightarrow ZnO(s) + H_2O(l) + 2e^- \qquad [E° = -1.28 \text{ V}]$$
$$2MnO_2(s) + H_2O(l) + 2e^- \rightarrow Mn_2O_3(s) + 2OH^-(aq) \qquad [E° = +0.15 \text{ V}]$$

The e^- represents electrons. Thus the oxidation reaction at the top actually generates electrons which flows out of the cell through the terminal to the rest of the circuit. The reduction reaction at the bottom simply returns the electrons from the circuit, thus completing the circuit and allowing current flow. The E° provided inside the square bracket to the right represents the **standard cell potential** generated by that half reaction. Each oxidation or reduction reaction generates a **cell potential**, with the *standard* cell potential (indicated by the °) measured when the the reagent is at 1 molar concentration and at 25 C. When the reagent starts to get exhausted, the potential would drop. Likewise when the temperature drops.

In our particular case, assuming that the concentration is 1 M, then the actual voltage generated by the alkaline cell can be calculated thus:

$$E°_{cell} = E°_{cathode} - E°_{anode} = 0.15 - (-1.28) = 1.43 \text{ V}$$

The chemical concentration and side reactions are designed so that at factory output, the voltage of an alkaline battery would be nominally 1.5 V.

The classical electric cells all have *two* terminals, because there are two half-reactions and thus two terminals are needed. One may notice that some batteries for cell phones or cameras may have three or four or even more terminals. The extra terminals are for computerized queries so that the cell phone or camera can know what kind of battery was inserted, and how much "juice" is left, and so forth. In other words, such batteries already have a "chip" inside. (OK, there is a third reason why the chip is added. It allows the original manufacturer to extract more profit out of the battery, as it makes it more difficult for someone to supply a generic replacement battery. In this sense it is the same reason why most ink cartridges now have a chip in it.)

If potential difference is between *two* points, how can the E° shown on the right hand side of the equation above be for only *one* electrode?
Answer: Glad you were thinking through it. E° is defined between *two* terminals: that of the half-reaction stated, and that of a standard hydrogen electrode (SHE). **Potential difference is always between two points**.

Potential Difference

The electronics student does *not* need to know or learn the chemistry indicated above. However, she would be wise to notice that there are **two cell potentials**, one for each half-reaction, or one exerted on the anode and another on the cathode, and that the resulting voltage generated by the cell is exactly the difference between the two potentials. That is why it is called **potential difference**.

Since we measure voltage between two points, we are really measuring the potential difference between the two points. It is important to recognize voltage and potential difference always deal with two points. Occasionally, we provide or measure the voltage "at a certain point" in the circuit. When only one point is indicated in the measurement, the other point would by default imply "ground" which is the default reference point, and assumed to be zero or 0 V. When measuring under such conditions, one of the lead (typically the black or negative lead) would be grounded, and the other (read or positive) lead would be used to measure that particular point. The measurement is still a potential difference between *two* points.

Electromotive Force

You may be curious why the cell potential has an abbreviation of E. It came from the historical understanding of an electromotive force. At one time it was believed so, and the term plus its abbreviation stuck with us until now. Of course we now know it is *not* really a force, at least not so in the case of the electrochemical cell. (There is an issue that energy is also represented as E in physics; in such situation, voltage is usually represented by V to avoid confusion.)

Electrochemical Cells in Series

Once we understand the concept of potential difference from above, it is easy to see that when we have cells in series, if they are connected in tandem (i.e. anode to cathode), then the resultant output is simply the **addition** of the individual voltages.

Now if we connected the cells in the wrong way (either unintentionally or deliberately, with anode to anode or cathode to cathode), then the resultant output is the **subtraction** of one from the other. This can be easily seen from the discussion on how the cell was constructed - when you reverse the cell, the potential difference is also reversed.

These cases are illustrated in the circuits below.

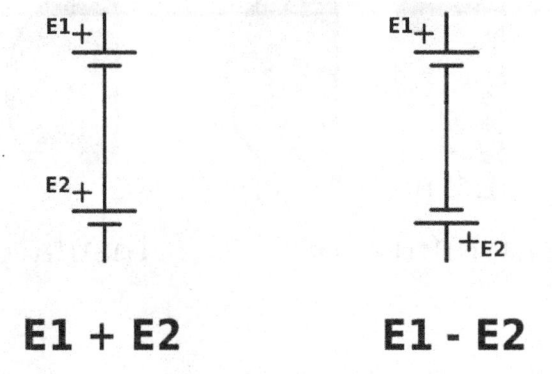

E1 + E2   **E1 - E2**

Battery

Supposedly a battery is the plural for cell. However, pop culture usage for battery is anywhere from one to many cells. Regardless, the symbol for cell and battery is different, and some people maintain the pop culture concept that a battery can be anywhere from one to many cells.

What does a cell store?

A cell does *not* store *charges*. Think about that. It stores *energy*. Batteries, supercapacitors, antiferroelectrics and electrostatic (dielectric) capacitors all can store energies. The latter two can handle *fluctuations* well while the former three can handle large energy *capacity*.

3.3 Current, Conductivity, Resistivity, Resistor, Resistance, Conductance

An electric current is often visualized and quantified as a flow of electric charge. Its unit is the ampere, abbreviated A Its symbol is I. The ampere is defined as the flow of electric charge across a surface at the rate of one coulomb per second.

Re-Cap: Electronic Band Structure

Just to be sure, we are not talking about electronic music bands and their structure, haha. Instead, we are applying what have learned in quantum mechanics, delving into the band theory for electrons in materials. This is a very useful application, explaining conductors, semiconductors, and insulators in one swoop, and clearly. (Prior to this, we *could* explain conductors and insulators using other *analogies*, somewhat vaguely; but will *have* to resort to the band theory for semiconductors regardless, although sometimes not showing the band structure *visually*, but just *verbally* mentioning the band gap. I believe visual learning is a powerful friend.)

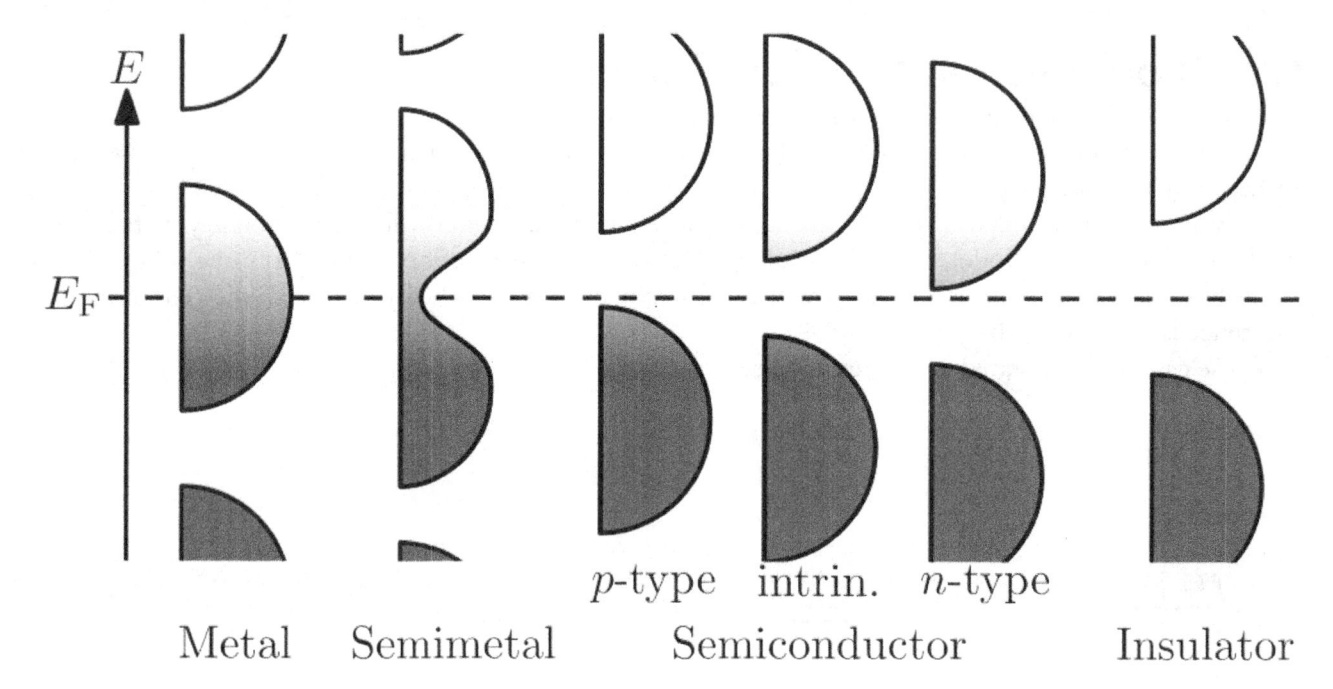

Metal   Semimetal   *p*-type intrin. *n*-type   Insulator
                              Semiconductor

Conductivity and Resistivity

Resistivity is commonly represented by the Greek letter ϱ (rho). Its unit is $\Omega \cdot m$. It quantifies how strongly a material opposes the flow of electricity.

Conductivity is the inverse (reciprocal) of resistivity. Be aware that we often switch between the two even in the midst of a sentence, because they are so related. For consistency and simplicity, we'll show numbers using only resistivity.

Charge Carriers

Charge carrier is a general term for a particle that is free to move that carries an electric charge. It can be an electron, atomic ion, or a "hole" which is a fancy concept for the absence of an electron in a semiconductor. Yes, that hole can move, thus becoming a charge carrier of opposite polarity to the electron. We'll illustrate the concept of charge carriers in the following discussions, starting with electrons that can be explained by the band theory mentioned earlier.

Conductivity in Insulators

As seen visually on the diagram in the previous (adjacent) page, at the rightmost column, the Fermi level of an insulator lies in between two energy bands. Some may say (verbally) that it lies in a forbidden band, energy interval that contains no energy levels. In insulators, the electrons happen to fill up the lower energy bands. The Fermi level falls within a band gap. There are no energy states available near the Fermi level, which means electrons are not available to move freely (or more specifically, require a lot of excitation energy to jump to the higher energy band in order to move), resulting in a very low conductivity. The wider the band gap, the more difficult it is to conduct.

Summary: Insulators: electrons fill up lower band; Fermi level lies in between bands with a wide gap.

Conductivity in Conductors

As seen visually on the diagram in the previous (adjacent) page, at the leftmost column, the Fermi level of a conductor lies within one energy band.

Conductivity in Semiconductors

As seen visually on the diagram in the previous (adjacent) page, at the middle columns, the Fermi level of a conductor lies within a band gap, although in contrast to an insulator it is close enough to be filled with holes or electrons thermally.

Conductivity in Electrolytes

In electrolytes the charge carriers are ions. Conductivity decreases with decrease in concentration. However, "molar conductivity" increases with dilution. Batteries contain electrolytes.

Conductivity in Plasma

Plasma displays function through plasma, which explains why there is high conductivity despite the vacuum.

Plasma is one of the fundamental states of matter (the others being solid, liquid, gas).

Conductivity in Superconductors

In metals like copper, as it is cooled, the conductivity increases, and the resistivity decreases. But the resistivity does not go to zero even in extreme cold. But in some materials known as superconductors, the resistivity drops to zero abruptly when cooled below a critical temperature. That means an electric current flowing through a superconducting wire will perpetuate with no power source to regenerate the current (but with power required to maintain the temperature below critical of course). Superconductivity is destroyed when the temperature goes above the critical temperature, or when the current density goes above the critical current density level which also depends on the material.

Superconductivity is a quantum mechanical phenomenon.

Thermionic Emission of Electrons

Originally known as the "edison effect," thermionic emission explains the hot filament of the incandescent light bulb releasing electrons into the vacuum tube. A second electrode will attract those electrons if it is at a more positive voltage. The result can be considered a good conductor at one direction only, but requiring power to operate.

Resistivity of Common Material

Material	Resistivity in $\Omega \cdot m$	Classification	Note
Graphene (carbon)	1.00×10^{-8}	GOOD CONDUCTOR	carbon has multiple allotropic forms;, can be good conductor to good insulator
Silver	1.58×10^{-8}		best conducting metal; but seldom used for that purpose as it tarnishes in air
Copper	1.68×10^{-8}		the go-to metal for conductors
Gold	2.44×10^{-8}		commonly plated on electrical contacts as it does not corrode
Aluminum	2.82×10^{-8}		commonly used for high voltage long-distance overhead power line: lighter for the same conductance as copper
Iron	9.71×10^{-8}		
Mercury	9.80×10^{-7}		conducting *liquid* metal
Graphite (carbon)	$\sim 5 \times 10^{-6}$		graphite is anisotropic: conductivity varies by direction, here it is the basal plane
Indium Tin Oxide	$< 1 \times 10^{-1}$		most used *transparent* conducting film
PEDOT:PSS	2×10^{-6} to 1×10^{-1} treatment dependent		*transparent* conducting *polymer* used as antistatic agent, conducting electrode, electrolyte in polymer electrolytic caps
Nichrome (nickel-iron-chromium alloy)	1.10×10^{-6}		**RESISTOR!** commonly used in heating elements
Germanium	$\sim 4.6 \times 10^{-1}$	BAD CONDUCTOR	semiconductor conductivity depends on doping
Sea water	$\sim 2 \times 10^{-1}$		considered borderline conducting; while deionized water is not; drinking water lies somewhere in between the two
Deionized water	1.80×10^{5}		actual conductivity depends on amount of gas dissolved
Diamond (carbon)	$\sim 1 \times 10^{12}$		too expensive as insulator commercially!
Glass	$1 \times 10^{13} - 1 \times 10^{15}$		
Hard rubber	$\sim 1 \times 10^{13}$		rubber is used as insulator early on before arrival of synthetic plastic
Dry Air	$\sim 2 \times 10^{14}$	INSULATOR	air is used as insulator in variable capacitors
Teflon	$\sim 1 \times 10^{24}$		teflon and other plastics used as insulators, as dielectric in capacitors

Chapter 3 – DC Analysis

Thus a very rough idea is that when resistivity is greater than one Ω·m, we usually say it is an insulator; good insulators have resistivity higher than 10^8 Ω·m. (I picked this number as it makes it easy to remember together with metals..) If it is less than one Ω·m, it is a conductor; better (metallic) conductors usually have resistivity in the 10^{-8} Ω·m range. A resistor has to be within the bad conductor range.

Classification via Charge Carriers

We had previously discussed charge carriers and so forth. Here is how their resistivity relate:

Material	Charge Carriers	Resistivity in Ω·m
Insulators	non-mobile electrons	$> 10^8$
Electrolytes	mobile atomic/molecular ions	variable depending on concentration
Semiconductors	mobile electrons and "holes"	variable depending on doping
Conducting Oxides and Polymers	mobile electrons and ions	variable depending on doping/treatment
Plasma	long-range collective-interaction of electrons and ions through EM field	variable depending on degree of ionization
Metallic Conductors	long-range collective-interaction of electrons through EM field	10^{-8}
Superconductors	"Cooper pairs" of electrons (quantum-mechanical)	0

The point of this table is to illustrate that there is no one single explanation for the conductivity of materials. Charge carriers vary, and actually becoming more and more exotic and strange as conductivity increases.

Do Electrons Actually Flow All the Way as Current?

You may have heard many times, or you may have said it yourself, that when charges flow it becomes current. In fact, current is often defined as charge flow per second.

When we switch on the light, the light turns on immediately. When lightning strikes, we can see that the lightning trail travels miles within fractions of a second. (More precisely, the negative-charged *step leader* can travel downward up to 60 miles a second, while the positive-charged *streamer* can travel upward up to 50 000 miles per second.) They seem to travel fast. But wait....

Electron Drift Velocity

In a copper wire of 2 mm diameter, with 1 A flowing through it, the electron drift velocity is 23 µm/s. (A better term is **electron drift speed**, since there is no direction involved, as "velocity" would imply. The formula used is in fact the basis for Ohm's Law, which we shall study shortly.) Remember a micrometer is 10^{-6} m. 23 µm is the width of a thin human hair.

In other words, if assuming the current is DC, the electrons don't even have time to leave the light switch, and yet we can already see the light turned on. If it is 60 Hz AC, then the electrons won't even have moved more than 0.2 µm, *ever*, if it is really electron particles that actually moved.

Speed of Electricity

Signals or energy travel down a wire (made of metal in the classification above), or lightning strikes (in the above classification, plasma), at the speed of light for that medium, as we can literally see. That is up to 300,000 kilometers per second. It reflects the speed of the EM wave.

Electron Mean Free Path

In some of the better conductors, electrons are relatively free to move around, able to roam around about 150 times the size of the atom away. This ability to roam around is call the **mean free path**. That is the basis for why copper and silver are good conductors.

The bulk electron mean free path of copper is ~39 nm. (The atomic radius of copper is about 140 pm. Remember 1 pm is 10^{-12} m. 39 nm is equivalent to $39000/(140*2) = 139$ diameters of copper atoms.)

The bulk electron mean free path of silver is ~53 nm. (The atomic radius of silver is about 172 pm. Remember 1 pm is 10^{-12} m. 53 nm is equivalent to $53000/(172*2) = 154$ diameters of silver atoms.)

Again we can see that the electron drift velocity is only about 1000x its mean free path per second. *Not fast.*

How Can a Slow Drift Velocity Support High Current?

If the drift velocity is only 23 µm/s, and current is the rate of charge flow, how can high current be supported? The answer is simply **there are lots of copper atoms around**. How much? Assuming arbitrarily that we have a 23 µm thin slice (remember that is the width of a thin human hair) of 14-gauge copper wire, we have roughly 8×10^{19} copper atoms in that slice. The electronic configuration for copper is an exception to the Aufbau Principle, which explains why different experts will give different answers to the question of how many valence electrons do copper have that can be used to conduct electricity. First, the actual *observed* electronic configuration is:

$$Cu = [Ar]\ 4s^1\ 3d^{10}$$

Second, the outermost orbital is 3d, and it is filled; however, 4s is very close to 3d. In fact, that is the reason why 4s is not filled, but rather 3d, the outermost, is, because it lowers the overall energy. That explains why some would answer copper as having 1 valence electron (by looking as 4s, and maybe assuming 4s is the valence shell since s usually is), other may say it has 2 (because they follow the rules and come up with $[Ar]\ 4s^2\ 3d^9$), and some may say it has 11 (because 1 from 4s and 10 from 3d). I'm going to defer that question of valence electrons to the experts; however, I'd say that because of this special configuration, up to 11 electrons are theoretically available for conductivity. On the other hand, even though all 11 are capable, they won't become charge carriers all at once. So although Cu(II) is more stable than Cu(I), and Cu(III) and Cu(IV) has been observed, let us be conservative, and assume only one electron per copper atom.

That would still give us >12 coulombs of charges, supporting up to 12 A of current within its drift velocity and within that thin slice. Of course, with that much current flowing in a 14 gauge wire, it may heat up, acting somewhat like a resistor.

The Nature of Charge Flow

It is like water flowing in a pipe from the water utility to a faucet. When the faucet is turned on, water starts pouring immediately. There is no need for water to travel all the way from the water utility to the spigot before water starts flowing. It is water *already* in the faucet that pours out from the spigot. When water flows, a **circuit** is established, and all water in the circuit from beginning to end starts motion at the same time. There is **long-range collective interaction** in a circuit.

Likewise, there is no need for electrons to travel all the way from the light switch to the light, It is the electrons already at the light that lights up the light. When the switch is turned on, all mobile electrons start moving at the same time. Because a circuit is formed, electrons that leave the light is replaced by electrons that enter. There is **long-range collective interaction** in a circuit.

There is danger in analogies. In the water pipe analogy, the water *actually* moves a long distance *eventually*. In the case of electron circuits, the electrons only *appear* to move a long distance. But none of the electrons *actually* move a long distance. We say the electrons in metals (and some polymers) are **delocalized** but **not completely free**. *Delocalized* because they can roam outside the region of their own atom. *Not completely free* because on average, they can roam around for about 150 times the size of the atom.

We know there is wave-particle duality for particles, where sometimes a physical phenomenon can be explained that an electron is wave, and sometimes it is particle. It is *not* our choice. Nature chose it for us.

Likewise, something as simple as electricity cannot be explained by *one* simple concept. Most of our power, voltage, and current calculations can be explained by electrons moving as particles *at a particular locality*. But when we are dealing with speed of electricity, we have to explain it as long-range *collective* interaction at the speed of EM wave.

Our culture tends to promote pick-and-choose mentality. In science, nature picked and chose for us. In other words, it is a bad idea to try to be creative and over-simplify, especially during an exam. Hint hint. :)

> While charge carriers in a circuit may move *slowly*,
> their motion *starts collectively*;
> that signal that *informs* them to proceed travels at the speed of light.
> Electrons in metals are delocalized but not completely free.

Collective Interaction Illustrated

Illustration is not proof, but it helps us understand how current electricity works. It is like two thousand fire **bucket brigade** members lining up over one mile, delivering a bucket of water to the next member each time they hear a whistle, say every second. How long does it take to get the first bucket of water to the fire? One second. That is fairly fast. How fast does the water travel? Only at about three feet per second, fairly slow. But it is collective interaction that made it all work, since every brigade member can hear the same signal and act accordingly. How fast did the signal travel? At the speed of sound, which is pretty fast.

In this case, the brigade members do not have to move much. Yet and only the water buckets are moving. However, in the case of current electricity, the electrons themselves do not move much, yet current is still moving.

Conventional Current Flow

As we had previously discussed, Ben Franklin defined what is considered positive and negative charges, and we had continued that same convention ever since. As a result, by convention, **the electric current direction is the direction which positive charge would move**. This is exactly the opposite of the purported direction of flow of electrons. Some people made a big deal out of this, but since charge carriers can be of many kinds, as we have learned, it is simply an arbitrary decision. And humans have traditionally prioritized positive over negative....

Static versus Current Electricity Revisited

I hope you realize by now, that even though electrons are involved in *both* static and current electricity, and conductors and insulators act essentially the same in *both*, the behaviors in many aspects are different. One does not see the typical charge repulsion or attraction from current electricity, for example. Many of these behavioral differences can be attributed to the collective interaction of current electricity.

Lumped Abstraction: Resistance and Conductance

Up to this point we had talked about resistivity and conductivity - they are **intrinsic properties** of a material - independent of the quantity, size, area, or length of the material.

$$R = \varrho L / A \qquad \text{(Resistance = resistivity * length / area)}$$

What if we have 3 meters of a copper wire with 1 mm² cross sectional area, which has resistivity of 1.68×10^{-8} Ω·m? Well we just use the formula: 3 m x 1.68×10^{-8} Ω·m / 10^6 m² = 5.04×10^{-2} Ω.

What we had just performed is a **lumped abstraction**. We lumped all of the distributed resistance implied by the resistivity into one lumped sum. In audio electronics, we usually deal with the lumped abstraction, like resistance and conductance.

As you may have guessed, conductance(G) is the inverse of resistance (R).

$$G = 1/R \qquad \text{(Conductance = 1/resistance)}$$
$$R = 1/G \qquad \text{(Resistance = 1/conductance)}$$

Chapter 3 – DC Analysis

♪ 3.4 Ohm's Law Explained & Clarified, Linearity, Noise of 1st & 2nd Kind

The historicity and empirical nature of Ohm's Law is first examined. Then the various options for its expression is explored, followed by the implication of the empirical nature and the impact of quantum theory. Noise of the first kind will then be discussed as it is important for audio..

An Empirical Ohm's Law and Linearity

In 1827, Georg Ohm published his **law of proportionality between voltage and current**, where the voltage was controlled by varying the temperature of a thermocouple, the resistance was varied by a length of test wire, and the current was measured by a galvanometer. Nowadays, the **constant of proportionality is the resistance** of the circuit. Thus we have $E = I R$, where E is the voltage, I is the current, and R is the resistance.

Note that Ohm's Law is an **empirical** formula - it was arrived at empirically through extensive experimentation, and it is applicable *only* to those materials for which it applies. We called materials **ohmic** if they follow ohm's Law behavior over a reasonable common range of conditions. Some materials are **non-ohmic** under certain circumstances.

For ohmic materials under conditions where Ohm's Law holds, we have $E = I R$, which is a **linear** response. Those materials or elements are known as **linear circuit elements**. Specifically, linearity demands that R is independent of E or I, that is, the resistance is constant regardless of the voltage or current. We shall use the term **linear** and **linearity** quite a bit further in this book, and it simply means that a constant of proportionality can be maintained independent of the other parameters being controlled and observed. It is best visualized as a graph where the response can be diagramed as a straight **line** (which we shall do shortly). Most of audio electronics circuits are considered **linear circuits**.

Reconstructing the modern Ohm's Law

It is instructive to reconstruct Ohm's Law of the modern form based on Georg Ohm's original observations:
1) **Voltage drives current**, in other words, $I \propto E$ (this is a shorthand saying I is proportional to E)
2) **Resistance impedes current flow**, in other words, $I \propto 1/R$
3) Combining the two together yields $I \propto E/R$
4) We can further *define* resistance R so that the constant of proportionality in 3) is simply 1, thus yielding $I = E/R$, where R is independent of E or I
5) Rearranging to solve for E, we get $E = I R$, which tells us that **the potential drop E across a resistor R is produced by the current flow I**
6) Rearranging to solve for R, we get $R = E / I$ for ohmic devices, where **resistance is the ratio of E / I which is a constant independent of E**

Please re-read the above statements in bold; together they express various implications of Ohm's Law.

Other Famous Empirical Laws: - Boyle's Gas Laws - Kepler's Law of Planetary Motion	Non-empirical formulas used in this book: - practically everything else :)
The key point is that empirical laws work *only* under certain conditions, like "only for ideal gas" or "*not* for quantum-sized or universe-sized objects".	For example, the Power formula is true by **definition**! We need to go through this level of understanding *only* for Ohm's Law. Be of good cheer!

Various Ways to Express Ohm's Law

Recall what we learned in chapter 1, that relationships can be expressed as a formula, a graph, a table, or in words. It is instructive to do that *once* here, for the famous Ohm's Law. (We shall not repeat that feat again as it becomes tedious very quickly.)

Table

Voltage E	0	1	2	3
Current I	0	0.2	0.4	0.6

Graph

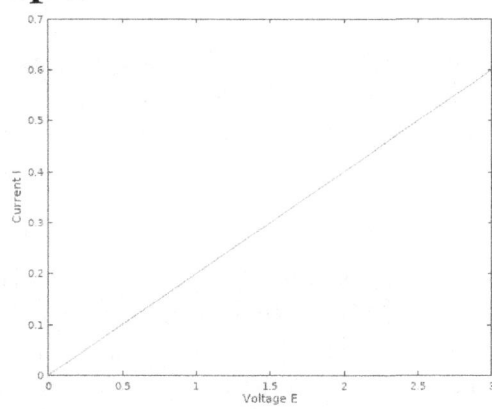

Formula

$I = E / R$
$E = I R$
$R = E / I$

Note that some people will accept *one* of the above as Ohm's Law, while others require *all three*. To be safe, you may consider learning all *three*.

Why some may say three?

3) Maybe your professor is sadistic. (Just kidding)

2) Because in practical applications you need to use all three.

1) Last but certainly not least, Ohm's Law is *empirical*. (I'm repeating because it is important.) Specifically, that means we read the equations *differently* from non-empirical laws. For example, having $I = E / R$ does *not* imply that $E = I R$ is also true. Thus all three are necessary to support Ohm's Law in words, and ditto in equation form.

In Words

Ohm's Law is an *empirical* law applicable to *ohmic* components only: voltage drives current; resistance impedes current flow; the current through a conductor between two points is directly proportional to the voltage across the two points, and inversely proportional to the resistance; the potential drop across a resistor is produced by the current flow; and resistance is the ratio of voltage and current, which is a constant independent of voltage and current.

Some students abhor formulas like nature abhors vacuum. I hope you can see that the formula approach is the most concise, the easiest to remember, and takes the least amount of time to write in an exam.

Non-Ohmic Expressions

There are a huge list of materials which do not obey Ohm's Law. Generalized categories:
- diodes (including LED)
- gas discharge and vacuum tubes
- electrolytes
- superconductors
- incandescent filament light bulbs
- thermistors

Likewise, Ohm's Law cannot be applied to a battery, or to non-linear devices in general, which may include active devices (transistors, MOSFETs, thyristors, etc.).

It is too difficult to express non-ohmic responses using equations, but it is instructive to show the responses as graphs:

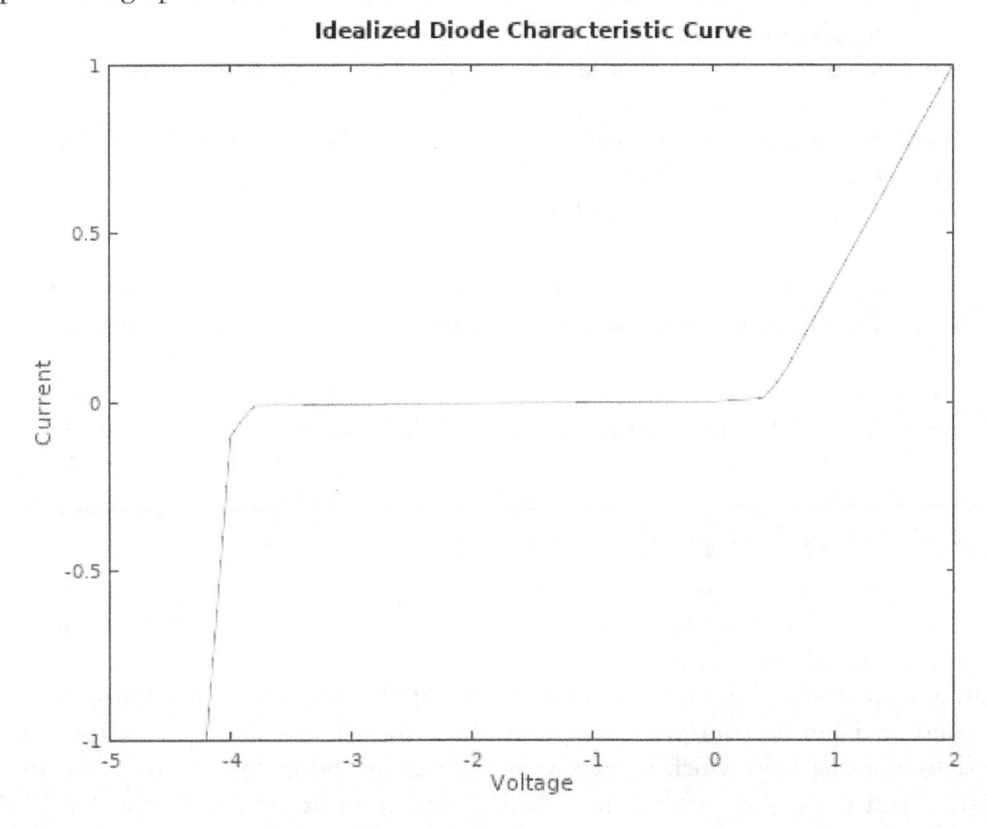

Idealized Diode Characteristic Curve

Here notice that starting at about 0.7 V, the current starts to vary linearly with the voltage applied. But below that, it does not conduct much (does not carry much current), until when it reaches about -4 V (for this particular diode) when it "avalanches" and a lot of current starts to flow (in the opposite direction). General small-signal diodes usually operate in the first region mentioned, while zener diodes and TVS diodes are designed to specifically operate in the avalanche region.

Regardless, the point is: the diode is an example of a non-ohmic device; it is *not* linear across a wide range of voltages - thus it is considered **non-linear** device.

♪ Noise of the First Kind - a Quantum Consequence

Because of the **discrete nature of charge** explained earlier when we presented quantum theory, and the probabilistic nature of it, the *instantaneous* current through a practical resistor actually has **statistical fluctuations**, which varies with temperature, even when voltage and resistance are kept constant, which Ohm's Law would have implied. This will show up in audio electronic circuits as noise. This was first measured by John B. Johnson at Bell Labs in 1926. He then described it to his colleague Harry Nyquist (of Nyquist frequency fame) who was able to explain it as thermal agitation of the charge carriers inside an electrical conductor regardless of *any* applied voltage, thus this thermal noise is frequently known as Johnson noise, Nyquist noise, or Johnson-Nyquist noise. In this book we shall simply refer to it as the **noise of the first kind**; other kinds of noise will be mentioned later on.

Noise of the first kind is **broadband**, where all frequencies are engaged. It can be described as approximately ***white*** noise, where the power spectral density is nearly constant throughout the frequency spectrum. This thermal noise will be averaged out over a longer period of time, thus ohm's law is accurate for current *averaged* over a long enough time.

It is important to note that noise *power* of the first kind is independent of applied voltage and current, thus it is also independent of resistance value. People often thought (incorrectly) that it is a sign of poor design or manufacture of resistors, but it should be noted that thermal noise is intrinsic to *all* resistors of *any* value. Noise power is dependent on temperature and frequency bandwidth, or $P \propto T \Delta f$. (And that means, when we apply Ohm's Law and power equation when we learn later, that noise *voltage* and *current* is dependent on resistance value.)

Examining the relationship above, we can see that we can reduce noise power by operating at a cooler temperature (at absolute zero, there will be no noise of the first kind), and by examining signal at a reduced frequency bandwidth.

We just said earlier that noise power is independent on resistance, but noise voltage is dependent. However, for an RC circuit, which we shall study later, the resistance value also drops out of the filter.

♪ Noise of the Second Kind - a flicker noise

In contrast, noise of the second kind is ***pink***. In other words, it is proportional to $1/f$, or as frequency increases, the noise decreases. It can be due to many factors; for resistors, it is mainly due to impurities in the conductive channel. It is always related to direct current.

Noise of the first kind is unavoidable, because it is independent of the material or resistance value. In contrast, noise of the second kind can be controlled, it is sometimes known as *excess* noise, where it is the excess on top of the noise of the first kind which is present in all resistors and is unavoidable. Specifically, carbon-composition resistors and thick-film resistors have high levels of noise of the second kind, while wire-wound resistors have the least amount, and thus is preferred in audio electronics sections where it matters most.

Flicker noise was first described by Aldert van der Ziel in the 1950s.

Other Controllable Noise types

Another kind of noise, called **shot noise**, is also a consequence of discrete quanta of charge carriers, but in this case caused by *fluctuations of the arrival times* of the charge carriers, especially when they are traversing a gap. In 1918 Walter Schottky (of Schottky diode fame) discovered vacuum tubes have shot noise because electrons randomly leave the cathode and arrive at the anode (plate); space charge tends to smooth out the arrival times and thus reduce the randomness of the current, reducing shot noise. In most other audio electronic devices, shot noise effect is less than noise of the first two kinds.

Burst noise, or **popcorn noise**, is another that may affect audio electronics, especially early op-amps.

3.5 FUNDAMENTAL NETWORK THEOREMS

There are a lot of network theorems. Surprisingly, most of them are "shortcuts." So do you want to know them? Most students would prefer to learn as little as possible, and so it would seem good *not* to learn them. But then why do they exists? Because people prefers to use short-cuts! So that is your dilemma - do you want to learn as little as possible, and in doing so end up with a lot of complications, or to learn shortcuts to eliminate the complexities, but in doing so has to spend time learning a little more?

While you are pondering that important decision, let us introduce the fundamental network theorems. These are needed regardless whether you learn shortcuts or not. In that sense they are the fundamentals. All other so-called network theorems are based on these fundamental theorems. (The presence of shortcuts of course implies these fundamental theorems are more difficult to understand and to use; so please take the time to understand them...)

1) Ohm's Laws
We had already studied the fundamental aspect of Ohm's Law.

2) Kirchhoff's Laws
Let us use this circuit as example to illustrate how to use Kirchhoff's Laws in network analysis:

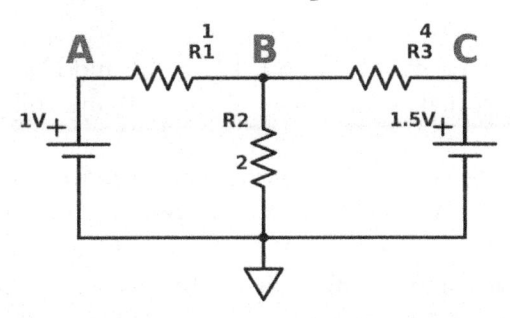

Kirchoff's laws require us to *first* identify **nodes** and **loops**. Node is like a branch on a road or a stream. This circuit has one non-trivial node B. The "ground" node would also be an obvious merging of streams, but it is usually considered a trivial node; as we can exercise our right *not* to handle one node, and that is usually the one taken. It has two small loops: the left one containing R1, R2 and the 1V battery; the right one containing R2, R3 and the 1.5V battery. There is also a bigger loop containing R1, R3, and the 1V and 1.5V batteries; as we can exercise our right *not* to handle one loop, and that is usually the one taken.

Notice that Kirchoff's Laws are plural. The **first law** is also called **Kirchhoff's Current Law**; the **second law** is also called **Kirchhoff's Voltage Law**. They are "duals" of one another. (See supplement on duality.)

Kirchoff's Current Law (KCL)

This is used often in derivations of formulas, and in such circumstances, the acronym KCL is often used.

> **Kirchoff's Current Law: the algebraic sum of currents entering and exiting a node must equal zero.**
> 1. Select a **node**.
> 2. **Label all currents** flowing in and out of that node with a direction. This can be arbitrary and does not need to reflect actual flow.
> 3. **Sum the currents to zero**, using a convention like all currents entering a node is positive, while all current leaving the node is negative. This **results in one equation**.

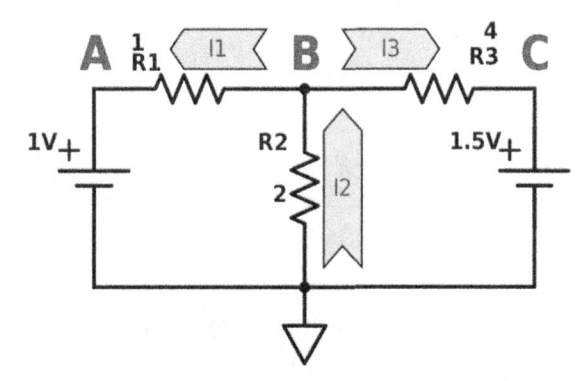

There is one non-trivial node B in the circuit, and we labeled it with I1 for R1, I2 for R2, and I3 for R3, all operating at node B. Note although the labeling is arbitrary, it makes sense too: I2 enters B and split into I1 and I3.

Using the convention mentioned, attach positive sign to I2 since it is entering, and negative signs to I1 and I3 since they are leaving, and sum them: $-I_1 + I_2 - I_3 = 0$. We got our first equation!

The above is how "left-brained" people likes to define it using a sum to zero. But for "right-brained" people, I suggest an alternative...

Alternative to the "convention": I usually *visualize* KCL with **current(s) entering = current(s) leaving**, or:

$$I_2 = I_1 + I_3$$

(Note we get the *same* result as before! Hopefully to some this alternative is less intimidating...)

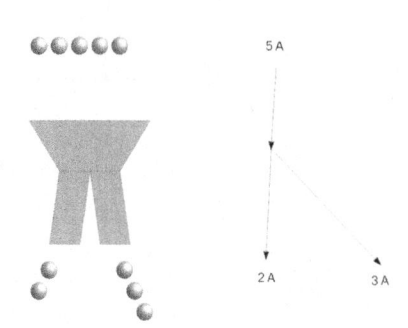

KCL can be illustrated as in the diagram on the left. Five marbles pass a funnel with two drains, two going to the left. How many marbles go to the right? Three of course, according to the conservation of mass.

Likewise, 5 A pass through a wire which is split into two. The left branch carries 2 A. How much current goes to the right branch? 3 A of course. Isn't it intuitive? We just applied KCL!

Kirchhoff's Voltage Law (KVL)

This is used often in derivations of formulas, and in such circumstances, often the acronym KVL is used. KVL is a "dual" of KCL, when we substitute voltage for current and loop for node.

> **Kirchoff's Voltage Law**: the **algebraic sum of voltages at a loop must equal zero**.
> 1. Select a **loop**.
> 2. **Label all voltages** operating within that loop with a direction. This can be arbitrary and does not need to reflect actual flow.
> 3. **Sum the voltages to zero**, using a convention like all voltages one way is positive, while all voltages the other way is negative. This **results in one equation**.

There are two loops in this circuit. First step is we select the *left* loop. Second step we already labeled the components in the left loop. Third step we sum them together, applying Ohm's Law to get the voltages:

$$I_1R_1 + I_2R_2 - V_1 = 0$$

Repeat using the *right* loop: $I_2R_2 + I_3R_3 - V_2 = 0$

Alternative to the "convention": I usually *visualize* KVL with **potential difference across two points in a loop = the same regardless which way (going left or right) you calculate the potential difference**, or:

Looking at the left loop, and the terminals across V1 battery in particular, we see:

$$V_1 = I_1R_1 + I_2R_2$$

Looking at the right loop, and the terminals across V2 battery in particular, we see:

$$V_2 = I_2R_2 + I_3R_3$$

(Note we get the *same* results as before! And to me this alternative is less intimidating…)

Many textbooks cover Kirchhoff's laws and their applications as late as possible, because that involves systems of equations and get complicated really fast. We also cover complicated examples at the end of this chapter. However, we introduce the concepts as early as possible, in the hopes that the reader will develop an intuition towards these fundamental theorems.

With that said, please ponder the material we covered in batteries in series earlier. How did we calculate them in series? With KVL of course! We deliberately use the definition of potential difference to enable us to calculate the potential differences in series, as it is intuitive, just to prove that KVL is intuitive.

Likewise, when we do voltage divider, the reader should think intuitively and visualize the whole thing as a KVL application.

3) Superposition Theorem

The **superposition principle** states that the net response at a given time and place caused by two or more stimuli is the sum of the responses that *would* have been caused by each stimulus individually. In plain words, that means we can evaluate only one stimulus (or signal) at a time, and then sum up all of responses together. It turns out this principle is valid for *all* linear systems.

In circuits, we apply this principle in the form of the **superposition theorem**: The total current (or voltage) in any part of a linear circuit equals the algebraic sum of the currents (or voltages) produced by each source separately. Of course we are assuming that the individual sources are not dependable on other sources (which would have complicated the problem). To evaluate the separate currents or voltages to be combined, replace all other voltage sources by short circuits and all other current sources by open circuits.

> The total current (or voltage) in any part of a bilateral linear circuit equals the *algebraic sum* of the currents (or voltages) produced by each independent source separately.

In other words, we "turn on" each independent sources **one at a time**, the rest are "turned off" (voltage sources replaced by short circuits, and current sources replaced by open circuits), and we sum all the responses *algebraically* to produce the total response. (Note again the emphasis on "algebraically".)

Using the example circuit shown earlier, we first do the analysis by turning on the 1 V and turning off the 1.5 V:

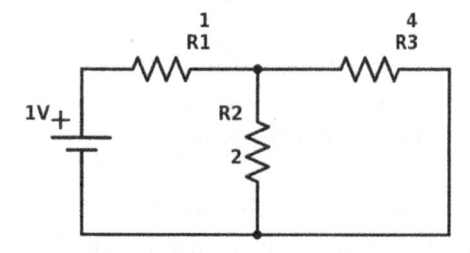

Next we turn off the 1 V and turn on the 1.5 V, as follows:

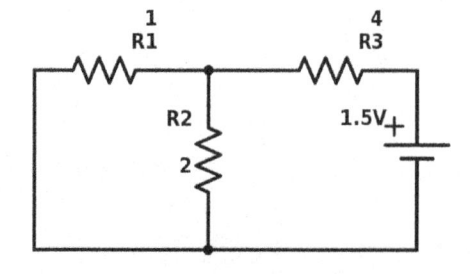

Then we sum the two responses together. This example will be tackled fully in section 3.13.

3-20

Chapter 3 – DC Analysis

3.6 Ohm's Law in Practice: an IER to an EIR

The mission, should you choose to accept, is to find the remaining one of the **three** (R, E, and I) when given two of them. E stands for Voltage; I stands for current; R stands for Resistance. This is a standard ratio formula application. Two common approaches are first described, and then examples given.

Algebraic Approach

Ohm's Law states:

| E = I R | (Ohm's Law) |

To find E, we simply substitute I and R into the formula.

To find I, we divide both sides by R so that I stands alone:
 E/R = I * R/R = I
Thus:

| I = E / R | (Alternative Form 2) |

To find R, we similarly divide both sides of the power equation by I so that R stands alone:
 E/I = I/I * R = R
Thus:

| R = E / I | (Alternative Form 3) |

Ohm's Law Memory Aid

Some call this the Ohm's Law Circle (when drawn as a circle instead of the outer rectangle shown here) or Triangle (when drawn as a triangle). I'd simply write the inner T and ignore the circle or triangle completely. The memory aid is defined thus:

E
I	R

Cover or circle or bracket the variable to be calculated, and the result will be exactly as visually shown.

To find E, cover it, and the remaining items are in a row:

[E]
I	R

Thus: E = I * R as indicated by the last row.

To find I, similarly cover I, and the remainder appears as a numerator E over denominator R:

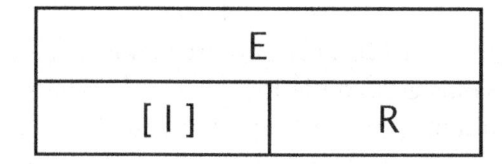

E
[I]	R

Thus: I = E/R as indicated by the fraction.

To find R, similarly cover R, and the remainder appears as a numerator E over denominator I:

E
I	[R]

Thus: R = E/I as indicated visually.

Unsurprisingly and without fanfare, the results are exactly the same as using the algebraic approach.

Examples
1) R = 5 Ω, I = 1 A. Find E.
Using Ohm's law directly, E = R * I = 5 * 1 = 5 V.
Notice that the unit of voltage (V) is given at the end. **Never forget to supply the unit!**

2) E = 100 mV, R = 10 mΩ. Find I.
Using the ratio memory aid, we have:

E = 100 mV
[I]	R = 10 mΩ

Covering I, we obtain: I = E/R = 100 m / 10 m = 10 A.

3) Let's assume a soldering iron with a current of 10 A running at a voltage of 100 V. Find the resistance of the soldering iron, assuming Ohm's law holds.
Using the algebraic approach, E = 100 V, I = 10 A. Find R. Divide both sides by I, E = I * R leads to E/I = I/I * R, Thus R = E/I = 100/10 = 10 Ω.

Chapter 3 – DC Analysis

3.7 Power Equation in Practice: to PEI or not PIE

The mission, should you choose to accept, is to find the remaining one of **three** (P, E, and I) when given two of them. P stands for Power; E stands for Voltage; I stands for current. This is a standard ratio formula application. Two common approaches are first described, and then examples given.

Algebraic Approach

The power equation states:

$$P = E I \quad \text{(Power Equation)}$$

To find P, we simply substitute E and I into the formula.

To find E, we divide both sides by I so that E stands alone:
  P/I = E * I/I = E
Thus:

$$E = P / I \quad \text{(Alternative Form 2)}$$

To find I, we similarly divide both sides of the power equation by E so that I stands alone:
  P/E = E/E * I = I
Thus:

$$I = P / E \quad \text{(Alternative Form 3)}$$

Power Equation Memory Aid

Some call this the Ohm's Law (when drawn as a circle instead of the outer rectangle shown here) or Triangle (when drawn as a triangle). The memory aid is defined thus:

P
E	I

Cover or circle or bracket the variable to be calculated, and the result will be exactly as visually shown.

To find P, cover it, and the remaining items are in a row:

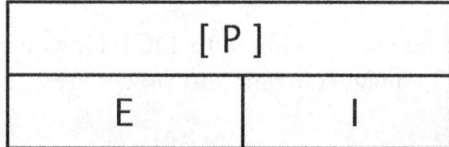

[P]
E	I

Thus: P = E * I as indicated by the last row.

3-23

To find E, similarly cover E, and the remainder appears as a numerator P over denominator I:

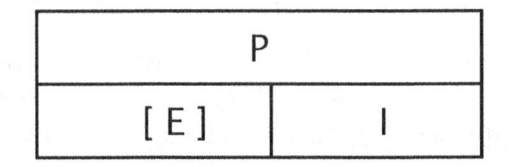

P	
[E]	I

Thus: E = P/I as indicated by the fraction.

To find I, similarly cover I, and the remainder appears as a numerator P over denominator E:

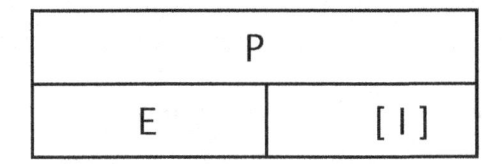

P	
E	[I]

Thus: I = P/E as indicated visually.

Unsurprisingly and without fanfare, the results are exactly the same as using the algebraic approach.

Examples
1) E = 5 V, I = 1 A. Find P.
Using the power law directly, P = E * I = 5 * 1 = 5 W.
Notice that the unit of watts (W) is given at the end. Never forget to supply the unit!

2) P = 100 mW, I = 10 mA. Find E.
Using the ratio memory aid, we have:

P = 100 mW	
[E]	I = 10 mA

Covering E, we obtain: E = P/I = 100 m / 10 m = 10 V.
Notice I used "m" instead of the scientific notation of 10^{-3} which is more cumbersome, since the two cancels out one another anyway. If they don't cancel, then it is better to put in the scientific notation to ascertain no clerical mistakes in the calculation.

3) Let's assume a DC soldering iron of 10 W at a voltage of 100 V. Find current, assuming the power law holds.
Using the algebraic approach, P = 10 W, E = 100 V. Find I. Divide both sides by E, P = E * I leads to P/E = E/E * I, Thus I = P/E = 10/100 = 0.1 A.

In case you wonder, so far we are working with DC. However, the power equation and ohm's law also works for AC when "effective" voltage, current and power are used. We'll discuss RMS further in the next chapter.

3.8 DOUBLE THE LAWS, DOUBLE THE FUN: PEIR ON A PIER

This more advanced mission, should you choose to accept, is to find the remaining one of the **four** (P, E, I and R) when given two of them. P stands for Power; E stands for Voltage; I stands for current; R stands for Resistance. Three approaches are described, with examples given after each one..

Gilbert's PIER (or PEIR) Method has no peers

The formulation is as follows; first draw the PEIR:

Power equation			
P	E	I	R
	Ohm's Law		

Then put in the values of the two known variables. Then circle variables to be calculated. There are several cases to consider:
1) All three are within the power equation range: you got it - follow the power equation approach in 3.7.
2) All three are within the Ohm's Law range: you got it - follow the Ohm's Law approach in 3.6.
3) All three are NOT within either range; however, the two given variables are all within one of the ranges: you will first apply the formula in the range the two variables are given, to get the variable you were *not* asked to calculate. Once that is done, you'll then have another variable which will allow you to use the other law to finally get to the variable you are to obtain. Yes, this is a two-step process, involving two laws. This visual approach makes it clear that that's exactly what you need to do.
4) Only one variable is given in each range - in other words, you were given P and R. In this case, just remember these two equations to find E and I:

$I = \sqrt{(P/R)}$
$E = \sqrt{(P*R)}$

These two formulas will be derived in the algebraic approach following, at the end of this section.

As you can see from the visual, you can remember the power formula as PEI and ohm's law as EIR.

If you don't like the word PEIR, you can substitute the word PIER, as in the foundation, like this:

Power equation			
P	I	E	R
	Ohm's Law		

In this case, you can remember power equation as easy as PIE, and lend an IER to Ohm's Law.

Examples

1) P = 100 mW, I = 10 mA. Find E.

P = 100 mW	[E]	I = 10 mA	R

This fits the power formula, and so E = P/I = 100 m / 10 m = 10 V.

2) E = 100 mV, R = 10 mΩ. Find I.

P	E	[I]	R

This fits Ohm's law, and so I = E/R = 100m/10m = 10 A

3) P = 100 mW, E = 10 V. Find R.

P = 100 mW	E = 10 V	I	[R]

In this case, the three parameters do not fit either Ohm's or power law. Since we have 2 parameters fitting power law, we first apply it to obtain the missing parameter I. Thus I = P/E = 100m/10 = 10 mA. Then we have two parameters available for Ohm's law: R = E/I = 10/10m = 1 kΩ. Notice this is the case where we have to apply *both* formulas.

4) P = 10 mW, R = 100 Ω. Find E.

P = 10 mW	[E]	I	R = 100 Ω

This is the case where we are given the two extreme blocks, so we just use the memorized formula:
E = √(P*R) = √(10m*100) = √1 = 1 V

The DC Formula Wheel

If you have this book with you, then applying the formula wheel is the easiest. It is a simple matter of looking up the wheel to determine which formula to apply. For those with excellent memory, this is also the approach to take. Many students taking the GCSE exam in the British Commonwealth (equivalent to a high school matriculation test) are known to memorize the whole wheel and write it down before a DC analysis question is to be answered.

Here is the formula wheel (OK, my formulation is slightly unique; I am a little "retro" making the wheel into a rectangle):

$P = E * I$	$P = E^2 / R$	$E = P/I$	$E = \sqrt{(P*R)}$
$P = I^2 R$	P	E	$E = I * R$
$I = P/E$	I	R	$R = E^2 / P$
$I = \sqrt{(P/R)}$	$I = E/R$	$R = P/I^2$	$R = E/I$

The wheel is constructed as follows: All cells from the upper left starts with P=; all cells on the upper right, E=; lower left, I=; lower right, R=. The two variables on the other side of the equation is picked from the closest *two* variables of PEIR. And then of course we have to derive the actual equation either using Gilbert's PIER method of the algebraic method given next. This is a quick analysis of the cells of this rectangular formula wheel. Three cells for power law; three cells for Ohm's Law; four cells for a combination, which means that squares are the result; and then two special cases that calls for memorization in Gilbert's PIER method, where square roots are the result, also using both laws.

Power law	Both² laws	Power law	Both √ Laws
Both² laws	P	E	Ohm's Law
Power law	I	R	Both² laws
Both √ Laws	Ohm's Law	Both² laws	Ohm's Law

Example

P = 10 mW, R = 100 Ω. Find E.

We have P and R, wishing to find E. We look up the wheel towards the direction of E and see the formula as $E = \sqrt{(P*R)}$, thus we have $E = \sqrt{(P*R)} = \sqrt{(10m*100)} = \sqrt{1} = 1$ V

Algebraic Approach

If you are good at algebra, this approach can always give you results, albeit it may take extra time. So in a sense this is the last resort if you are allowed to use any approach to solve the problem.

Examples

Let's derive the formulas for case 4 in Gilbert's PIER approach earlier. To re-cap, we are given P and R, and required to find E and I.

We know $E = I * R$, ... (1)

and $P = E * I$ ..... (2)

Let us find I first. Substituting (1) into (2), we get $P = (I * R) * I = I^2 * R$.

Thus, $I^2 = P/R$, and $I = \sqrt{(P/R)}$.

Let us now find E. From (1), $I = E/R$. Substituting into (2), we get $P = E * (E/R) = E^2/R$.

Thus, $E^2 = P * R$, and $E = \sqrt{(P*R)}$.

Likewise we can derive all the formulas shown in the formula wheel, which will be left to the reader.

Suggestion

1) If you can look up the formula wheel, then use the formula wheel. That is the foolproof method, unlikely to make clerical errors during an exam.
2) If not, but you have perfect memory, also use the formula wheel, for the same reason that it is least likely to fall into the trap of clerical errors.
3) If your memory is not that good, as in the case of Gilbert, he has created an approach that requires far less memorization. Just memorize how to do Ohm's Law and power equation, and the PIER or PEIR method, plus the two additional formulas. (Please note that most of us believe our memory is good, but during duress of exam, most of us we behave differently.)
4) If you can't even do that, then you'll have to brush up and practice your algebraic skills. Remember, under exam conditions and panic modes, the algebraic approach may get you stuck easily. If you have to depend on that approach, be sure to practice enough to make it perfect.

Two PIERs

There are occasions when you may need to apply the PIER approach twice. Simply carefully set out the PIERs, record what are known and what are to be calculated, and then find a correlation point between the two PIERs, solving one PIER after another. Remember there are no short-cuts. So don't guess for it, as most likely you'll get a wrong answer by attempting any short-cuts. Rather, work consistently and carefully, using the visual aids to help reduce chances of clerical errors. This approach is best illustrated with an example.

Example

I connected a cell phone to a USB charger which is plugged in to the cigarette lighter socket in a car. The cell phone is drawing 2 A at 5 V; how much power is it drawing from the USB charger? The charger itself is drawing that same power from the car battery at 12 V; how much current was the USB charger drawing from the battery?

First, identify the two problems using PEIR.
Cell Phone:

	Power Equation		
[P]	E = 5V	I = 2A	R

USB Charger:

	Power Equation		
P = above	E = 12V	[I]	R

Both PEIRs use the power equation. The first one gives P = E * I = 5 * 2 = 10 W.

Plugging in 10 W into the second PEIR based on the question (which by the way assumes the efficiency of power conversion is at 100% for simplicity's sake), we get I = P/E = 10/12 = 5/6 A.

Notice that I have provided the answer as a fraction rather than a recurring decimal. Use whatever your boss or professor requires.

3.9 Series & Parallel Topologies: Be Simplified

The reason why we study unique topologies is *not* because we want complications, but rather because we want short-cuts and simplified workflow. Interestingly, simplicity in real life often occurs because we are willing to accept *necessary* complexity. (In our case, the necessary complexity is that we need to recognize four standard topologies.) It turns out that a supermajority of circuit problems we run into in audio electronics revolves around these four topologies, and solving problems using these topologies are easy, compared with using more general techniques like Kirchhoff's Laws which are far more complicated and take much more time and effort. (It may be interesting to note that the formulas given for these four topologies can be derived from Kirchhoff's Laws. So we are perfectly safe to go there.)

In other words, we are using visual identification and pattern recognition, part of language skills, to help us navigate circuit analysis quickly and effectively.

By the way, if you follow the duality principle earlier, series and parallel are duals when we consider resistance and conductance as duals.

Series Topology

So let us dive right in. Resistors (and in the general case, impedances which we will study in future chapters) in series can be recognized and represented thus:

The formula for the equivalent resistance (or impedance) of a series circuit is simple: it is the sum of all the resistances. One way to represent it is:

$R_s = R_1 + R_2 + ...$	(Resistances in series)
$Z_s = Z_1 + Z_2 + ...$	(Impedances in series)

Example

Two 4 ohm loudspeakers connected in series is equivalent to what load?

The equivalent load is simply the sum of the two, which is $4 + 4 = 8 \ \Omega$. (*This is DC-style analysis.*)

Parallel Topology

Resistors (and in the general case, impedances which we will study in future chapters) in series can be recognized and represented thus:

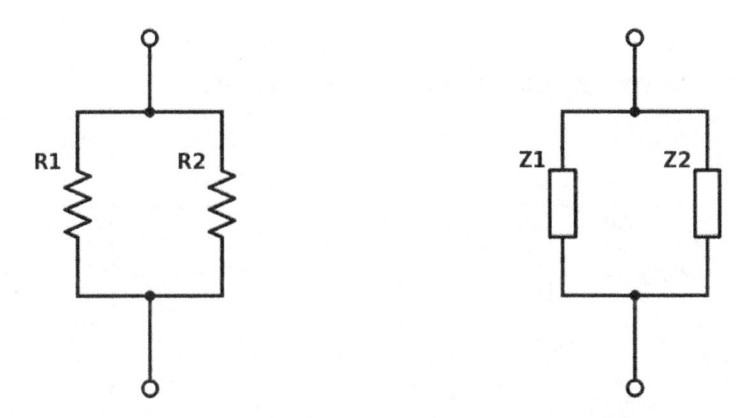

Remember how we defined conductance? We have a similar generalized concept called admittance:

$G = 1/R$	(Definition of Conductance)
$Y = 1/Z$	(Definition of Admittance)

The formula for the equivalent conductance (or admittance) of a parallel circuit is simple: it is the sum of all the conductances (or admittances). One way to represent it is:

$G_p = G_1 + G_2 + ...$	(Resistances in parallel)
$Y_p = Y_1 + Y_2 + ...$	(Impedances in parallel)

Working with resistance values, then the formula becomes, after applying the conductance values:

$/R_p = 1/R_1 + 1/R_2 + ...$	(Resistances in parallel)
$1/Z_p = 1/Z_1 + 1/Z_2 + ...$	(Impedances in parallel)

Example

Two 4 ohm loudspeakers connected in parallel is equivalent to what load?
Using the resistance formula, the equivalent load is simply the reciprocal of sum of the two reciprocals:
$1/((1/4) + (1/4)) = 1/(2/4) = 1/(½) = 2\ \Omega$.

The inverse of an inverse may be confusing to some students, many of which may commit clerical errors and may miss the correct answer. Let us instead compute using the conductance or admittance approach.

The admittance of each loudspeaker is ¼ = 0.25
The admittance of the load = The sum of the admittances = 0.25 + 0.25 = 0.5
Thus the impedance of the load = 1/0.5 = 2
If the last step is done (or checked) with a calculator, clerical error is more than likely avoided.

Network Simplification

We can use the series and parallel formulas in sequence to simplify a network into a single equivalent impedance.

Example
What is the equivalent circuit for the network shown below at upper left?

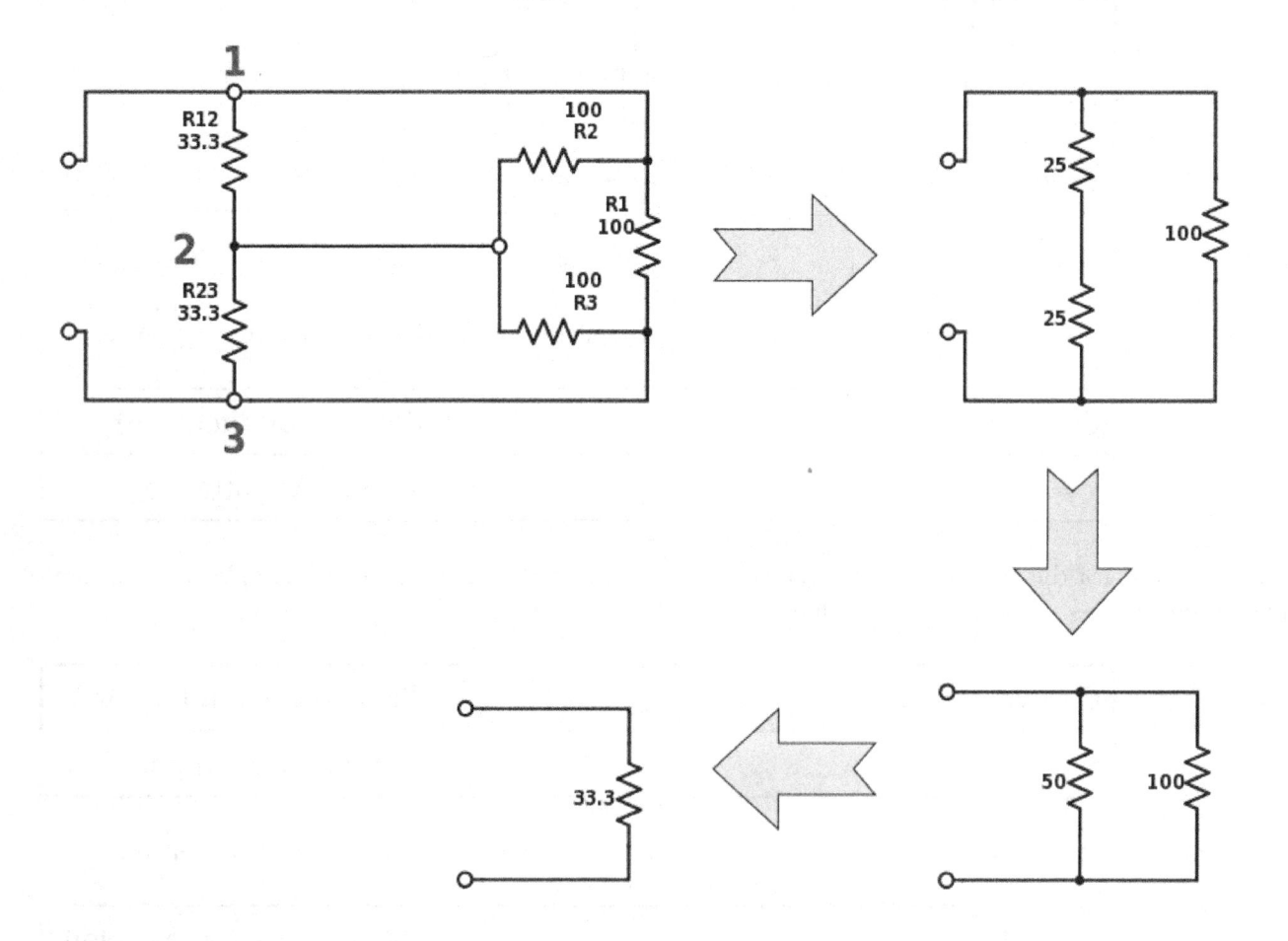

We basically would be looking for opportunities - anywhere in the circuit where we can apply the series or parallel transforms. Notice that R12 and R2 are in parallel, and so are R23 and R3. Not only that, but they mirror each other, so we can do the calculations *once* for both networks!
33.3 || 100 = 1/ (1/33.3) + (1/100) = 25 Ω

Thus we obtain the network on the upper right, with the 25 Ω equivalent resistances shown. Then we notice the two 25 Ω equivalent resistances are in series, so applying the series transform we obtain the network on the lower right, with the equivalent resistance being 25+25 = 50 Ω.

Then we notice that the result is a parallel network, so applying the parallel transform, we obtain the final equialent circuit on the lower left: 50 || 100 = 1/ (1/50) + (1/100) = 33.3 Ω

You can see from this example that the series-parallel transforms are very powerful - learn them well!.

3.10 Voltage and Current Divider Topologies: Be United

I'll cut to the chase: the voltage divider topology is based on the series topology, and the current divider topology is based on the parallel topology (as we can also deduce from duality). The difference lies in the *viewpoint* and *usage*:
- for series and parallel topologies we calculate the equivalent **impedance** (or resistance),
- for voltage and current dividers we calculate the **voltage or current ratio**.

Voltage Divider

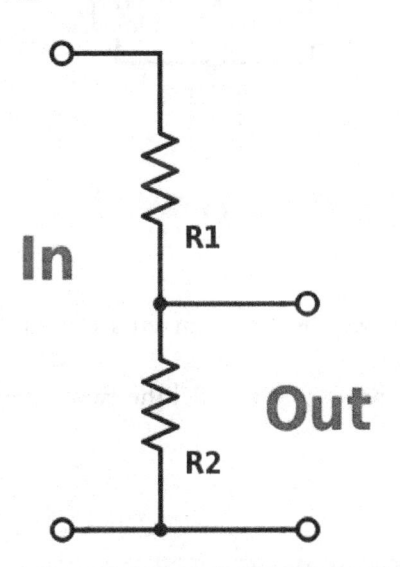

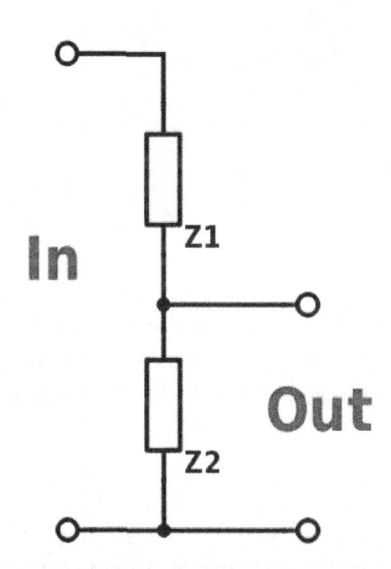

Notice again the circuit is the *same* as the series circuit. The equivalent impedance would be R1+R2 and Z1+Z2 respectively. Let us assume that current I flows through the series circuit. Then applying Ohm's law, we know the input voltage: In = I · (R1+R2) and In = I · (Z1+Z2) respectively.

Since the same current I flows through R2 and Z2, we can apply Ohm's law to that part of the circuit to obtain the output voltage: Out = I · R2 and Out = I · Z2 respectively.

To find the **voltage ratio**, which we usually call the **transfer function**, we get:
   Out/In = (I · R2) / [I · (R1+R2)] = R2 / (R1+R2)
   Similarly, Out/In = (I · Z2) / [I · (Z1+Z2)] = Z2 / (Z1+Z2)
Other than Ohm's law and power law, this will be the formula most commonly used in audio electronics. Thus it is essential to get that committed by heart.

| Out/In = R2 / (R1+R2) | (Voltage Divider, Resistance Formula) |
| Out/In = Z2 / (Z1+Z2) | (Voltage Divider, Impedance Formula) |

Example
One volt is applied to two equal resistors in series. What is the voltage on *one* of them?
   Assume resistors have value R. Using the formula, Out = In · R2 / (R1+R2) = In · R / (R+R) = 1 · ½ = ½ V

Current Divider

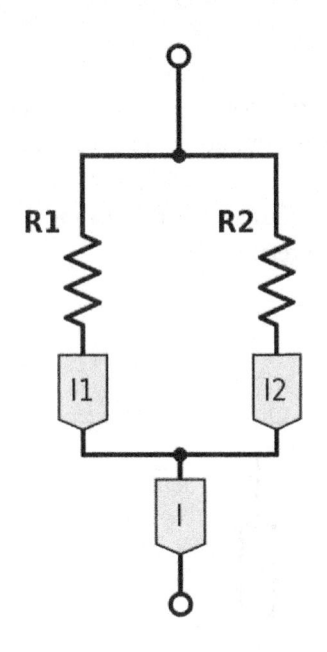

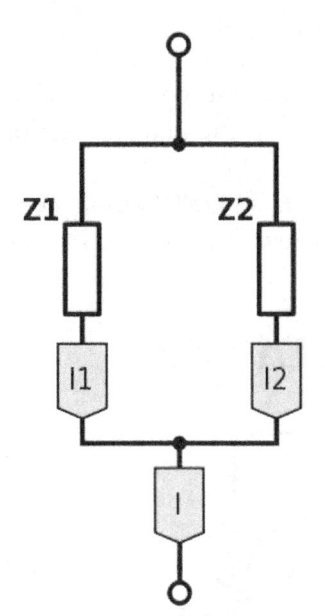

The current divider topology is the same as the parallel, except here we show the currents I1 and I2 flowing through R1 and R2 or Z1 and Z2 respectively. I is the total current.

The voltage E across R1 is I1R1, and across R2 is I2R2. Since they are in parallel the two voltages **must** be the same.
I1 = E/R1
I2 = E/R2
The total current = I1 + I2 = E/R1 + E/R2
Current ratio I2/I = (E/R2) / (E/R1 + E/R2) = (1/R2) / (1/R1 + 1/R2)

If we use conductance the formula will look much easier:
I1 = E · G1
I2 = E · G2
The total current = I1 + I2 = E · G1 + E · G2
Current ratio I2/I = (E · G2) / (E · G1 + E · G2) = G2 / (G1 + G2)

Thus we arrive at a formula very similar to that of the voltage divider, except it is a current ratio for conductances applying to parallel circuits, versus voltage ratio for resistors applying to series circuits. Duality in action.

Out/In = G2 / (G1+G2)	(Current Divider, Conductance Formula)
Out/In = Y2 / (Y1+Y2)	(Current Divider, Admittance Formula)

Gilbert's Visual Method

This method is so intuitive that we may as well try a more complicated example to illustrate its value...

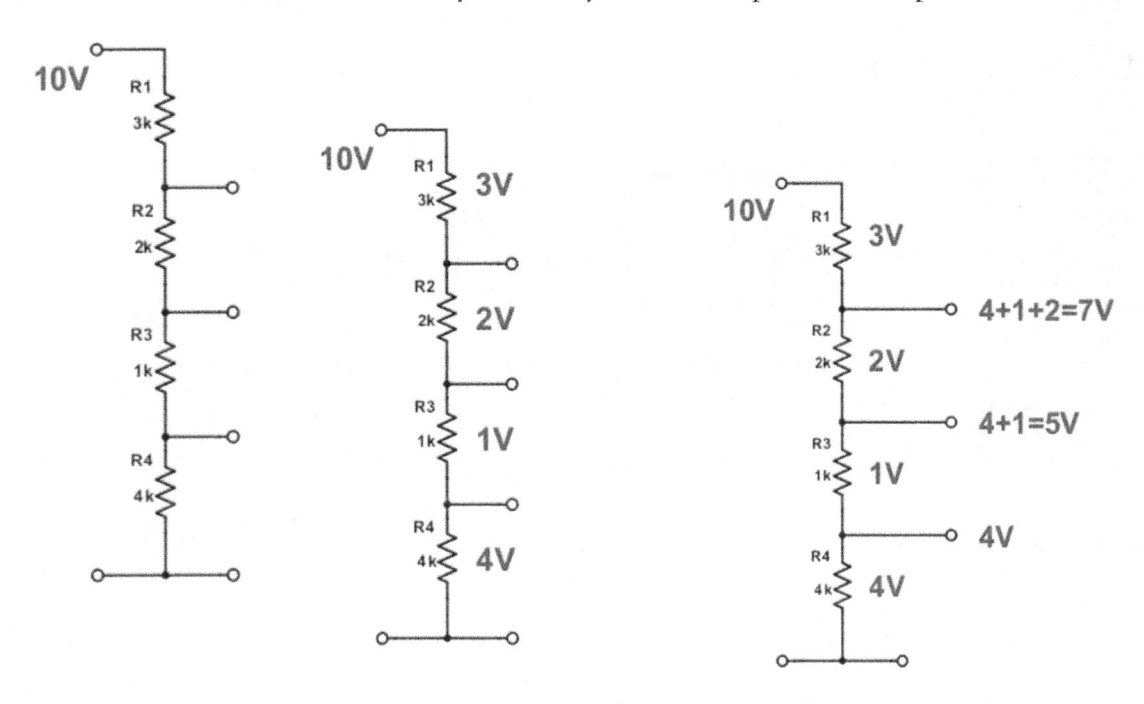

How do we calculate the voltages in the leftmost circuit? *First*, we need to recognize this as a voltage divider. *Second*, we sum the resistances, which is 3k + 2k + 1k + 4k = 10k. *Third*, we divide the voltage per unit resistance - this is where the term voltage-divider comes into place. In other words, we have 10V divided into 10k, or 10V/10k or simply 1V/1k. That means for each 1k resistance we allocate 1V of voltage. *Fourth*, we allocate such potential differences across the individual resistors: R1 has 3k and so gets 3V, R2 has 2k and so gets 2V, R3 has 1 k and so gets 1V, and R4 has 4k and so gets 4V. We label the circuit with such voltages, as shown in the middle diagram. Note that we had divided or distributed the potential difference of 10V across the resistors, and such voltages are also potential differences *across the individual resistors*. If we want to calculate the voltages across the terminals, we simply add the voltages in series. As shown in the rightmost diagram, the voltage at the top of R4 is simply the potential difference across R4, which is 4V. The voltage at the top of R3 is a summation of the potential differences across R4 and R3, which is 4+1 = 5V. And the voltage at the top of R2 is 4+1+2 = 7V. And just as a double-check, the voltage at the top of R1 should be 4+1+2+3 = 10V, which is the same as the supply voltage.

If you think this is simply intellectual exercise and that you'll never encounter such in practice, think again. This is actually one of the most commonly encountered circuit in audio electronics! It is used in biasing of active circuits. We'll be performing this type of calculations many times, so we may as well practice it and do it well. For such a simple example as above, notice that some of us could have done all of the calculations in our head!

Now let's turn to another example that occurs commonly, this time a dual power supply is used. A **dual power supply** is one which supplies two voltages, typically +Vcc and -Vcc, both referenced to ground.

Example

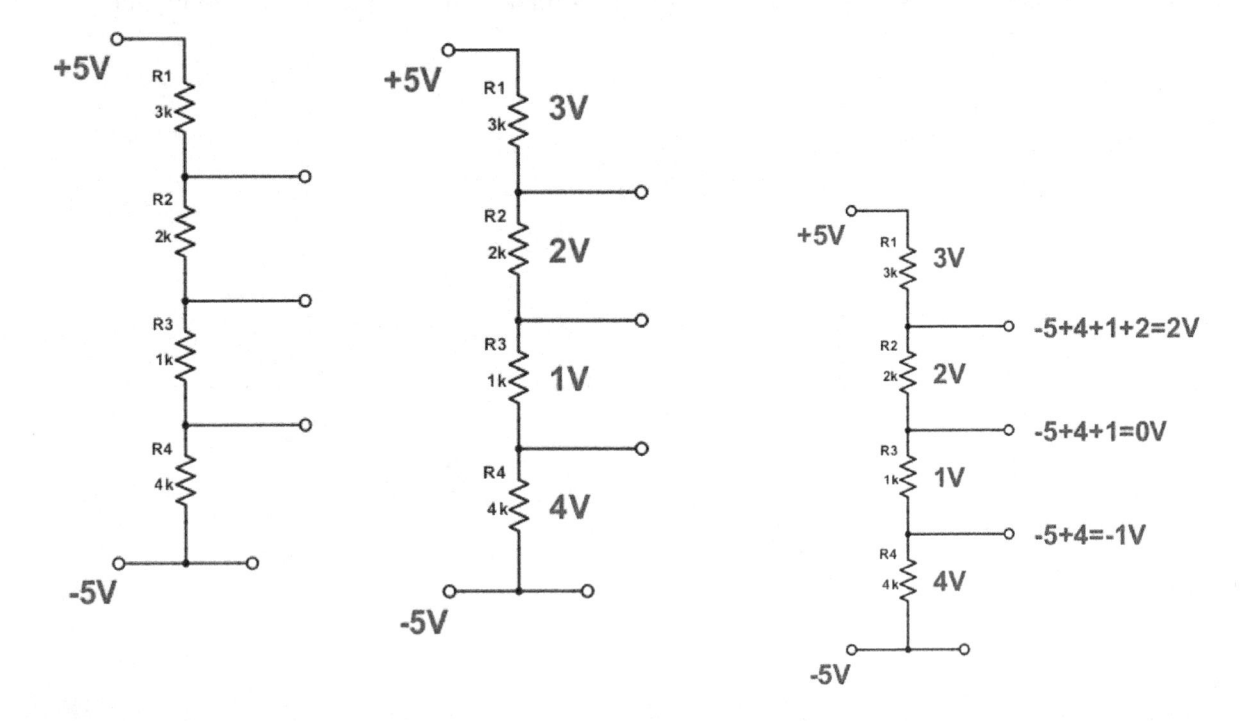

Notice the circuit is the same as before, other than that instead of a 10V single power supply, we have a +5/-5 V dual power supply.

The potential difference between the two power supplies are +5 - (-5) = 10 V, which is the same as before. So we can divide the potential difference of 10V the same way as before across the resistors, as shown in the middle diagram.

Then in the final step when we calculate the actual potential, instead of adding from the previous bottom which is ground or 0 V potential, we now add from the bottom of -5 V. Thus all the calculations start with -5V, as shown in the rightmost diagram. The answer for the three terminals would be -1 V, 0 V, and +2 V respectively. As a double check, we can also total up to the top power supply: -5 + 4 + 1 + 2 + 3 = +5 V, exactly as the top power supply.

Note that the same circuit can be analyzed using KVL and Ohm's Law, and in fact that is how we derive the formula. The equivalent circuit is one resistor R = R1 + R2 + R3 + R4 = 3k + 2k + 1k + 4k = 10 kΩ, and so the current flowing through the equivalent circuit is I = (5 - (-5))/10 k = 10/10k = 1 mA. That is also the current flowing through the original circuit, due to equivalence. So the potential difference across each resistors are IR1, IR2, IR3, and IR4 respectively, or 1m*3k = 3 V, 1m*2k = 2 V, 1m*1k = 1 V, and 1m*4k = 4 V, *exactly* as visualized above. Finally we use KVL one more time to calculate the actual voltages at the terminals as indicated earlier, adding potential differences to get the final answer.

3.11 Combining Series-Parallel and Dividers: A Wall of PIER or PEIR

When an analysis involves combining series-parallel and voltage-current divider topologies, it is often useful to use another aid: by filling in a table of P-E-I-R versus each of the circuit elements and equivalent circuit elements. It is basically an extension of the basic Gilbert's PEIR approach as we had discussed previously, making it into a **wall of PEIR**. An example is best to illustrate the approach.

Example

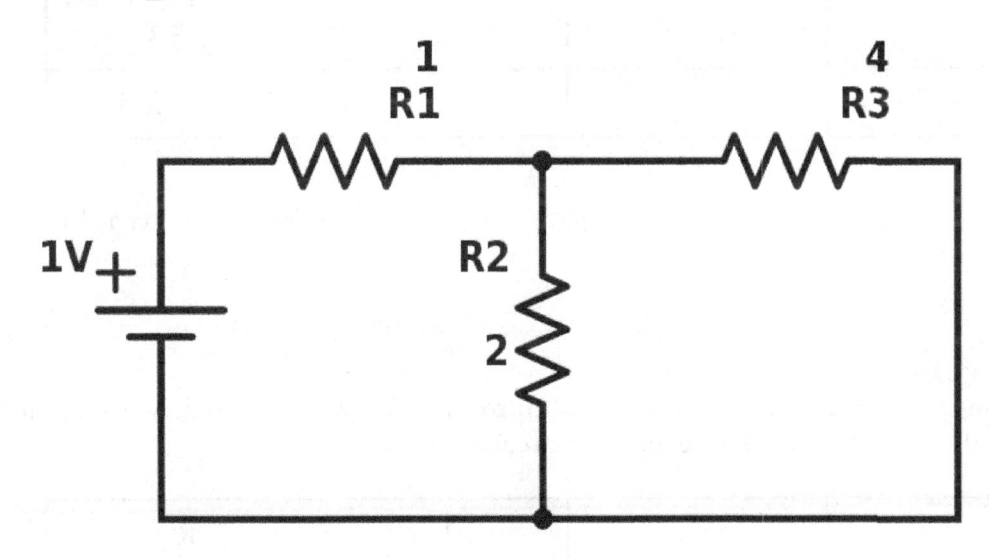

How do we find all the voltages and currents in this circuit?

Step 1 - Listing What's Known

First we create the basic elements of Gilbert's Wall of PEIR: four columns labeled P, E, I, R (or PIER), and then on the left the column we list all the elements in the circuit, then we list already-known values:

	P	E	I	R
R1				1
R2				2
R3				4

Step 2 - Series-Parallel Transforms

Next we go through the steps of finding the equivalent resistance, adding rows as we go along: R2||R3 is the parallel topology, then it is in series with R1, yielding R1 + (R2||R3) as the final answer:

	P	E	I	R
R1				1
R2				2
R3				4
R2‖R3				1/[(1/R2)+(1/R3)] = 1.33
R1 + (R2‖R3)		1		R1 + ↑ = 2.33

I used ↑ represent what was there in the cell above. With the completion of the total equivalent circuit, I can also notate E=1 with it.

Step 3 - Initial Current

Then we can proceed using Gilbert's PEIR approach - **look for rows with two cells filled**, which occurs at the last row, and calculate the current through the equivalent circuit.

	P	E	I	R
R1				1
R2				2
R3				4
R2‖R3				1/[(1/R2)+(1/R3)] = 1.33
R1 + (R2‖R3)		1	←/→ = 0.429	R1 + ↑ = 2.33

3-38

Step 4 - Continue Network Analysis on Current, Divider Transforms

Then we know that current is the same current through R1, or R2||R3 as well:

	P	E	I	R
R1			↓ = 0.429	1
R2				2
R3				4
R2‖R3			↓ = 0.429	1/[(1/R2)+(1/R3)] = 1.33
R1 + (R2‖R3)		1	←/→ = 0.429	R1 + ↑ = 2.33

From that we can apply the current divider formula to get the current flowing through R2 and R3:
 I2 = I · (½)/(½)+(¼) = 0.429 · (2/3) = 0.286 A
 I3 = I · (¼)/(½)+(¼) = 0.429 · (1/3) = 0.143 A

	P	E	I	R
R1			↓ = 0.429	1
R2			I2 = 0.286	2
R3			I3 = 0.143	4
R2‖R3			↓ = 0.429	1/[(1/R2)+(1/R3)] = 1.33
R1 + (R2‖R3)		1	←/→ = 0.429	R1 + ↑ = 2.33

The equations for I2 and I3 are left with the table, thus making clear documentation of the process.

Step 5 - Apply Ohm's Law

With currents known, we can calculate all the voltages using Ohm's law, E=IR. Basically we scan every row again to see where there are two elements filled and then complete the calculations:

	P	E	I	R
R1		I1R1 = 0.429	↓ = 0.429	1
R2		I2R2 = 0.572	I2 = 0.286	2
R3		I3R3 = 0.572	I3 = 0.143	4
Rx=R2||R3		IxRx = 0.572	↓ = 0.429	1/[(1/R2)+(1/R3)] = 1.33
R1 + (R2||R3)		1	←/→ = 0.429	R1 + ↑ = 2.33

$I2 = I \cdot (½)/(½)+(¼) = 0.429 \cdot (2/3) = 0.286$ A
$I3 = I \cdot (¼)/(½)+(¼) = 0.429 \cdot (1/3) = 0.143$ A

I introduced Rx notation here to simplify the table.

Note that E2=E3 which should be the case since R2 and R3 are in parallel. Likewise they should be the same as their lumped circuit of R2||R3, which is the case. This is a good check.

To learn a consistent technique, after all the voltages and currents have been calculated, and summed if using the superposition theorem (explained in next section), *then* power can be calculated from E, I, R. Do not calculate power earlier, as it may mess up everything.

In sum:
1) We begin by performing **series-parallel** equivalent circuits until we compute all impedance values.
2) We then compute either all currents or voltages, whichever is easier, using the **divider** approach. In this case, it is the current divider.
3) We then compute the remaining entries, in this case voltages, through **Ohm's Law**.
4) We may add up voltages and currents through **superposition** (next section) as necessary.
5) For a final step we then compute the **power dissipation** for individual elements. (In other words, we go backwards from steps based on the letter P-E-I-R or P-I-E-R, starting from R...)

One may complain that this is tedious. But what is the alternative? We will later be shown Kirchoff's laws, which is the alternative usually bandied around by others. But that requires solving a system of simultaneous equations, and the math is even *more* tedious... Most students saw Kirchoff and left, so they never learned network analysis; while with this method based on series/parallel and dividers, they may have a chance...

In other words, *this method* is probably the easiest and fastest shortcut known to engineers to obtain a correct answer. We just need to set our reference for comparison to realize its benefits. (We need to resist coming up with *even faster* methods, as that probably will give wrong answers!) Fortunately you'll probably *not* be asked to do such a problem in your exams, unless you have a long time limit for the exam. Still it is important to understand the process, and how having a systematic method (using the Wall of PIER) makes it more structured and more tolerable, with potentially fewer mistakes along the way.

3.12 Star, Delta, and Bridge Topologies: Reaching for the Moon...

The four topologies presented so far covers the majority of what we'd normally see in audio electronics - and those are what we'll concentrate on in this book. We now cover remaining likely to be encountered occasionally in the rest of the cases in audio electronics.

T-Π or Star-Delta Transformation

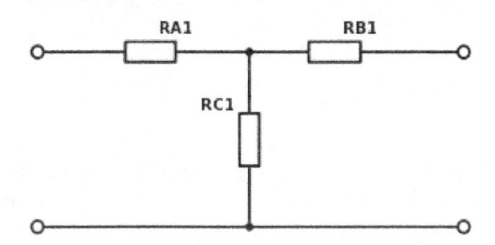

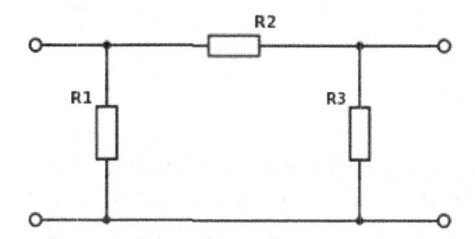

On the left is what we call a T- or Y- (or Wye-) or star- circuit. The T should be obvious; the Y can be visualized when we draw it like a Y, with RA1 and RB1 raised up slightly. In that way it also looks like a star.

On the right is what we call a Π (or spelled out as pi) or delta- circuit or mesh or triangle. The delta or triangle can be visualized when we draw it with the lower terminals of R1 and R2 converged to one point.

You just need to **learn to visually identify them** by recognizing their special shapes, and realize that they have to be treated differently from the other topologies.

After recognizing the topology, if the elements cannot work out using the previously known methods, then we can execute a **T-Π or Star-Delta Transformation** - *that* should then allow us to continue our analysis. That is the concept.

From Π to T
$RA1 = R1 \cdot R2 / (R1+R2+R3)$
$RB1 = R2 \cdot R3 / (R1+R2+R3)$
$RC1 = R1 \cdot R3 / (R1+R2+R3)$

From T to Π
$Y1 = YA1 \cdot YC1 / (YA1+YB1+YC1)$
$Y2 = YA1 \cdot YB1 / (YA1+YB1+YC1)$
$Y3 = YB1 \cdot YC1 / (YA1+YB1+YC1)$

Here again it is easier to work with conductance rather than resistance..

Bridge Topology

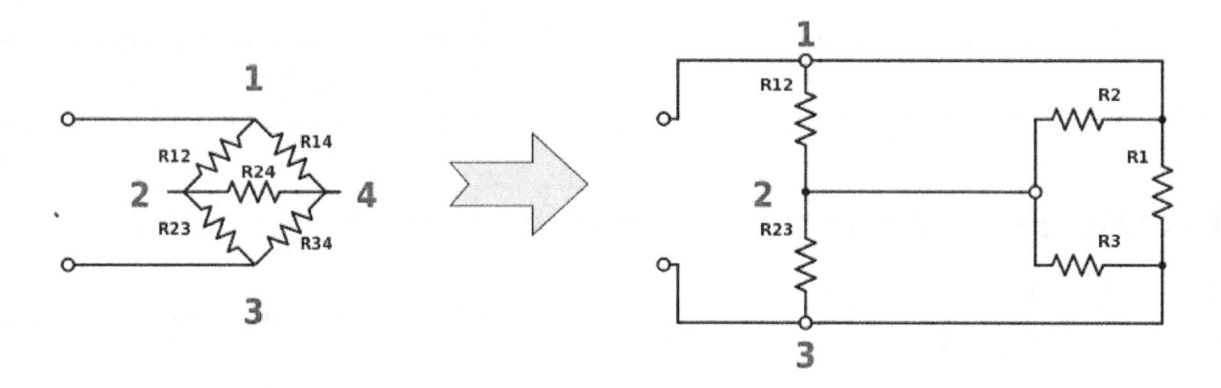

The bridge topology is shown on the left. Four elements labeled R12 (across nodes 1 and 2), R23 (across nodes 2 and 3), R34 (across nodes 3 and 4), and R14 (across nodes 1 and 4) are arranged in the form of a rhombus or diamond. Then a fifth element R24 (across nodes 2 and 4) is overlaid across the two nodes (2 and 4) that are not connected to the port.

We cannot simplify the network using series parallel transformations like before. However, we can remove node 4 by transforming the three elements connected to node 4, which forms a T-network, into a Π-network, as shown on the right. Then immediately we can merge R12 and R2 as parallel, and R23 and R3 as parallel, and then those two resulting elements in series, which can then be finally merged with a parallel transformation. Review the network transformation example earlier in section 3.9: it shows how this circuit on the right can be simplified.

An alternative is to transform the lower delta in the bridge into a Y, and then proceed with series-parallel transformations, as shown below:

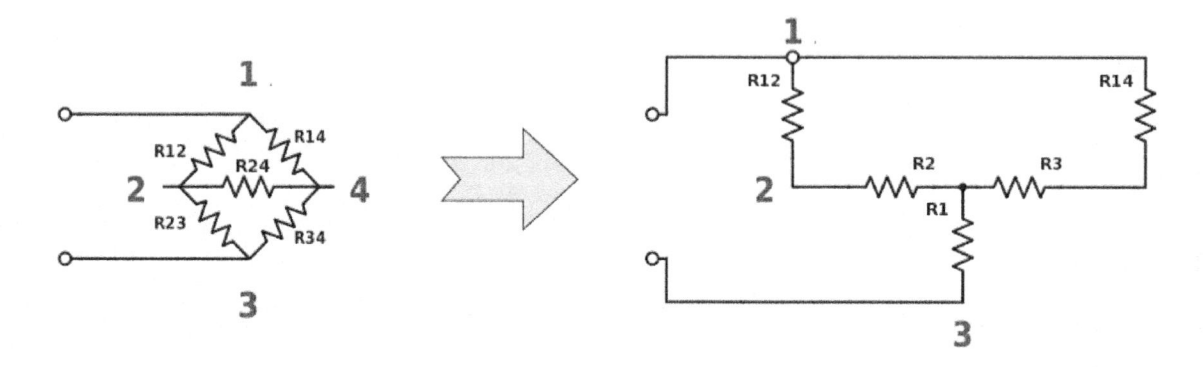

After the delta of R23, R24, and R34 is transformed into a T of R1, R2, R3, from there we can see that R1 and R2 are in series, and R3 and R14 are in series. The two resulting elements are in parallel, which are then in series with R1, so that we finally can obtain the equivalent impedance. The calculations will be left for the reader.

3.13 DC Analysis for General Audio Circuits

Up to this point we have analyzed only resistors. Many textbooks stop right there, and students felt unequipped when questioned about how to handle general circuits in practice. This section covers the rest of the components, presenting a general strategy. First we'll cover other passive components, then diodes, then op-amps when active.

DC Analysis involving Capacitors and Inductors

Capacitors blocks DC. Thus we can simply **replace it with an open circuit** (that is, remove the component).

Inductors do *not* block DC. In fact, it appears as a short circuit under ideal conditions (or a low serial resistance in practice). Thus we can simply **replace it with a short circuit**.

DC Analysis involving Diodes

There are many specialty diodes, too many to discuss individually here; however, they all have one thing in common with the general-purpose diode: in at least a certain range of voltages and currents, current flows readily in one direction but not the other. In other words, the resistance is high in one direction, and low in the opposite direction. In fact, the ideal diode has zero resistance for the forward bias polarity, and infinite resistance for the reverse bias polarity.

Thus, for DC analysis involving diodes, we first try to determine the voltage across the diode. If we can be sure it is either conducting or blocking, then we use that in our analysis. If we cannot, then we make *two* determinations, one assuming it is conducting (zero resistance), and another assuming it is blocking (infinite resistance).

DC Analysis involving Op-amps

When active, the op-amp maintains a virtual ground between the differential input. In other words, we can assume that there is a **short circuit between the differential input**. The output impedance of the op amp is low, so we can assume that it is essentially shorted to signal ground.

DC Analysis involving Other Components

In general, find out the equivalent circuit for the component, and substitute them accordingly. For example, for loudspeakers, we substitute its "nominal" resistance.

Example DC Analysis for Sample Audio Circuits

Can we handle general audio circuits with what we have covered so far?. This is where we want you to start reading schematics and visually identify the key configurations we had discussed so that you get the experience to analyse circuits you'll meet in the future.

Before we begin, let us make clear an important and fundamental principle:

> "If you are a hammer, everything looks like a nail."
> If you are DC analysis, everything is a resistor.

In other words, once you realize only resistors are what you have to deal with (everything else converted properly - not just ignored), then things would be dandy.

LED Current Limiting Resistor

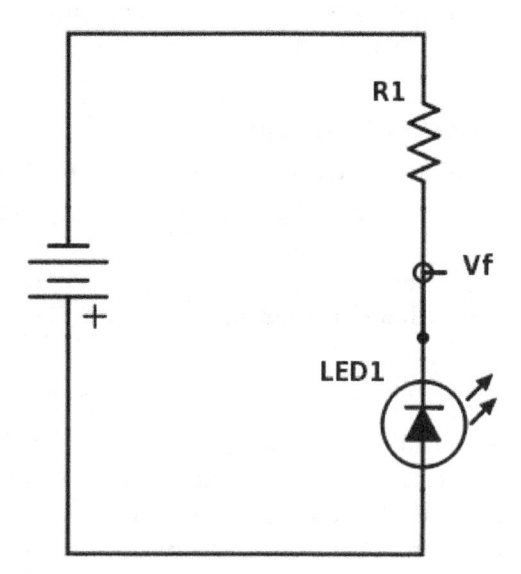

We look up the LED specification sheet and find that the forward voltage Vf for the LED is 3.1 V, and the forward current is 30 mA. If we use a 5V power supply, then we have R1 = (5 - 3.1)V / 30 mA = 63.3 kΩ. Using E6 series (20% tolerance) resistor value, the next up would be 68 kΩ.

The voltage across the resistor is (5 - 3.1) = 1.9 V. The current is 30 mA. So the power dissipated would be 30m · 1.9 = 57 mW. The standard ⅛ W resistor (or 0.125 W) would satisfy this requirement.

DC Response of High Shelving Filter

Again remember the principle of *not* to be scared by big words, until you realize you really should. In this case, the words "high-pass shelving filter" is irrelevant. You simply look at the circuit and analyze!

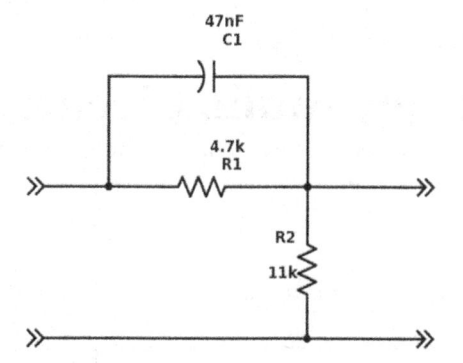

The DC bias of the input is 5V. The capacitor can be assumed to be open circuit for DC analysis. Thus the circuit can be analyzed as a voltage divider. The output DC bias level would be: 5 V · 11k / (4.7+11k) = 3.5 V.

DC Response of James Network

Again remember the principle of *not* to be scared by big words - no need to worry about what is James network!

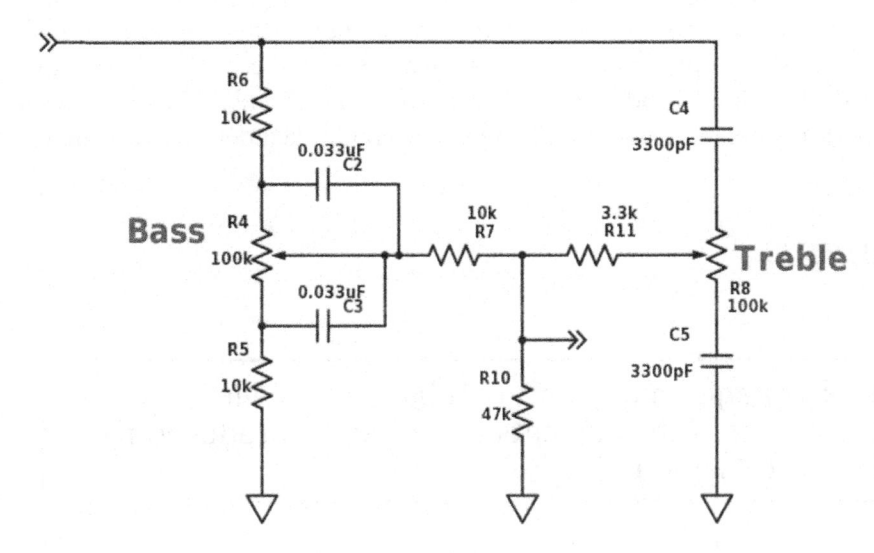

But this circuit does look complicated! Do we have to be scared? No! Remember for DC analysis, we change the capacitors to open circuits? This results:

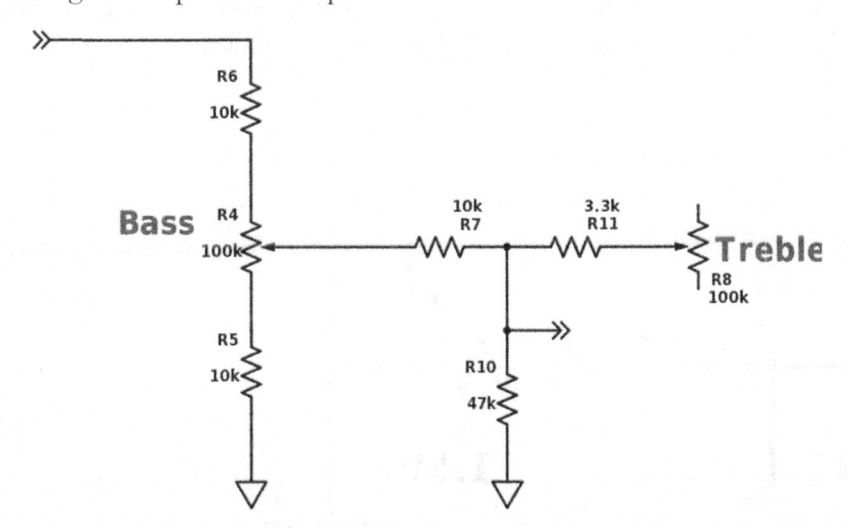

Now this is far less scary. Then notice the "bridge to nowhere" on the right? R11 and R8 can be removed from consideration! So now this is simply a voltage divider (R6, R4, R5) feeding another voltage divider (R7, R10)...

For the purpose of this analysis, let us say R4 consists of R4A above the tap, and R4B below the tap. Then we have a current divider with two branches, one consisting of R4B and R5 in series, another of R7 and R10 in series. Above the current divider we have R6 and R4A in series.

The equivalent impedance seen from the input is thus [(R4B+R5)||(R7+R10)] + (R6+R4A). If the input DC voltage is Vin, then the current flowing through the input would be Vin / {[(R4B+R5)||(R7+R10)] + (R6+R4A)}. The current flowing through the R7+R10 branch would be divider according to the current divider. Then we can obtain the voltage on the R7+R10 branch. Applying voltage divider rule, we can get the voltage at R10 which is the output. The details is left to the reader to finish the calculations.

3.14 MULTI-SOURCED, CHANGING LOAD: MORE NETWORK THEOREMS

One may think Ohm's Law is the key fundamental law, and it is understandable given its emphasis in high school and in many textbooks including this one. But remember it is *only* an empirical law. There are other more fundamental principles that needs be understood, although may not be as widely referenced or recognized. For example, how do we perform network analysis when we have multiple independent sources?

Superposition Theorem

> The total current (or voltage) in any part of a bilateral linear circuit equals the *algebraic sum* of the currents (or voltages) produced by each independent source separately.

In other words, we "turn on" each independent sources **one at a time**, the rest are "turned off" (voltage sources replaced by short circuits, and current sources replaced by open circuits), and we sum all the responses *algebraically* to produce the total response. (Note again the emphasis on "algebraically".)

Example

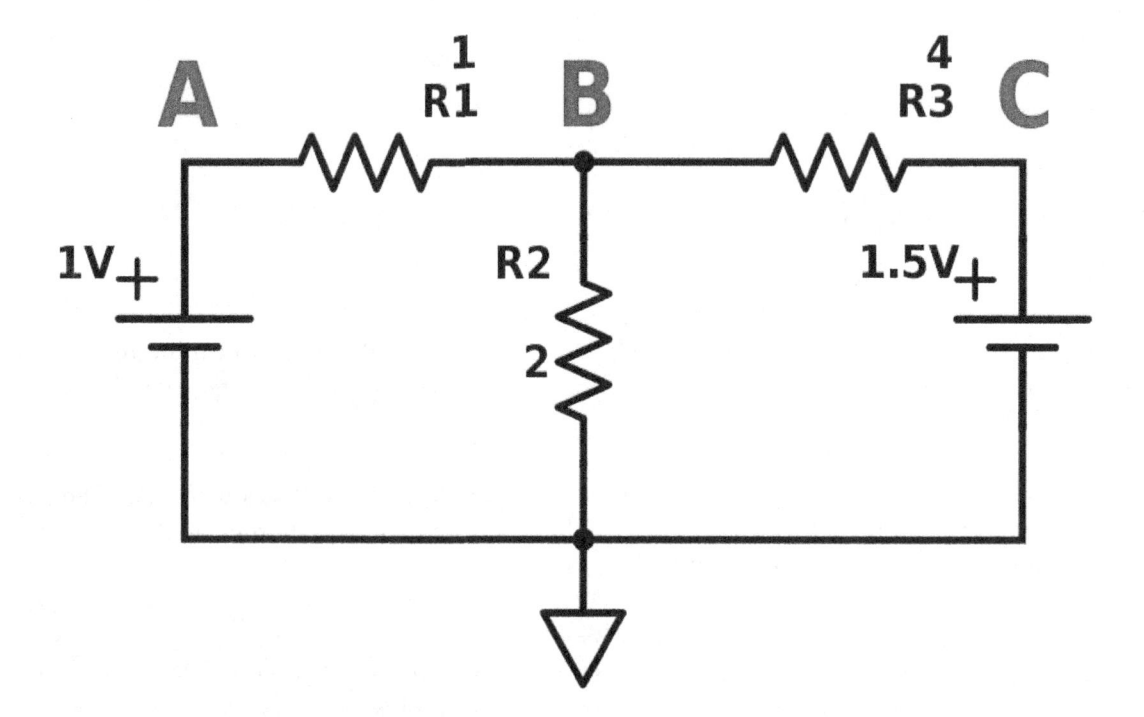

How do we perform the DC analysis?

Chapter 3 – DC Analysis

Step A = Source A (1V)
We first do the analysis by turning on the 1 V and turning off the 1.5 V:

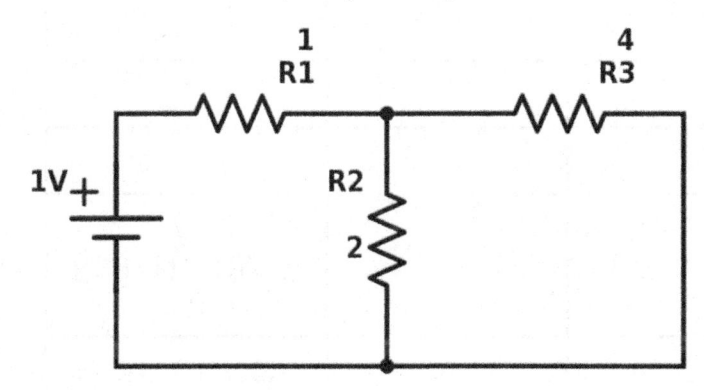

We had already done this as example in the previous section 3.11, with this result:

	P	E	I	R
R1		I1R1 = 0.429	↓ = 0.429	1
R2		I2R2 = 0.572	I2 = 0.286	2
R3		I3R3 = 0.572	I3 = 0.143	4
Rx=R2||R3		IxRx = 0.572	↓ = 0.429	1/[(1/R2)+(1/R3)] = 1.33
R1 + (R2||R3)		1	←/→ = 0.429	R1 + ↑ = 2.33

I2 = I · (½)/(½)+(¼) = 0.429 · (2/3) = 0.286 A
I3 = I · (¼)/(½)+(¼) = 0.429 · (1/3) = 0.143 A

Step B = Source B (1.5V)
Next we turn off the 1 V and turn on the 1.5 V, as follows:

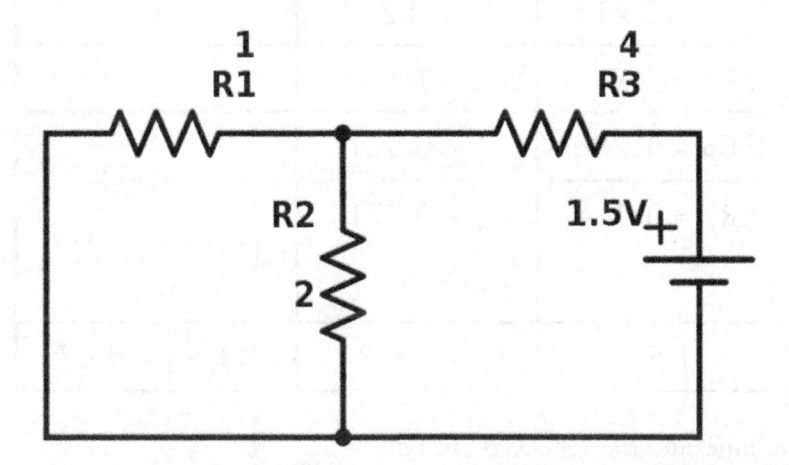

Notice the equivalent resistances are *different* from before;

3-47

	P	E	I	R
R1				1
R2				2
R3				4
Ry=R1||R2				1/[(1/R1)+(1/R2)] = 0.67
R3+(R1||R2)				R3 + ↑ = 4.67

Next we calculate the current through the battery, or R3, or R1||R2, and then the current divider:

	P	E	I	R
R1			I1 = 0.214	1
R2			I2 = 0.107	2
R3			↓ = 0.321	4
Ry=R1||R2			↓ = 0.321	1/[(1/R1)+(1/R2)] = 0.67
R3+(R1||R2)		1.5	←/→ = 0.321	R3 + ↑ = 4.67

After that we apply Ohm's law to complete the table:

	P	E	I	R
R1		I1R1 = 0.214	I1 = 0.214	1
R2		I2R2 = 0.214	I2 = 0.107	2
R3		I3R3 = 1.286	↓ = 0.321	4
Ry=R1||R2		IyRy = 0.214	↓ = 0.321	1/[(1/R1)+(1/R2)] = 0.67
R3+(R1||R2)		1.5	←/→ = 0.321	R3 + ↑ = 4.67

Notice E1 and E2 and Ey should be the same, and they checked out fine.

Step C = Algebraic Sum of Effects from All Sources

Next we sum the two E values collected from steps A and B, noting whether current flows from each source in the same polarity for each source or not:

	E from Step A	E from Step B	Opposite?	E Total
R1	0.429	0.214	Opposite	0.429 - 0.214 = 0.215 V
R2	0.572	0.214	Same	0.572 + 0.214 = 0.786 V
R3	0.572	1.286	Opposite	1.286 - 0.572 = 0.714 V

If you want to the more systematic, then define a direction for each circuit element. The current or voltage will show a sign, and then the E total will likewise show a sign. That way is *necessary* when we have more than two sources. But when there are *only* two sources, we can use this simpler treatment of just figuring out if they are opposite or not.

	I from Step A	I from Step B	Opposite?	I Total
R1	0.429	0.214	Opposite	0.429 - 0.214 = 0.215 A
R2	0.286	0.107	Same	0.286 + 0.107 = 0.393 A
R3	0.143	0.321	Opposite	0.321 - 0.143 = 0.178 A

Notice that *power* is not a linear function, and so should *not* be added using superposition *directly*. However, since current and voltage are *linear* functions, they can be added through superposition, and then power can be calculated *indirectly* from superposed currents and voltages.

	E Total	I Total	Total Power
R1	0.429 - 0.214 = 0.215 V	0.429 - 0.214 = 0.215 A	0.215 x 0.215 = 0.0462 W
R2	0.572 + 0.214 = 0.786 V	0.286 + 0.107 = 0.393 A	0.786 x 0.393 = 0.309 W
R3	1.286 - 0.572 = 0.714 V	0.321 - 0.143 = 0.178 A	0.714 x 0.178 = 0.127 W

This fundamental principle of superposition will be useful when performing AC analysis, combining results from multiple AC sources of multiple frequencies, or combining DC and AC analysis.

Thevenin's Theorem

What if I now want to change the value of R2 in the previous circuit? Oops... I'll have to go through the whole series of calculations again... What if *all* I want is to know the voltage and current for R2 and *nothing else*?

Thankfully there is a short cut. I did not lie earlier - if we need *all* the voltages and currents the earlier method is probably the fastest shortcut, but if we change the requirement and need *only* voltage and current for R2, then there is a shortcut for that. Sort of like that commercial (there's an app for that) - there's a theorem for that too! By now you probably realize that **network theorems are shortcuts**! Every theorem teaches us a shortcut that we can use...

> *Any* linear circuit with voltage and current sources and resistances only (no matter how complex!) can be replaced at a terminal of interest by an equivalent circuit consisting of a voltage source in series with a resistance.

Example

Let us show the circuit again (and solve it with Thevenin's Theorem, treating R2 as "the load"):

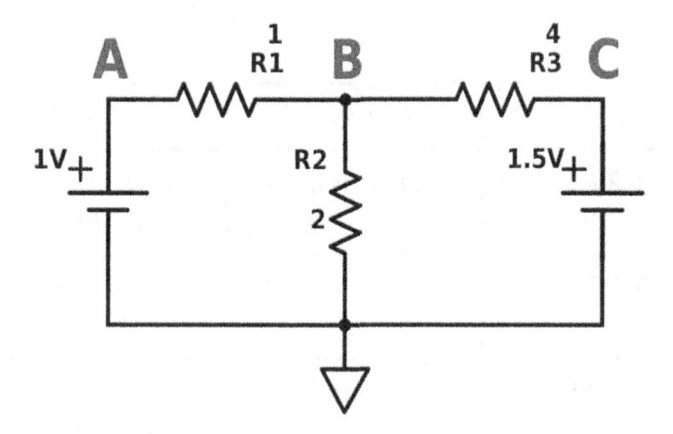

Step 1: Remove Circuit Element of Interest

Let us remove R2 from the circuit. Look how much simpler the circuit is now!

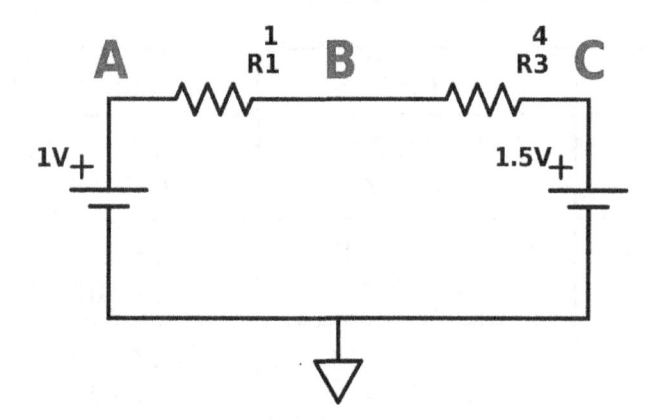

Step 2: Find Thevenin Equivalent Impedance

Let us find the equivalent impedance looking from B and ground. We do that by replacing the voltage sources with shorts, and current sources with opens, just like for Superposition Theorem. Look even how much simpler this circuit is!

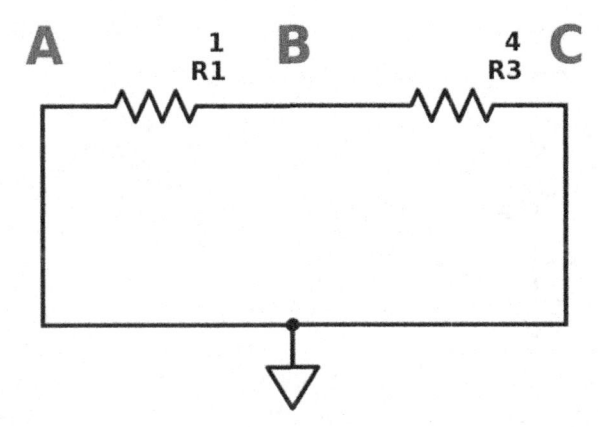

The Thevenin equivalent impedance is simply $R1 || R3 = 1/[(1/1)+(1/4)] = 0.8\ \Omega$.

Step 3: Find Thevenin Equivalent Voltage

Let us find the equivalent potential difference between B and ground.

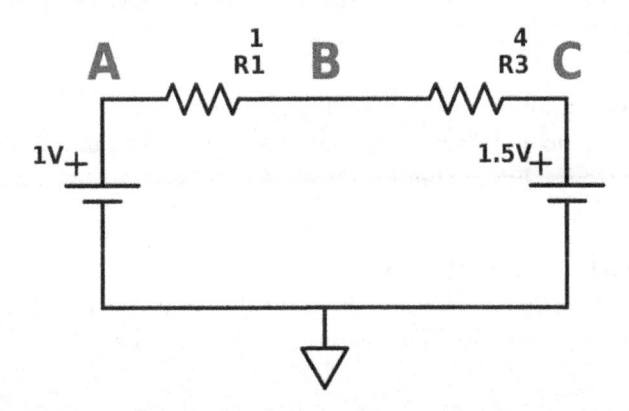

This is a simple series circuit; the equivalent resistance is $1+4 = 5\ \Omega$.
The equivalent voltage source is $1.5 - 1 = 0.5$ V.
The current flowing through that circuit is thus $I = E/R = 0.5 / 5 = 0.1$ A.
The voltage drop across R3 is $E = IR = 0.1\ 4 = 0.4$ V.
So voltage at B is $1.5 - 0.4 = 1.1$ V.

Alternately we can calculate the voltage drop across R1 is $E = IR = 0.1\ 1 = 0.1$ V.
So voltage at B is $1 + 0.1 = 1.1$ V. It checked out.

Alternatively we can use the voltage divider to get the voltage too. All roads lead to Rome.

Step 4: Put Equivalent Circuit Back Together with the Load

Now we just need to put the Thevenin Equivalent Circuit together with the circuit element of interest. Look how simple it is!

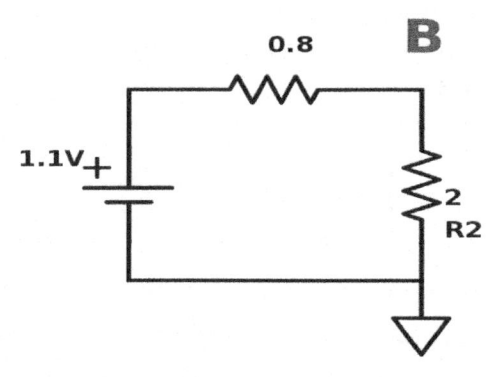

Even more important: look how easy it is to calculate the voltage and current for R2!
The current flows through a series circuit of $0.8 + 2 = 2.8\ \Omega$. So current = $E/R = 1.1/2.8 = 0.393$ A.
The voltage at B is $E = IR = 0.393 \cdot 2 = 0.786$ V.

Yeah! It matches our previous result. And the calculations are much simpler than using superposition! But watch this: let us change R2 to $1\ \Omega$...

The current flows through a series circuit of $0.8 + 1 = 1.8\ \Omega$. So current = $E/R = 1.1/1.8 = 0.611$ A.
The voltage at B is $E = IR = 0.611 \cdot 1 = 0.611$ V.

See how simple this is? Thevenin's Theorem is an incredibly powerful technique, isn't it?
We have to thank Hermann von Helmholtz (who independently derived it in 1853) and Léon Charles Thévenin (who published it in 1883) for this shortcut... hm... theorem.

Alternate: Doing Step 3 using Superposition

We can also do that by using Superposition Theorem. We are showing this only for didactic purpose, because obviously you want a faster shortcut. We have two relatively simple circuits to analyze!

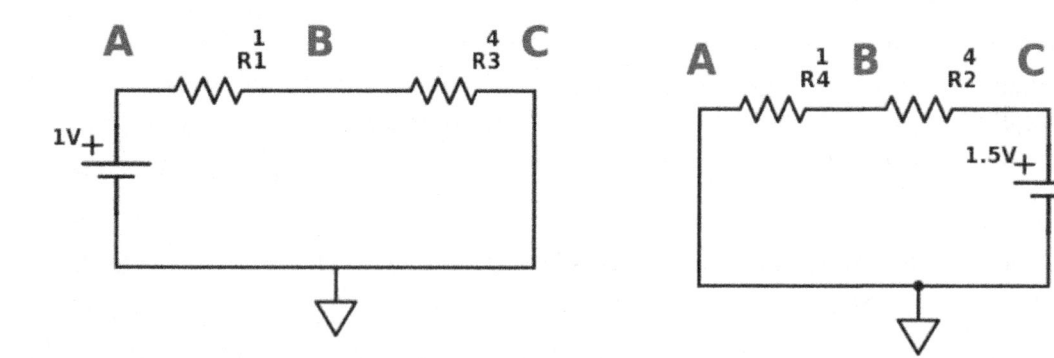

The circuit on the left is a voltage divider at point B. The voltage at B is $1V \cdot 4/(1+4) = 0.8$ V.
The circuit on the right is also a voltage divider at B. The voltage at B is $1.5V \cdot 1/(1+4) = 0.3$ V.

Beware: when we sum them together by superposition, we need to sum them algebraically! Since in both cases the voltages are additive, the summed voltage is $0.8 + 0.3 = 1.1$ V. This matches our earlier result.

Kirchhoff's Laws

Finally! The two fundamentals laws of circuit analysis are Ohm's Law and Kirchoff's Laws (plural - there are *two*). (The rest we studied, remember, are theorems, i.e., shortcuts. Power equation or formula is just a definition.) We started with Ohm's Law and will end with Kirchoff's Laws. That is because using Kirchoff's Laws nearly always lead to a **system of simultaneous equations**. Solving them are messy and error-prone manually - unless you have, and know how to operate, a calculator or computer software that can handle them well.

Let us use again the same circuit as example to illustrate how to use Kirchhoff's Laws in network analysis:

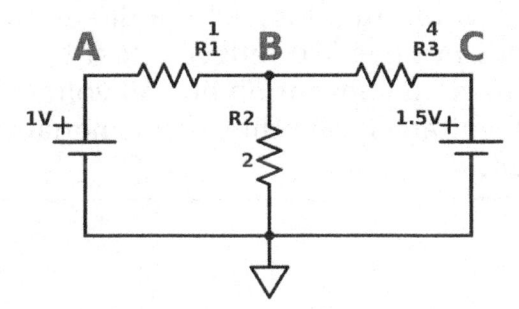

Kirchoff's Current Law (KCL)

This is used often in derivations of formulas, and in such circumstances, the acronym KCL is often used.

> **Kirchoff's Current Law: the algebraic sum of currents entering and exiting a node must equal zero.**
> 4. Select a **node**.
> 5. **Label all currents** flowing in and out of that node with a direction. This can be arbitrary and does not need to reflect actual flow.
> 6. **Sum the currents to zero**, using a convention like all currents entering a node is positive, while all current leaving the node is negative. This **results in one equation.**

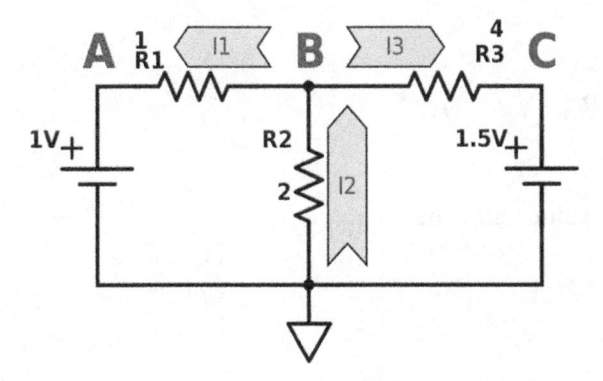

We have selected node B, and labeled it with I1 for R1, I2 for R2, and I3 for R3, all operating at node B. Note although the labeling is arbitrary, it makes sense too: I2 enters B and split into I1 and I3.

Using the convention mentioned, attach positive sign to I2 since it is entering, and negative signs to I1 and I3 since they are leaving, and sum them: -I1 + I2 - I3 = 0. We got our first equation!

Alternative to the "convention": I usually use KCL *literally*, thus current(s) entering = current(s) leaving, or:

   I2 = I1 + I3   (Note we get the *same* result as before! And to me this alternative is less intimidating...)

Kirchhoff's Voltage Law (KVL)

This is used often in derivations of formulas, and in such circumstances, often the acronym KVL is used. KVL is a "dual" of KCL, when we substitute voltage for current and loop for node.

> **Kirchoff's Voltage Law**: the **algebraic sum of voltages at a loop must equal zero**.
> 4. Select a **loop**.
> 5. **Label all voltages** operating within that loop with a direction. This can be arbitrary and does not need to reflect actual flow.
> 6. **Sum the voltages to zero**, using a convention like all voltages one way is positive, while all voltages the other way is negative. This **results in one equation**.

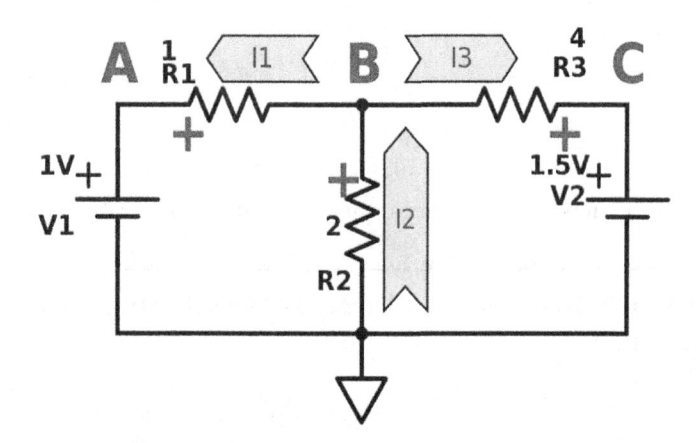

I have labeled all passive devices with polarity signs, using a convention that follows the current drawn earlier: the current enters the component at negative label and leaves at positive label (remember this can be arbitrary). Usually for illustration it is drawn one loop at a time; but in practice we usually label all at once.

First step is we select the *left* loop. Second step we already labeled the components in the left loop. Third step we sum them together, applying Ohm's Law to get the voltages:

$$I_1 R_1 + I_2 R_2 - V_1 = 0$$

We arrive at *another* equation!

Then let us repeat using the *right* loop: $I_2 R_2 + I_3 R_3 - V_2 = 0$
We got our *third* equation!

Now we can substitute resistor and voltage source values into the 3 equations:

$$-I_1 + I_2 - I_3 = 0 \quad (1)$$
$$I_1 \cdot 1 + I_2 \cdot 2 - 1 = 0 \quad (2)$$
$$I_2 \cdot 2 + I_3 \cdot 4 - 1.5 = 0 \quad (3)$$

How do you proceed to solve the system of equations is up to you, since there are so many ways. Here is one method:

Substitute (1) into (2) to eliminate I_1: $\quad I_2 - I_3 + I_2 \cdot 2 - 1 = 0$
Simplifying: $\quad I_2 \cdot 3 - I_3 - 1 = 0$
or, $\quad I_3 = I_2 \cdot 3 - 1 \quad (4)$
Substitute (4) into (3):: $\quad I_2 \cdot 2 + (I_2 \cdot 3 - 1) \cdot 4 - 1.5 = 0$
or, $\quad I_2 \cdot 2 + I_2 \cdot 12 - 4 - 1.5 = 0$

or, $\qquad$ I2 · 14 - 5.5 = 0

or, $\qquad$ I2 = 5.5 / 14 = 0.393 A $\qquad$ (5)

Now that we got I2, everything is downhill from here. Substitute (5) into (4):
$$I3 = I2 \cdot 3 - 1 = (5.5 / 14) \cdot 3 - 1 = 0.17857 \text{ A} \qquad (6)$$
Substitute (6) and (5) into (1):
$$I1 = I2 - I3 = 0.393 - 0.17857 = 0.2144 \text{ A} \qquad (7)$$

The results are the same as that derived earlier using superposition (note rounding errors may occur at the last significant digit). Voltages may likewise be calculated using Ohm's law as before.

Some who likes to solve algebraic equations may like Kirchhoff's Law better than superposition. (But note: if you choose the wrong path to solve the system of algebraic equations, it may be very tedious, and maybe won't get you anywhere even.) The key thing to remember is we need to **systematically label the nodes and loops** and then proceed from there, or else everything falls apart.

Other Theorems?

There are plenty of other network theorems... For example, the duality for Thevenin's Theorem is called Norton's Theorem, and we can similarly arrive at Norton's Equivalent Circuit, which we shall leave for the reader's enjoyment... (As reminder: **duality means substitute voltage for current, series for parallel, and resistance for conductance - or vice versa**...)

The point is, treat all network theorems as shortcuts - use them only when it is necessary. So we shall not indulge you with more theorems. We are now ready to analyze real audio electronics circuits, IRL!

3.16 Review Exercises

Chapter 3 Review Guide

Theory
- **concepts** of matter, energy and electricity
- **differentiation** between charge, electron, charge carrier, static electricity, current, voltage, work, energy, heat, power, electron mobility, speed of electricity transmission, force, field
- **indivisibility** versus **quantization** of matter, energy and electricity
- **collective interaction** versus **mobility** of electrons
- **units**, **symbols**, **abbreviations** of basic **quantities** in electronics (P, E, I, R, energy, work, charge, force, pressure, electronvolt)
- **avoidance of wrong concepts**

Practice
- ability to **read** most simple audio electronics **schematics**
- ability to perform **DC analysis** on such circuits
- ability to **calculate power, voltage, current, resistance** given *any* two of those values

We cannot wait until the whole chapter is complete before reviewing the material. So the exercises are split into parts A (after 3.1 and supplements), B (after 3.3), C (after 3.6) and D (after 3.11).

Exercise A

1. Differentiate between charge, electron, and charge carrier. Give three occasions where the word electron can't be used but charge is OK.

2. Differentiate between energy and work. Give one example each that the word energy cannot be used but work can, and vice versa.

3. Differentiate between force and power.

4. Differentiate between power and energy.

5. Differentiate between current and voltage.

6. Your exercise consists of pushing yourself hard against the wall, without moving. According to physics, are you working hard? Or are you hardly working? Explain.

Exercise B

Does your local electric utility supply you with billions and billions of electrons every day when you turn on your electric appliances? Explain why or why not.

True or False? "The alkaline cell can be a source of charge in a circuit. The charge that flows through the circuit originates in the cell." Explain.

When an alkaline cell no longer works, is it out of charge? How do you recharge it?

Explain if we can construct an electric cell with only *one* terminal instead of the two terminals commonly seen.

Explain which of the following is/are true?
a) the *energy* of an electron is quantized
b) the *charge* of an electron is quantized
c) the *charge* of an electron is the smallest charge theorized
d) an electron in a copper atom moves very quickly across millions of atoms when a voltage is applied, at near the *speed* of light
e) an electron in a copper atom is free to roam across millions of atoms when a voltage is *not* applied
f) an electron in a rubber molecule is free to roam across millions of molecules when a voltage is *not* applied
g) the *charge* inside an electrochemical battery contributes to its *voltage*
h) the *charge* inside an electrochemical battery contributes to its *current* when the circuit is closed
i) an *atom* is *not* divisible
j) the atomic *nucleus* is *not* divisible
k) a *proton* is *not* divisible
l) a *neutron* is *not* divisible
m) an *electron* is *not* divisible
n) a *photon* is *not* divisible
o) the *elementary charge* (a fundamental physical constant, $1.602\ 176\ 621 \times 10^{-19}$ coulombs) is *not* divisible
p) an object can be *charged with static electricity* through a conductor when connected to an object with more static charges
q) the *static electricity* in an object can be *discharged* through a conductor when connected to "earth" or "ground"
r) a rechargeable battery can be charged with *static* electricity through a conductor when connected to a rechargeable battery charger

2. Explain if you can measure the voltage at *one* point of a circuit.

An battery case holds four 1.5 V cells. What voltage does it produce? Would you expect the cells to be in series or in parallel? What would be the problem if it is connected in the other configuration?

Give examples of five different classes of material that conducts electricity.

Explain the speed of electron movement in a metal. Explain the speed at which the light bulb is lit when its switch is turned on. Explain the conundrum.

Explain the direction of the conventional current flow.

Explain if electricity is quantized. Explain both static and current electricity.

How do you explain "analog" current if electricity is quantized?

Exercise C

Explain whether an electric current *can* flow in an object when there is no voltage applied. State Ohm's Law and apply it to calculate the resistance of this object.

1. If the current is 10 amps and the voltage is 24 volts, what is the total resistance in the circuit? (State which law is involved. Cite the law using equation circle or algebraic equation. Derive the formula to be used. Substitute values. Calculate the result. Normalize to appropriate units.)

2. If the resistance is 200 Ω and the voltage is 12 volts, what is the current flowing through the circuit?

3. If the power of a circuit is 480 watts and the current is 3 amps, what is the voltage?

4. How much current is flowing through a circuit if it produces 200 watts of power with 12 volts?

5. I connected a cell phone to a USB charger which is plugged into the cigarette lighter socket in a car. The cell phone is drawing 2 A at 5 V; how much power is it drawing from the USB charger? The charger itself is drawing that same power from the car battery at 12 V at 80% efficiency; how much current was the USB charger drawing from the battery? (Hint: there are two problems you need to solve, one after another.)

6. A soldering iron has an equivalent resistance of 240 ohms, and an equivalent of 1/2 amps of current is flowing through it. How much power is it dissipating? (Hint: you may have to apply two laws in sequence.)

7. What load does the amplifier see when two 8Ω speakers were connected in series and in parallel?
In series:

In parallel:

8. Two volts is applied to two equal resistors in series. What is the voltage on *each* of them? Draw a schematic and label the voltages.

Exercise D

1. Explain what happens when two cells, measuring 1.4 V and 1.1 V individually, are put in the three configurations below. After the cells are inserted, what are the voltages measured at the system terminals in each case? Use this example to explain why batteries are usually in series or parallel.

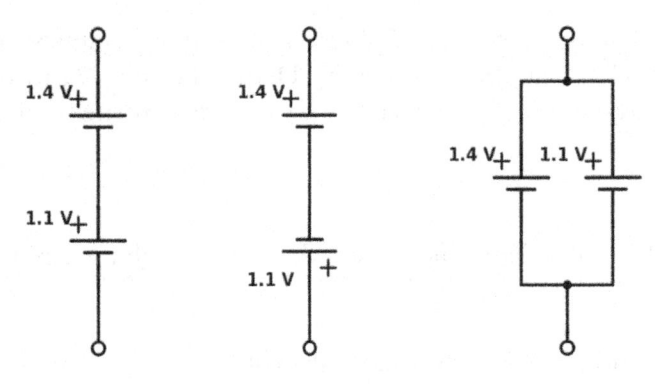

2. Calculate the current through the LED in the circuit below, using a 4V battery, with desired voltage drop of 2 V across the LED, and R1 is 100 Ω.

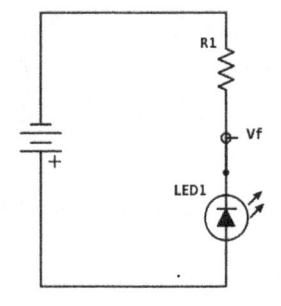

3. Calculate the equivalent DC resistance of the circuit below, when both pots are set to the middle (50 kΩ each side of the arm):

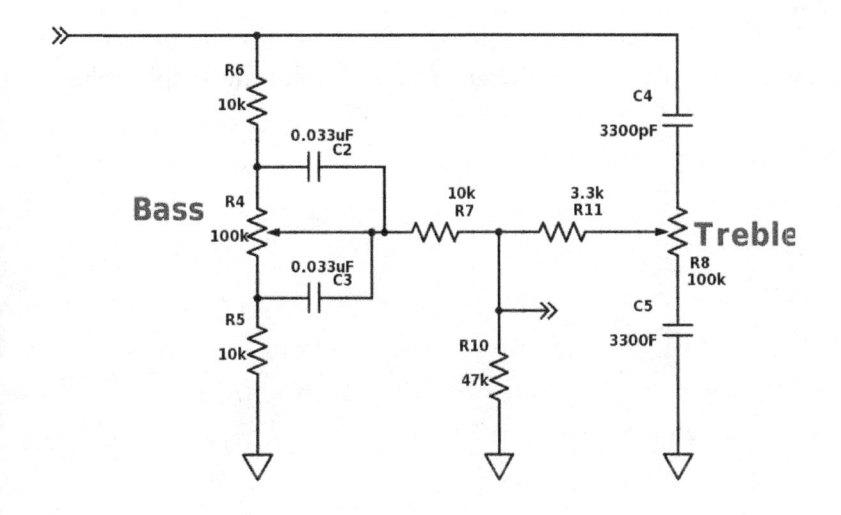

4. Calculate the equivalent DC resistance of the circuit below. Draw equivalent circuits to show each step of your simplification or transformation.

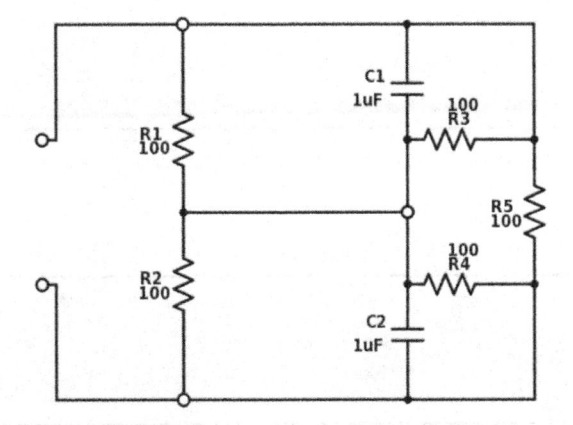

5. Locate the voltage divider and calculate the bias voltages at the base of the two transistors Q1 and Q2. You can ignore the rest of the circuit and concentrate only on the voltage divider. Notice the supply voltages are +12V on top and -12V at the bottom.

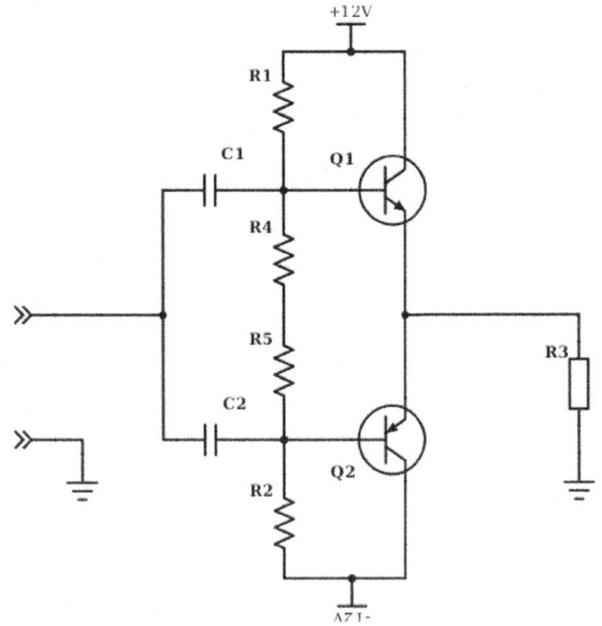

R1	R4	R5	R2	Calculation of DC Bias Voltage
9k	1k	1k	9k	at base of Q1:
9k	1k	1k	9k	at base of Q2:
9k	0	0	9k	at base of Q1:
9k	0	0	9k	at base of Q2:

Supplement N - Order Out of Chaos - From Matter to Current Electricity

Before receiving his Nobel Prize in 1965, professor Richard Feynman, "The Great Explainer", opened his legendary *Feynman Lectures on Physics* to freshmen at Caltech in 1961 thus:

> "If, in some cataclysm, all of scientific knowledge were to be destroyed, and only one sentence passed on to the next generations of creatures, what statement would contain the most information in the fewest words? I believe it is the **atomic hypothesis** (or the atomic fact, or whatever you wish to call it) that **all things are made of atoms—little particles that move around in perpetual motion, attracting each other when they are a little distance apart, but repelling upon being squeezed into one another**. In that one sentence, you will see, there is an enormous amount of information about the world, if just a little imagination and thinking are applied."

His approach differs from traditional physics professors, who'd start with "constant mass" law which is only approximate (and which he argued philosophically as completely wrong), and would wait until the third year in college to get into quantum mechanics, using differential equation approach. But he also recognized that "in its **real applications**—especially in its **more complex applications**, such as in electrical engineering and chemistry — **the full machinery of the differential equation approach is not actually used**." (bold added) We carry on his tradition (in the 1960's, electrical engineering is subdivided into *heavy* and *light* - the latter being *electronics*) by **not using differential equation**, yet starting off with the atomic hypothesis…

His vision was adopted by chemistry in recent decades; some chemistry textbooks had been completely rewritten, emphasizing the influence of, and starting with, **atoms**. Some organic chemistry textbooks even start with **electrons**. (*All* chemical reactions work through electrons! Chemical bonds are different ways of sharing electrons. The periodic table of the elements are organized by electronic structures. Acids and bases are now redefined as electron-pair acceptors and donors. Oxidation and reduction are now redefined by a change in the oxidation state, which in turn is determined by valence electrons. It is a surprise that they did not rename chemistry as electronics - just kidding.)

In contrast, it is a surprise to me that electronics textbooks are still typically written based on the classical atomic theory updated with subatomic particles, which is more than half a century old. Shouldn't we update our electronics textbooks to reflect the most current understanding of electrons, its namesake? Well, we need to be careful, and this attempt concentrates everything surrounding electrons, atoms, fields, energy, and so forth inside this section, so that it is easier to understand and teach, and easier to update as new information becomes available.

Science is not settled.

In music, one of the first things we need to learn is that music theory and practice is to be interpreted by **musical periods**. Likewise, theories of electronics need to be interpreted by specific **scientific model periods**. We will be presenting a quick overview of timelines for **matter**, **electricity**, and **energy**.

Learn to differentiate between when to use which term!
Please note that terms like charge, electron, charge carriers, electricity are *not* interchangeable. Do *not* use force when you mean work or field or energy or heat or power.

Just catch the drift!
The student is *not* expected to memorize all dates or names of all scientists. Instead, listen to the big picture and notice anything said that is different from what you thought or what you had been taught. The instructor can teach new things, but the student must first **unlearn** old things. Plus, try to catch the relationship between events discussed and their implications. When the student realizes that many old concepts were wrong, it is hoped that he will then be careful and *not* invent his own ideas or concepts, which we see all the time during exams.

Changing Views about Matter

As a reminder, the SI unit for mass is the **kilogram**, but we use the **gram**, abbreviated g, as the basis to connect with the SI prefixes.

But what is matter?

Matter matters much, not only in electronics; we shall learn some aspects that impact our understanding of electronics. Fundamentally, matter is mostly empty space, punctuated occasionally by some relatively massive subatomic particles that organized themselves into groups of recognizable units, and coordinating with one another through some relatively lightweight or even weightless particles. The smaller in dimension we look, the more chaotic it appears. And those particles sometimes behave as waves. We can't pick and choose their behavior - nature picks and chooses them.

Ancient Greek	Ancient Greeks ~400 BC • Building blocks of matter = "*atomos*"	[Artist's rendering by Gilbert assumed they weren't thinking of spheres yet but rather building blocks like bricks.] Around 4th century B.C., the ancient Greeks, famous for their philosophy, came up with the initial atomic theory, reasoning that matter were made up of discrete units which were indivisible. In fact, the word "atom" came from a Greek word which means "indivisible."
Early Atomic Theory	John Dalton 1808 Simple — weight 5 units Binary — weight 12 units Quaternary Ternary — weight 7 units Quintinary • Atomic Weights: "N=5, O=7" (inaccurate)	In 1789, Antoine Lavoisier discovered the law of conservation of mass, stating that the total mass of chemicals involved in a *chemical* reaction remained unchanged. (In contrast, we later discovered that in a *nuclear* reaction, the total mass does change to create energy.) In 1804, John Dalton expressed the Law of Multiple Proportions, generalizing on the Law of Definite Proportions discovered earlier in 1799 by Joseph Louis Proust, stating that the ratios of the masses of elements engaged in chemical reactions can be expressed in small whole numbers. These two fundamental laws of chemistry, together with the conservation of mass, laid the groundwork for the atomic theory Dalton proposed in 1803-1805, listing atomic weights for the elements for the first time. In 1869, Dmitri Mendeleev listed chemical elements then known in the form of a table, roughly in order of their atomic weight, and grouped with similar characteristics. He accounted for missing elements (which was later found) through this periodicity, and even switched elements occasionally to violate atomic weight orders when the grouping deemed appropriate. In 1871, he swapped columns and rows, leading to what we now called the periodic table. Germanium was the first semiconductor material we used to make transistors for audio electronics. Now we use Silicon, which lies in the same column, just above Germanium, in the periodic table.

Subatomic Particles	J.J. Thomson 1897 • Small lightweight negative electrons embedded in a sea of positive matter	In 1897, J.J. Thomson discovered the electron while working on cathode ray tubes. He discovered that the rays could be deflected by an electric field, in addition to magnetic field. (More on fields a little later.) He measured the charge-to-mass ratio, showing that the electron has a negative charge and a mass of 1/1800 that of hydrogen, the smallest atom. Thus, he inferred that the electron is a divisible part of an atom.
	Ernest Rutherford 1911 • Positive charge located within small compact central "nucleus" of atom	In 1909, Ernest Rutherford discovered in his famous gold-foil experiments that the main mass and charge of an atom reside in a small, compact area of the atom. We now call it the nucleus. He proposed that it consists of protons of positive charge.
	Neils Bohr 1913 • Electrons with quantized energy levels, in circular orbits around the nucleus	In 1913, Niels Bohr improved upon Rutherford's model, depicting the atom as a small positively charged nucleus surrounded by electrons in orbits just like planets surrounding the sun, but mediated by electrostatic forces rather than gravity. The Bohr model is the beginning of quantum theory which we shall discuss next. In 1932, James Chadwick discovered the possible existence of neutrons. (That was quickly followed by nuclear fission in 1938, the first self-sustaining nuclear reactor based on nuclear chain reaction in 1942, and the first nuclear weapon in 1945.)
Quantum Theory		Back in 1803, Thomas Young performed the famous double-slit experiment which led to the wave theory of light. In 1860, Gustav Kirchhoff posited his law of thermal radiation and defined a perfect black-body. (This Kirchhoff's law of thermal radiation should not be confused with Kirchhoff's laws of circuit which we shall study in chapter 3. Yes, Kirchhoff defined them all.) In 1900, Max Planck provided Planck's law of blackbody radiation, by assuming that energy was quantized (i.e., available in discrete steps, not continuous). He argued that each quantum energy element is proportional to its frequency, according to the simple formula $E = h\nu$. In 1905, Albert Einstein built on this and explained the photoelectric effect by proposing the quantization of electromagnetic radiation. The quanta in question here is specifically known as the photon. In 1924, Louis de Broglie proposed that moving particles exhibit wave-like behavior. In 1926, Erwin Schrödinger published his famous Schrödinger's equation, thus establishing the wave-particle duality: objects have characteristics of both particles and waves. In 1925, Werner Heisenberg presented quantum mechanics using matrices and probability theory. In 1927, he stated the famous uncertainty principle. (Yes, he was certain about that. And we are certain that that did happen.) That led to a can of worms, or maybe a box of cat that we don't know if dead or alive? In 1935, Erwin Schrödinger described a thought experiment, now affectionately known as Schrödinger 's cat...

In summary, there are three characteristics of quantum mechanics: 1) discrete values for energy, momentum, and other quantities; 2) wave-particle duality; 3) uncertainty principle.

Standard Model of Elementary Particles

In 2013, CERN scientists confirmed the Higgs Boson, which was proposed by Peter Higgs in 1964.

With that discovery, the latest confirmed fundamental particles are listed in the table below, and classified according to shades of color: the left three columns are **fermions**, with the upper two rows **quarks**, and the bottom two rows **leptons**; the right two columns are **gauge** and **scalar boson**s respectively.

Fermions			Bosons	
			Gauge Bosons	Scalar Boson
Up Quark	Charm Quark	Top Quark	Gluon	Higgs Boson
Down Quark	Strange Quark	Bottom Quark	Photon	
Electron	Muon	Tau	Z Boson	
Electron Neutrino	Muon Neutrino	Tau Neutrino	W Boson	

Together with the particles listed above, the Standard Model also explains three of the fundamental forces in nature: strong, electromagnetic, and weak. So far it has not been able to explain the fourth fundamental force: gravity. Graviton (as a boson) was proposed, but so far, no proof was obtained.

Relating Old Terms to the New

The traditional nucleus is now nucleon, and nucleons are **baryons** which are composites of fermions. Specifically, protons consist of two up quarks and one down quark, while neutrons consist of two down quarks and one up quark. The traditional electron, being a fundamental lepton, remains as an electron.

Old	New
Nuclei	Nucleons are baryons (composite fermions)
-- Protons	== up quark + up quark + down quark (*uud*)
-- Neutrons	== down quark + down quark + up quark (*ddu*)
Electrons	Electrons

Changing Views about Electricity

Just like matter, there is a history and difference in the model of what is electricity. It is more complicated than we think, and deserves careful understanding.

Static Electricity

Static electricity is observed when an object *has* a **non-zero and motionless net electric charge**.

(In contrast, the alternative is not called "dynamic electricity," but rather, current electricity, or simply, electric **current**, which is the **motion** or **flow** of electric charge *through* an object, which produces **no net gain or loss of electric charge** in the object. However, *both* are a consequence of the *same* electrons and protons appearing in opposite polarity.)

Around 600 BC, the Greeks noted that when fur is rubbed on amber, the latter is charged and it can attract light objects like hair. Sufficient charges could even cause a spark to spark. (By the way, the Greek word for amber is ἤλεκτρον, from which our "electron" came.)

In 1720, Stephen Gray, in his mid-fifties, stricken by poverty and living in a "home for destitute gentlemen", noticed that the cork at the end of his rubbed glass tube attracted small pieces of paper as the tube would. He then extended the cork with a fir stick, and then a wire, and realized that the same terminating ivory ball will continue to attract the small pieces of paper as the tube would. He also noticed that the wire can bend or even go up (against gravity), but it cannot touch the wall, or the effect will disappear - he used silk to isolate the wire from the wall, and everything still worked. With that type of experiment, he was the first to classify materials as conductors and insulators.

In 1733, C. F. du Fay proposed the **two-fluid theory**, with electricity coming in two variety that can cancel one another: "vitreous electricity" (when glass was rubbed with silk) and "resinous electricity" (when amber was rubbed with fur). Two pieces of rubbed glass repel each other; likewise, two pieces of amber; but rubbed glass attract resin.

Later in the 19th century, Benjamin Franklin proposed the **one-fluid theory**, and first identified the term "positive" with "vitreous electricity" and "negative" with "resinous electricity". Later, we adopted the convention for current polarity that is consistent with Franklin's definition of positive and negative. (That is how science usually works - we are the legacy of history.)

Many semiconductor devices used in electronics are sensitive to static electricity, and are often shipped in antistatic bags or shields. If so, you should return unused components back to the same container or shield. Careful operators also use an antistatic wrist strap while working with such components, and touching a solid grounded metal *before* starting work, to prevent damage to any components touched.

Piezoelectric & Pyroelectric Effects

. We had just discussed **contact-induced charge separation**.

. We can also have **pressure-induced charge separation**, where mechanical stress applied generates a separation of charge in certain types of crystals and ceramics molecules. This is called the **piezoelectric effect**, and is reversible, meaning that in those crystals, applying electricity can induce mechanical stress. That effect is utilized in the piezo microphone and the piezo pickup, which, if you follow so far, can be used as an piezo speaker! The piezo transducer is also common in making "new music" instruments.

. We can also have **heat-induced charge separation**, where heat applied generates a separation of charge in the molecules of certain materials. This is called the **pyroelectric effect**, and currently known pyroelectric materials are also piezoelectric

. Plus, we can have **charge-induced charge separation**, where a charged object brought close to a neutrally charged object induces a separation of charge *within* the object. This is called **electrostatic induction**, and is the principle behind the famous Van de Graaff generator.

Electric Charge

There are two types of electric charges: positive and negative. Elementary particles like **protons carry a positive electric charge of +e** while an electron carries a negative charge of **-e**.

The **electric charge is quantized**, i.e., it comes in integer multiples of a very small unit called the elementary charge, which was measured as 1.602×10^{-19} coulombs.

That it is called the elementary charge and that it is quantized does not imply it is the *smallest* charge accountable. Let me now expand on one small detail that I had discussed before: that protons and neutrons are triquarks consisting of up and down quarks. The **up quark (u) has a charge of +⅔ and the down quark (d) has a charge of -⅓**. That is the smallest charge known thus far, yet these fractional charges are *still* quantized.

This is how they are summed together: proton is **uud** = +⅔ +⅔ -⅓ = +1; neutron is **ddu** = -⅓ -⅓ + ⅔ = 0, exactly as we expected, and ending up with an integer elementary charge. This explains why even though fractional charge is integral to the standard model, it has never been observed.

The up quark (u) has a charge of +⅔ and the down quark (d) has a charge of -⅓.

Could the discovery of this number 3 popping up in unusual and unsuspecting places, associated with the fundamental quantum charges of the quark, renew interest in the fixation of the number 3 in music or pop culture? Who knows?

In the just intonation of music toning system, the perfect fifth is expressed as a simple integer ratio of frequencies of 3/2, while the perfect fourth is expressed as 4/3. They sound consonant, pleasant, and well-tuned. In fact, other than unison and octave, these two ratios with the number 3 are the most consonant intervals.

Back in the medieval period, triple meter (3/2 or 3/4) was the favorite in music, given the term *tempus perfectum* (perfect time), and indicated with a full circle for the time signature as a sign of perfection. This was because of the influence of numerology, with 3 representing the Trinity.

Trivia: The time signature symbol for what we now call "cut time" is not a "C representing cut" but rather a carry-over from the notational practice of that period up to Renaissance, where it signified *tempus imperfectum diminutum* (diminished imperfect time), represented by a broken circle with a line across it. And of course, "common time" wasn't common during that period, as it is considered "imperfect," represented by a broken circle (just looks like a C).

Aside: It should be noted that the scientists who developed the science of electronics philosophized a lot. For example, there is a philosophy of *duality*. It is wise to understand it, because duality states that equations stated in voltages have a corresponding one stated in currents, or resistance have a corresponding one in conductance. In other words, if you truly understand duality, you'll have to memorize only half of the equations needed. In the list of key equations, I have deliberately *not* used duality; although equally I've deliberately stated them in a way that it is easy to see the duality principle. See if you can figure out which one is part of a duality pair and thus reducing your memorization load.

Electric Force

Coulomb's law states that the **electrical force between two charged objects is directly proportional to the product of the quantity of charge on the objects and inversely proportional to the square of the separation distance between the two objects.**

$$\text{Force} = -k\, Q_1 Q_2 / d^2 \qquad \text{(Coulomb's Law)}$$

This embodies the inverse square law, and that like charges repel while dissimilar charges attract (because the charges are represented as positive or negative as previously described, leading to negative or positive forces). The coulomb's constant k is approximately $9.0 \times 10^9\ \text{N} \cdot \text{m}^2 / \text{C}^2$ in air.

Electric Field

The electric field strength is defined simply as the **force per charge** ratio. The electric field is caused by a **source charge**, while we imagine a **test charge** to obtain the force. The strange thing about a ratio is that *any* test charge can be used - the force simply increases proportional to the test charge, with the electric field remaining constant.

$$\text{Electric Field Intensity} = \text{Force} / \text{Charge} \qquad \text{(Electric Field Definition)}$$

Combining Coulomb's Law and the Electric Field definition, using a source charge Q and a test charge q:

$$\text{Force} = -k\, Q\, q / d^2$$
$$\text{Electric Field Intensity} = (-k\, Q\, q / d^2) / q$$

$$\text{Electric Field Intensity} = -k\, Q / d^2 \qquad \text{(Practical Electric Field Formula)}$$

For those curious, $k = 1/4\pi\varepsilon$ where ε is the permittivity.

Current Electricity

Current is the rate at which charge moves past a point on a circuit. $I = Q/t$. (If we use differential calculus, then it would be $I = dQ/dt$.)

> Current is the rate at which charge moves past a point on a circuit.
> $I \triangleq Q / t$ (Definition of Current)

To help with memorizing this as a *definition*, we can state it as $I \triangleq Q / t$ where the delta above the equal sign help us remember that it is a definition, and thus not an approximation like Ohm's Law that may not apply on occasions; it always applies!

The dimension symbol for current is I; the unit for current is the Ampere, abbreviated A. The unit for charge is Coulomb, abbreviated C. The unit for time is second, abbreviated s. Thus, A is equivalent to C/s.

Notice that the basis for SI units do *not* necessarily relate to how the *quantities* are *defined*. The ampere is a base SI unit. Thus, as far as SI units are concerned, the Coulomb is a derived unit, expressed in terms of **s·A**, and derived from the second and the ampere using $Q = t \cdot I$

One student asked, how come in high school they defined $I = E/R$? We'll examine Ohm's Law soon, but suffice to say, that was **not** a definition of current, and that was only *empirical* and *approximate*…

This will be further covered later in Chapter 3. Note that some of the charge separation mechanisms above can produce current electricity also.

S.I. Unit Measurement for Current

One may be inclined to think that the S.I. units are defined according to the definitions of the quantities. That is simply **not** true. Instead they are often defined according to what is easiest to measure accurately and stably and consistently. As a result, the measurement method may be significantly convoluted. The measurement methods defined for the S.I. unit may also change over time; whenever a better method is available, the challenge is to move to the better method…

As an example, the kilogram was originally defined as a cylindrical block of steel stored at multiple locations of the world. A sensitive balance is used to measure against that standard. Well, science is not settled. The definition for kilogram is expected to change in 2019, when it will be defined by the Planck constant, the speed of light, and the cesium hyperfine frequency - very convoluted, but less likely to be degraded by the environment.

Likewise, the SI unit of current of one ampere is currently defined as "the constant current which, if maintained in two straight parallel conductors of infinite length, of negligible circular cross-section, and placed one meter apart in vacuum, would produce between these conductors a force equal to 2×10^{-7} newton per meter of length." It is based on the electromagnetic wave behavior of a current.

In the future, the S.I. unit is expected to change so that it reverts to its definition, that an ampere will be approximately equivalent to 6.2415093×10^{18} elementary charges moving past a boundary in one second.

AC/DC

When the current is relatively steady, flowing in one direction, we call it **direct current**, or **DC**. When the current changes direction regularly, we call it **alternating current**, or **AC**. **Diodes** (in that form we often call it rectifiers) are used to convert AC to DC. **Transformers** allow AC to be stepped up in voltage for efficient long-distance transmission, and then stepped down in voltage for safety for home use.

> ### "War of Currents" and "Edison's Revenge"
>
> During electrification of the U.S. in 1880s and 1890s for the nation's grid, Thomas Edison proposed DC and George Westinghouse proposed AC (having licensed Nikola Tesla's polyphase AC and induction motor patents in 1888). Local cities preferred DC, but transporting electricity efficiently at that time requires AC using transformer to step up and down voltages, thus carrying electricity for long distances using thinner and less expensive cables. So, Edison lost "the War of Currents" in 1892, and three-phase AC has been the standard ever since, but...
>
> In 2017, the Plains and Eastern Line is the first American near- Ultra High Voltage Direct Current (UHVDC) super-grid being built to connect wind generators in Oklahoma to Tennessee (slated to be operational by 2020). At ultra-high voltages, DC has lower losses in transmission when compared with AC. (AC cables also can't go under water effectively.) The limit for AC is usually cited at around 400,000 volts. In this case, they operate at 600,000 V DC, barely under the 800-kV definition of UHVDC.
>
> UHVDC started in China in 2010, with 23 point-to-point UHVDC continental links expected to be operating by 2030; its success was exported to Brazil in 2015. Now India and Europe also is getting their UHVDC continental super-grids, with Europe really having never given up on DC transmission, having used it sporadically throughout the 20th century in specific locations, especially in undersea cables between countries. (By the way, the major suppliers of HVDC technology are from Europe: ABB in Swiss/Switzerland and Siemens in Germany.) The world record for longest distance, highest voltage, and largest transmission capacity is set in 2016 by an ABB contract in China to build a 3,000 km (1,900 miles) length, 1100 kV voltage, and 12 GW power links.
>
> In other news, data centers had also switched from AC to DC in the last decade, thanks to Google engineers discovering significant efficiency advantages; as they are the biggest user of data center, they in turn changed the whole industry. Next there are thoughts of using the same concept at the building level, with solar and wind generators and battery storage, and maybe with Tesla's and Priuses hooked up to it as well - DC is potentially more advantageous under such situations as well.
>
> Incidentally, the generator in an automobile used to be DC, since the battery is also DC, but it has changed to AC because the various speeds (or RPM of the generator) that the automobile has to run on.
>
> **There is no easy way to say whether AC or DC is "better" ...**
>
> As to what happened to the corporations, in 1892, at the conclusion of the War of Currents, Thomas Edison's electric company merged with Westinghouse's chief AC rival, the Thomson-Houston Electric Company, forming General Electric (GE). GE founded RCA in 1919, RCA co-founded NBC in 1926, and GE reacquired RCA in 1986. (*Trivia*: The pitches of the jingle of NBC spells G-E-C, highlighting General Electric Company's ownership at one time.) As of 2017, GE is the 13th largest company in the U.S. by gross revenue, and the oldest component of Dow Jones Industrial Average.
>
> Westinghouse founded Westinghouse Electric Company in 1886, entered into home appliance in 1914, broadcasting in 1920s, television in 1951, and attempted further diversification in 1990s after a financial catastrophe in 1990, buying and selling (Defense) Electronics Systems, nuclear systems, and CBS. Eventually Westinghouse has to sell off nearly everything else, and renamed itself CBS in 1997. Westinghouse Electric Company filed for bankruptcy in 2017.
>
> In retrospect, one may say Edison did really well, both in business legacy and in foresight on DC.

Magnetic Force & Magnetic Field

Magnetic poles produce forces in analogy to charges, and magnetic fields work somewhat similar to electric fields. **Maxwell's Equations** tie both magnetic and electric domains together. Einstein's **special theory of relativity** show how the magnetic and electric forces can be perceived differently depending on the frame of reference; in other words, they are *different* aspects of the *same* phenomenon. Together, we call them **electromagnetic (EM) field**.

Electromagnetic Waves

In 1862-1864, James Clerk Maxwell developed the (Electromagnetic) EM field equations, commonly known as Maxwell's equations. (Even though it is simple, because it requires vector differential operator, we will not show it here.)

In 1887, Heinrich Hertz produced the first deliberate radio waves. (As we now know, it is easy to *unintentionally* produce radio waves. We now name the unit of frequency as Hz after him.)

EM waves follow the inverse-square law. Its propagation speed is the speed of light.

Sound waves are not EM waves, and our ears cannot hear EM waves directly without aid, although they both use Hz as the unit of frequency.

SUPPLEMENT O - FROM ENERGY TO VOLTAGE

Big Picture Views about Energy, Work, Heat

First, the SI unit for energy, work, and heat is the **joule**, abbreviated J. Three birds, one stone - surprise!

Second, the fact that they share the same unit does not give us license to equate them or treat them as interchangeable; more so we need to carefully delineate and differentiate between them.

Work

When a constant force with a magnitude of F displaces a point object by a distance of s *in the direction of the force*, we say work (abbreviated in W) done is $W = F s$. (If the distance traversed is not exactly in the direction of the force, a cosine must be applied to calculate the displacement in that direction.)

The joule is the work expended by a force of one Newton through a displacement of one meter.

Heat

Energy must be transferred to an object to do work (heating the object is also work). Energy cannot be created or destroyed

Energy

Energy is the capacity to do work or to generate or remove heat.

Energy must be transferred to an object to do work (heating the object is also work). Energy cannot be created or destroyed, according to the Laws of Thermodynamics, and under such conditions. Of course, we have all also heard of $E = m c^2$.

Potential Energy

Energy must be transferred to an object to do work (heating the object is also work). Energy cannot be created or destroyed

Kinetic Energy

Energy must be transferred to an object to do work (heating the object is also work). Energy cannot be created or destroyed

Electric Potential Difference

The electric potential difference across the two ends of a circuit is the **potential energy difference per charge** between those two points.

$$\text{Potential Difference} = \Delta PE / Q \qquad \text{(Definition of Potential Difference)}$$

Let us examine the units. Potential difference is expressed in **volts (V)**, which is an SI *derived* unit. The unit for potential energy is **Joule (J)**, and the unit for charge is **Coulomb (C)**. Thus, volt can be derived from Joule per Coulomb, or J/C.

Work-Energy Principle
W = ΔKE - ΔPE

We have conservation of mass and conservation of momentum, likewise we have conservation of energy.

> Energy cannot be created or destroyed, but can be transformed or altered from one form to another.

Simple Views about Power

Once we understand energy, work, and heat, then we can tackle power. We can understand a lot by just taking note that the SI unit for power is Watt, abbreviated W, which is expressed in terms of other SI units as J/s. **Power is the rate of doing work, or the rate of generating or consuming energy, or the rate of producing or dissipating heat.**

> Power = change in energy / time   (Definition of Power)

Combining the definitions of current, potential difference, and power, we get:

> change in potential energy = potential difference * Q
> Power = (potential difference * Q) / time = potential difference * (Q/time)

> Power = potential difference * current   (Power Equation)

Again, let us examine the units. Power is expressed in **Watt (W)**, an SI *derived* unit. Potential difference is expressed in volt, another SI *derived* unit; and current in ampere, which is an SI *base* unit. Thus, from this equation, watt can be derived and expressed as V· A. Alternatively, we can also derive volts as W/A from this equation.

This will be applied in chapter 3 as the **power equation**, or power formula if you like.

Supplement P - V or E for Voltage?

Students often get confused over the character used to represent voltage in formulas. Some think it makes most sense when V stands for Voltage, and so when they see E, they often come to believe it stands for Energy or something else…

Electronics is a language - *both V and E are used!*

Again, it illustrates that electronics is a language. In fact, *sometimes* we see it represented by E, *sometimes* by V. There are dialects and different usages…

In some parts of physics, energy is used often, and often E is used to represent energy. But in beginning physics, velocity is used often, and represented by v, and so it makes even more sense *not* to use V to represent voltage…

However, in electronics, energy is seldom used in formulas, and thus E can often be used for voltage without creating problems. For DC analysis, about half the world of electronics use E for voltage, and another half use V.

For AC analysis and beyond, especially when we use the voltage transfer function for analysis, the majority tends to use V for voltage.

How did we get E for voltage in the first place? It probably dated back to the old days when EMF (Electro-Motive-Force) was the common name for voltage.

The Formula Changes

	The E form	The V form
Ohm's Law	$E = IR$	$V = IR$
Power Equation	$P = IE$	$P = IV$
Transfer Function	$H = E_{out} / E_{in}$	$H = V_{out} / V_{in}$

Take-away

One needs to be flexible in communications, in *understanding* what other meant, and not impose or dictate a fixed idea about what character *must* be used.

Supplement Q - XLR/TRS Cable Project

Your mission is to build a (contrived) cable with XLR connector on one side and TRS connector on the other side. One is not *supposed* to mix these types of connectors, although occasionally it is done. This gives a wider exposure to soldering different kinds of connectors as well as soldering cables, which is one of the more difficult soldering tasks.

The key to a quick and successful cable soldering project is to **plan ahead carefully** and **test incrementally** whenever you can before going to the next step. Here are some specifics:

1) **Insert heat shrink wrap early on,** *way* **before you shrink wrap**.
2) **Insert the connector covers early on, again** *way* **before you finally cover them up.**
3) **Insert the plastic wrap to the cable, again way before you use it to cover the solder joints and screw the TRS cover in.**
4) **Test if the cable can fit through the strain relief coils of the TRS plug** *before* **spending efforts and time stripping and soldering.**
5) **Test if the stripped lengths are correct** *before* **soldering begins by lining up the cable to the crimping position. Most likely you may have to strip the core further, or in some cases, strip the outer plastic further to make room.**

Which hole is Tip?

	XLR Male Connector	Cable	¼" TRS Male Connector
	pin 1	Shield	Sleeve
	pin 2	Red	Tip
	pin 3	Blue or other color	Ring

Relatively good soldering

Poor soldering

♪ SUPPLEMENT R - IMPEDANCE BRIDGING & MATCHING, DAMPING FACTOR

Many audio professionals like to tell others: "match your impedances." What they mean is "if you use balanced XLR connection on one side, you should match it on the other side." The reason why one may get the idea of "impedance matching" is because XLR is considered low impedance and 1/4" connection is considered high impedance, and they are *not* supposed to "match," plus there is a history to the term "impedance matching" ...

Impedance Matching

Once upon a time (early 20th century), when vacuum tube was the only modern active technology in town, it was difficult to produce more than 100 watts of power output. To maximize the power delivered to the loudspeakers so that the audience can hear louder speech or music, engineers took advantage of the **maximum power transfer theorem**. The theorem states that maximum power transfer occurs when the output impedance of the amplifier is equal to the impedance of the loudspeaker. At the time, it was common for loudspeakers to be specified at nominal 16 ohms, and so the amplifier was designed to have an output impedance of 16 ohms nominal as well.

Time has changed. Modern amplifiers can routinely produce 1000 watts of power output, and so priority shifted. The downside of maximum power transfer is that the power bandwidth becomes more restricted, and it was more difficult for the amplifier to control the movement of the loudspeaker voice coil when under impulse situations...

Damping Factor

The ratio of the load impedance to the amplifier internal output impedance is known as the **damping factor** (DF). The load impedance includes wire resistance in addition to the loudspeaker impedance. The reason why it is called damping factor is because the higher the factor, the more the amplifier can control the overshoot of the loudspeaker voice call, stopping it quickly with an impulse input. In other words, the system will get a better impulse response with a higher damping factor.

With this discovery, gone were the days of impedance matching for amplifiers. Modern amplifiers routinely have a DF of somewhere between 2 to 20.

Impedance Bridging

Even before the turn of events concerning impedance matching on the output of the power amplifiers, engineers already had been implementing impedance bridging on the pre-amplifiers. Engineers discovered that the lower the output impedance of a prior amplifier stage, and the higher the input impedance of the next amplifier stage, the less loss of signal results. With that lower loss of signal came also better frequency response and fidelity. (It is important to note that lower loss of signal does not equal to higher power transfer efficiency - the two occurs at different conditions.)

The modern standard is that the input impedance of the next stage should be a hundred times higher than the output impedance of the prior stage. This results in good impedance bridging.

In other words, in modern amplifiers and in audio studios, we use impedance bridging throughout. When we have an XLR cable connecting from a microphone to the pre-amp input, the microphone output impedance is far lower than the pre-amp input impedance. In fact, often *that* is the selection criteria for pre-amps - some vacuum tube and FET pre-amps have super-high input impedance of 1 MΩ or more.

Similarly, the line output impedance is typically a hundred times less than the line input impedances.

THAT - in a nutshell, is what cable matching implies - it is impedance bridging at work (and NOT impedance matching)!

♩ Supplement S - How to Really Turn Off Power

Long time ago, every electrical appliance has a "real" switch. When we turn off the switch, we turn off the power.

But with the arrival of the television about half a century ago, that assumption no longer rang true. To enable the CRT (cathode ray tube) in old televisions to turn on quickly, manufacturers secretly apply small amount of current to the filament *all* the time, so that it does not take as long to have a picture (when compared to cold start) when "turned on". **To *really* turn off power one must *unplug* the television set altogether**...

Those were the days, my friend... Now we have electronic devices that use so little power, plus there are ***no*** power switches at all. An example is my exercise heart-rate monitor. It is powered by a coin battery. When "not in use", the microprocessor "sleeps" in a super-low power mode. It sips so little electricity that is roughly equivalent to the battery power that would have been lost naturally under storage anyway. (At least that is the claim to remove the physical switch, which turns out to be not cheap if we want a reliable switch.) So, the only way we can turn off the device is to remove the battery. You can do the same with most android cell phones, but if you have an iPhone, the battery may not be replaceable (nor accessible), so good luck with *really* turning off power! (In general, if you do not have a real physical switch, but rather a capacitive sensor for power switch, you can assume that it is never powered off, as the capacitive sensor needs power to operate!)

For audio electronics in the studio, there is usually a certain specific order and protocol needed to turn on or off specific devices in the chain. Since most users can't be expected to follow the precise order specified, there is usually a **power sequencer** which sequences the power on and off cycles. Certain equipment need to be turned on earlier or later, and really **no equipment should be turned on simultaneously**, as the sudden rush of current may damage components. In no case should we attempt to turn *all* equipment on a power extension all at once, as that will cause a power surge due to inductance which may be high enough to damage some components. One should thus **only** use the sequencer to turn on or off the equipment cluster.

Chapter 4   WAVES, TIME & FREQUENCY DOMAINS, dB

> Waves, representation in space, time, & frequency, and calculations.
> dB and related calculations.
>
> *Expected Learning Outcome*: The student will be able to handle waves, time and frequency domain implications, and calculations relating to them, and handle them with ease and confidence if this is practiced sufficiently.

CONTENTS

Sine Waves in Space & Time Domains	4-3
Sine Waves in Time & Frequency Domains	4-6
Basic Periodic Waveforms	4-9
RMS, Peak, Peak-to-Peak, Average, and Conversions	4-11
Comparing Waves: Magnitude and Phase	4-13
Basics of Sound, ratio, dB, and Conversions	4-14
Review Exercises	4-17

This chapter prefcedes AC Analysis. It covers all the prerequisites in order to smooth AC analysis:
- Sine waves in space, time, and frequency.
- Basic periodic waveforms.
- Magnitude: RMS, peak, peak-to-peak, average
- Phase

- Ratios, dB
- Consumer and Pro audio

This chapter needs to be practiced, with a lot of calculations. Necessary before the student becomes comfortable to tackle AC analysis with ease and confidence. It is also a review of DC analysis to some extent – using logarithms. Again, do not skip or make light of this chapter!

4.1 SINE WAVES IN SPACE AND TIME DOMAINS

Waves in Time

Engineers often refer to the following waveforms as "sinusoidal waves", with "sine waves" as the shorthand. Unless otherwise stated, the usual understanding is that the waves shown are only a snapshot - that they actually are periodic from time past to the snapshot shown and continue thereafter. From that perspective, notice they all look alike; hence the grouping together as one category of waves. However, in terms of mathematics, they are generated differently: the first one uses a sine function (having an amplitude of zero at time zero), the second a cosine (one at time zero), the third a negative sine (moving towards negative at zero), and the fourth a negative cosine (going up at zero). The negative waves are simply the mirror image of the positive waves, reflected across the x-axis. Using the concept of electronics as a language, please understand that people can and often use the term "sine wave" loosely and interchangeably, unlike in mathematics. They are really the same wave when we look at them with different point of the wave as origin - assuming the wave is continuous, from everlasting to everlasting of course.

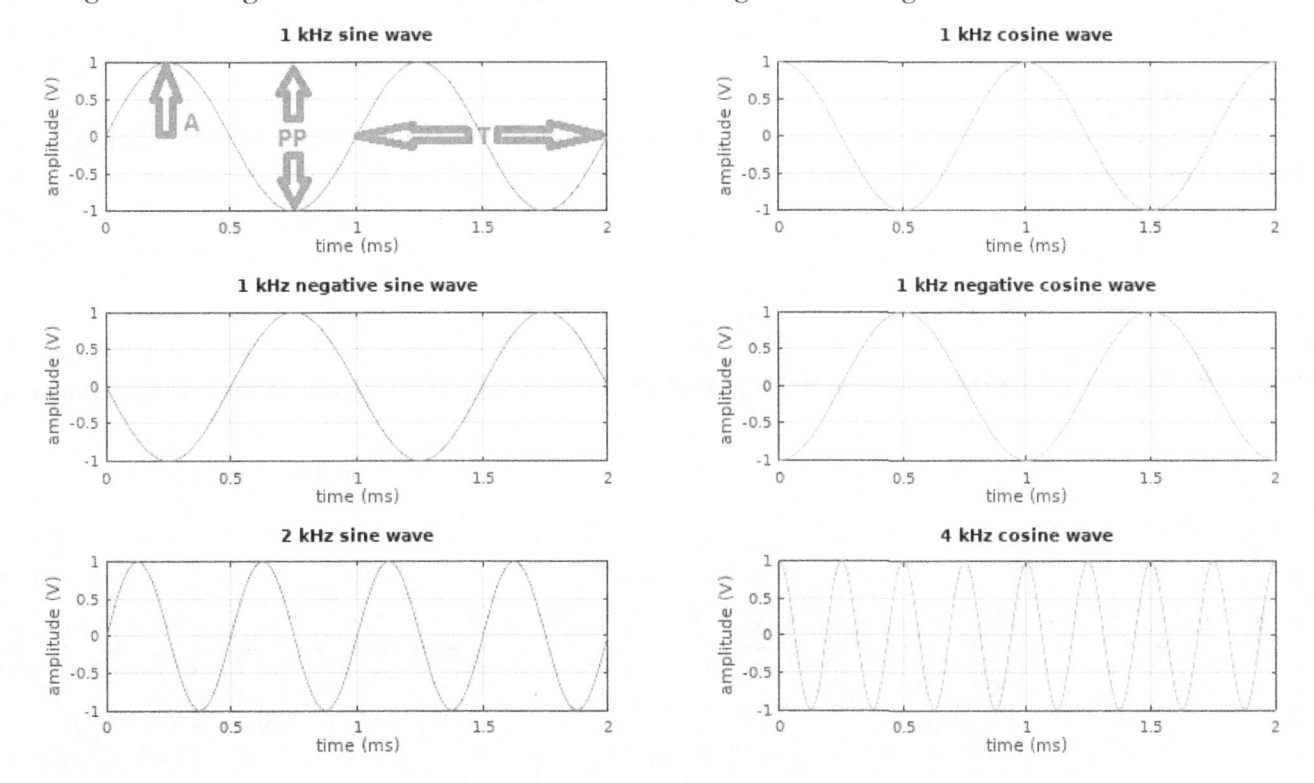

There are three characteristics we can associate with a sinusoidal wave: its **amplitude**, **period**, and **frequency**.

There are two ways engineers use the term amplitude: 1) The y-axis of all these plots are labeled amplitude (V), where the **instantaneous amplitude** is the value of y at a particular time. 2) The **amplitude of a wave**, though, some consider as the maximum excursion the waves make from its neutral point, while others consider as the maximum value. Since we use **bipolar waves** in all these example, which moves equally from zero to positive and negative peaks, the amplitude of the waves is the same by either definition, and is simply the distance from zero to either positive or negative peaks.

In the first plot, the amplitude is indicated as "A"; the **peak-to-peak** is indicated as "PP", and the **period** is indicated as "T". For all the plots, the amplitude is 1 V, and the peak-to-peak is 2 V (reminder: always note the unit). The period for the first plot is 2 - 1 = 1 ms. We can calculate period from any point to where the waveform repeats - in this case we use the zero-crossing point. We could have used the distance between peak to peak, and get 1.25 - 0.25 = 1 ms. Likewise, verify for yourself that the period is 1 ms for all first four plots. The period for the fifth plot is 0.5 ms, and for the last plot it is 0.25 ms.

> For waves, we can apply the formula:
> frequency = 1/period
> or, period = 1/frequency

Thus, the frequency of the first four waves are 1/1m = 1 kHz. (Notice we are using SI unit prefixes to our advantage in calculating the result.) The frequency for the fifth plot is 1/0.5m = 2 kHz. The frequency for the sixth plot is 1/0.25m = 4 kHz.

Waves in Space

Below are another set of plots of sine waves; they may look identical to the previous figure at first. But it is important to notice the axis labels - they are different!

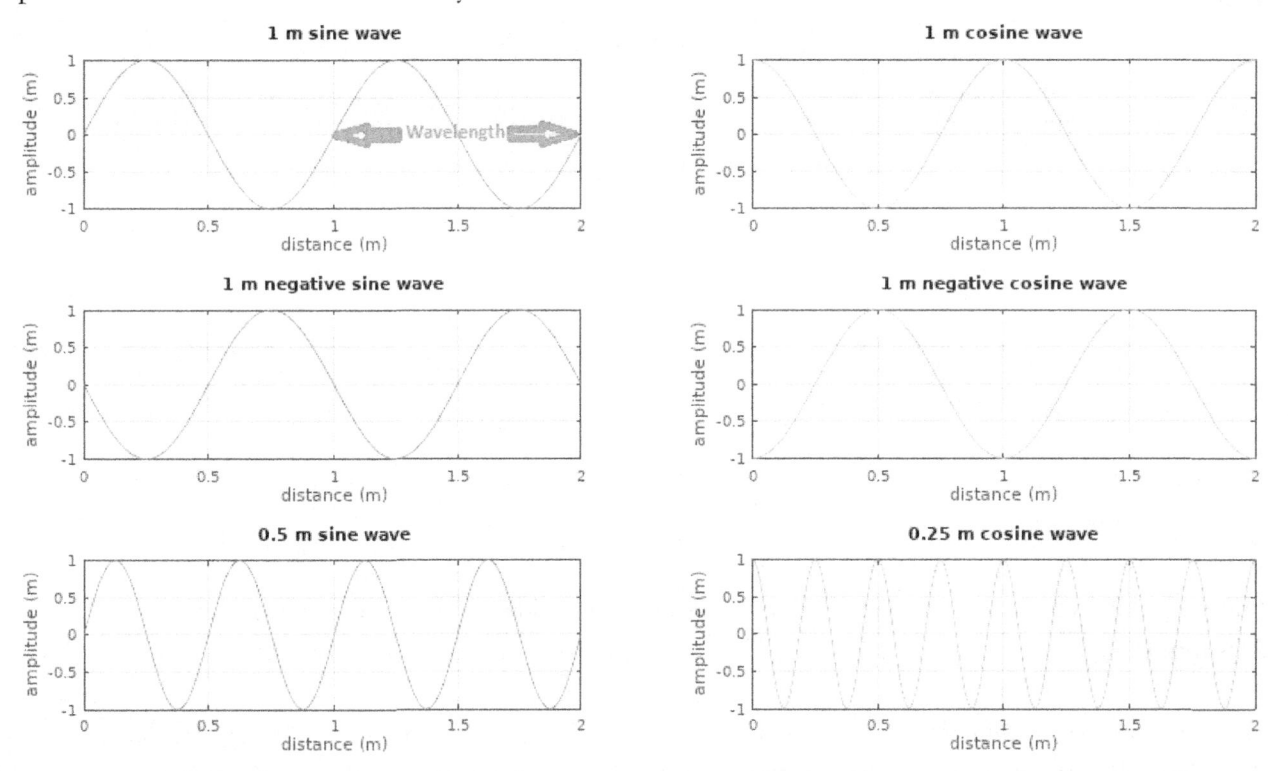

The x-axis was measured in time (ms) previously; now it is in distance (m). The former we say is in **time domain**, while the latter is in **space domain**. In space domain, it could be longitudinal or transverse waves, depending on how the standing waves travel, longitudinally or transversely, both of which can be represented as above. A guitar and violin strings produce transverse waves, while wind and long string instruments produce longitudinal waves. Sound waves in air and fluid media are longitudinal waves. Thus the guitar string vibrates transversely, but the music we hear comes longitudinally.

The amplitude for all the waves shown above are 1 m. The peak-to-peak are 2 m.

When in space domain, the **wavelength** is the distance between any repeating points in the wave; for the first four waves the wavelengths are all 1 m. The wavelength for the fifth wave is 0.5 m, and for the sixth wave, 0.25 m.

> For waves measured in space domain, we can apply the formula:
> frequency = speed of propagation / wavelength
> or, wavelength = speed of propagation / frequency

Please note that the formula for frequency is different for time and space domains. We first need to decide which domain we are concerned about, and then use the appropriate formula.

The speed of sound in dry air is about 340 m/s (at 0 °C). Thus the frequency for the first four waves would be 340/1 = 340 Hz, for the fifth wave 340/0.5 = 680 Hz, and for the sixth wave 340/0.25 = 1360 Hz.

The period for the wave in space is still governed by the formula period = 1/frequency. Thus the period for the first four waves would be 1/340 = 0.00294 s or roughly 3 ms. **Note that we do *not* see the period in space domain plot, rather only the wavelength!**

♫ Room Modes

The space in a room constraints waves, because there can be no physical movement of fluids at the boundaries of the room since that is in solid state. Because the room is in three dimensions, when excited by an acoustic source, a collection of resonances will exist, influenced by the room dimensions, and becoming the biggest obstacle to accurate sound reproduction.

Space domain is used to study room modes. A cubic room will have resonances from the three pairs of walls at the same frequency, exaggerating its effect. In less symmetric situations, of course, there are **axial**, **tangential**, and **oblique** room modes. In contrast, walls that are angular with irregular shapes will diffuse sounds and break up resonances by having far more resonances at varying frequencies, thus limiting their individual effects. Absorption through resonating chambers also helps; foam wedges a little less.

Room modes can be identified through three methods when they agree with one another: 1) peaks in the frequency response; 2) slow decay in the time domain; 3) physical measurements of room dimensions.

Examples

1) **The lowest frequency for axial room mode can be calculated as the room width being half a wavelength.** For a 10 feet cubic room, the wavelength will be **twice the width**, meaning 20 ft (so that half the wavelength is 10 feet, the room width). The speed of sound is roughly 1130 ft/s. Thus applying our formula, room mode frequency is 1130/20 = 56.5 Hz.

2) The rule of thumb for good room acoustics is to get the important room modes **below the range of hearing, which is 20 Hz**. The speed of sound is roughly 1130 ft/s. Thus applying our formula, minimal room dimension is 1130/20 = 56.5 ft! The ceiling in most buildings restrict the minimum dimensions. So only concert halls, Hollywood sound-stages, and ancient cathedrals would meet such requirement. Note that there are other room modes above the minimum, and the idea is to get them below hearing too. That explains why if we can get the minimum dimensions of the open space to be 100 ft. tall, that would be even better... :)

4.2 Sine Waves in Time and Frequency Domains

Adding Waves in Time

Waves can be mixed together linearly, in such a case, we simply add the amplitudes of the waves at each point in time to get the resulting wave. An example is shown below.

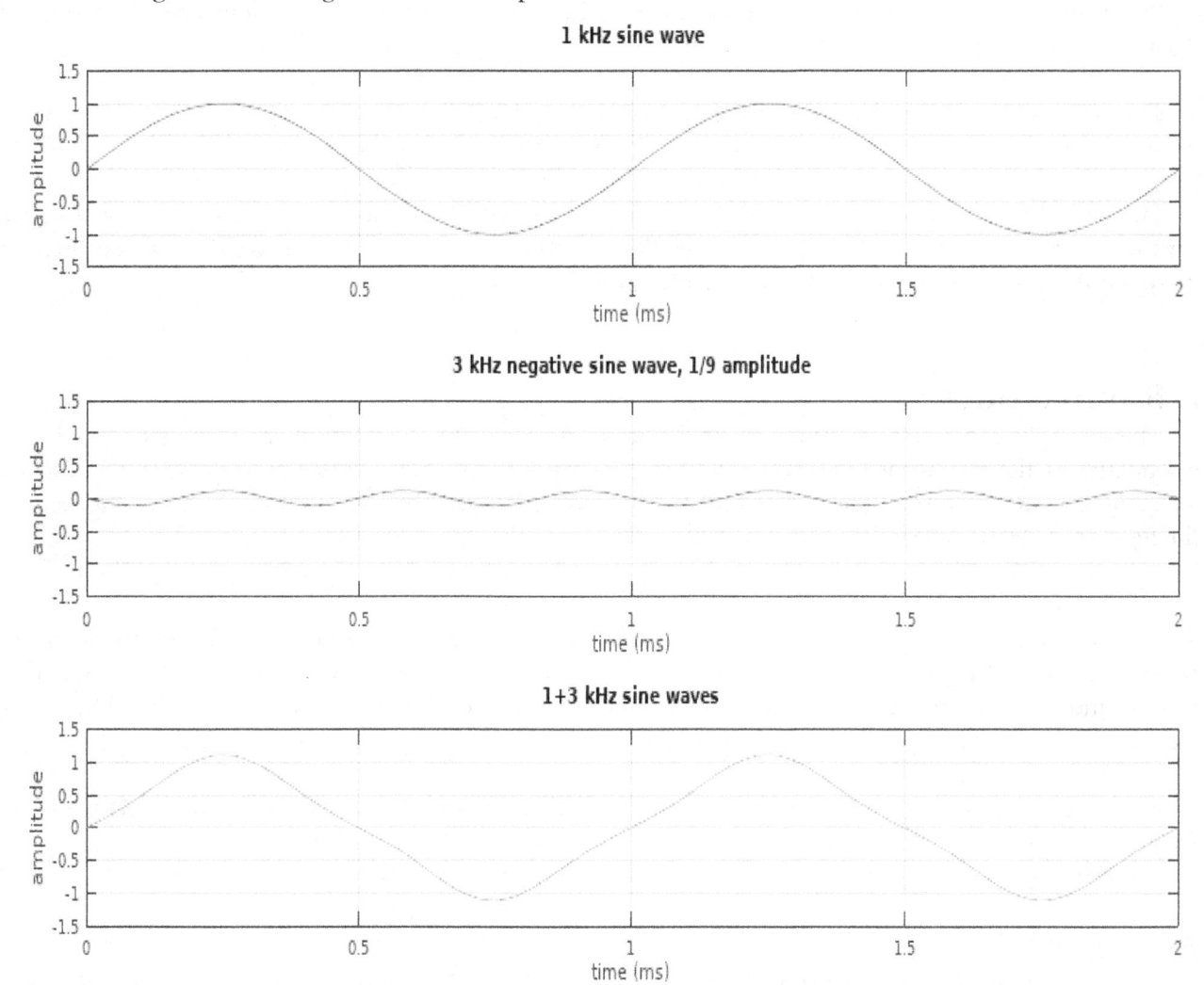

Note that the first wave has an amplitude of 1, a period of 1 ms, and a frequency of 1/1m = 1 kHz. The second wave has an amplitude of 1/9, or roughly 0.1, a period of ⅓ ms (obtained by noticing that there are three full cycles between 0 and 1 ms), and a frequency of 1/(⅓)m = 3 kHz.

It takes a lot of effort to add the two waves together at each point in time. The result is shown in the third plot. Notice that the amplitude of the resultant wave is slightly larger than 1. It already looks like a triangular wave…

From Time to Frequency Domain

What if we plot the behavior of the **same** sine waves as before, but using frequency as the x-axis (instead of time)? The results are shown to the right of the same time domain chart repeated here to the left:

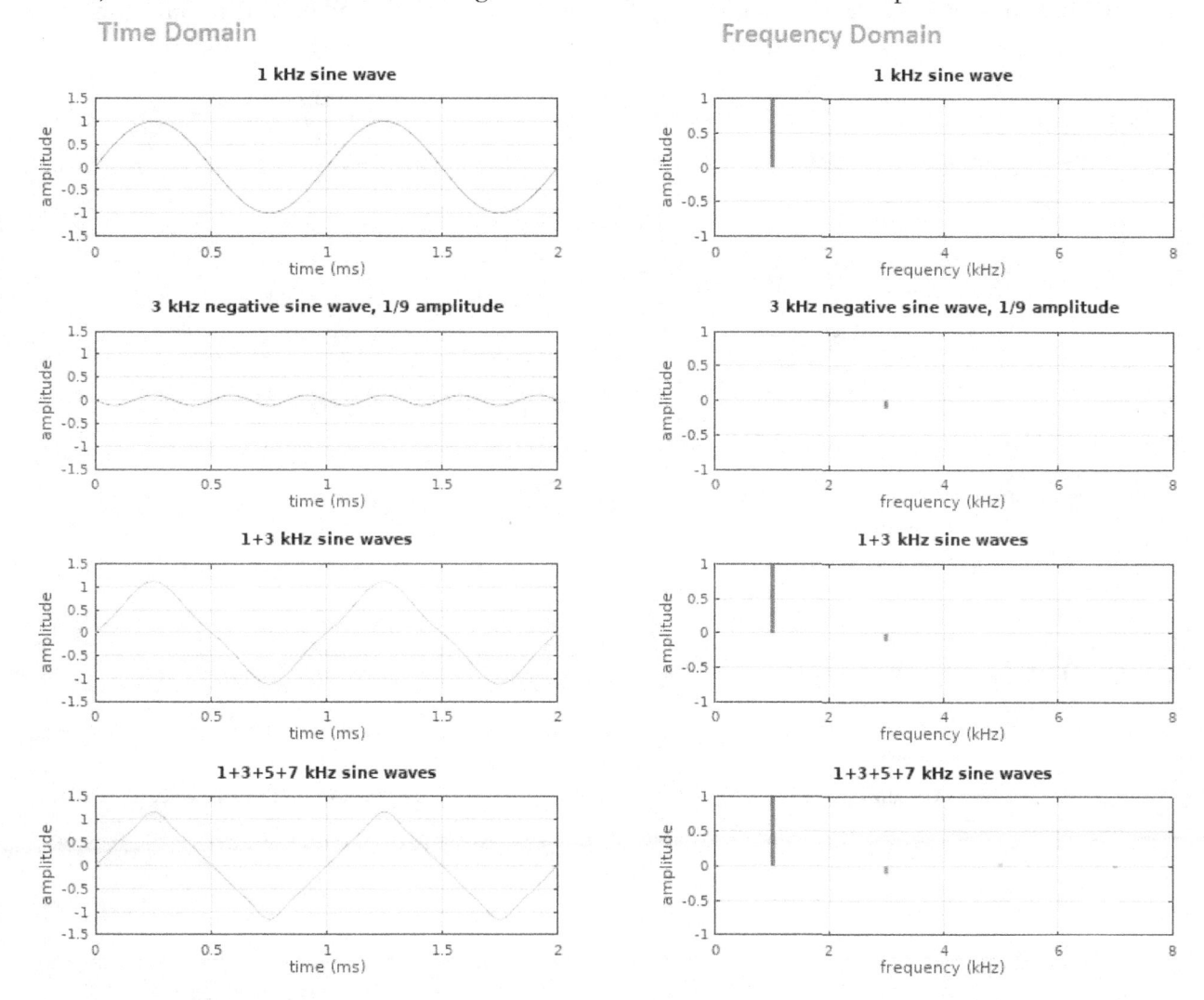

In the first "pane" to the right, there is a simple magnitude of 1 at frequency of 1 kHz, which was what we calculated earlier. In the second pane, there is a simple magnitude of 1/9 at frequency of 3 kHz, which was again what we calculated. Note that the amplitude is negative to denote a negative sine wave. We simply draw those two parameters as vertical bars in the proper axis. In the third pane, we simply add the two waves together, just like in the time domain, except that this time it is very simple - that is often the advantage of doing things in frequency domain. Notice that one cannot easily predict the amplitude when two waves are added together - in the frequency domain one has magnitude of 1 and another has a magnitude of roughly -0.1, so one may expect the resulting wave has an amplitude of less than 1. In reality, as we know in time domain, the resulting peak is larger than 1.

We have added two more waves in the bottom left and right traces, at 5 kHz and 7 kHz respectively. The resulting wave already looks fairly much like a triangular wave. In fact, **triangular waves can be created by adding odd harmonics in alternating negative and positive small and smaller amounts**, as illustrated with the final plot.

From Frequency to Time Domain

Let us attempt another example, this time starting from frequency domain and moving to time domain.

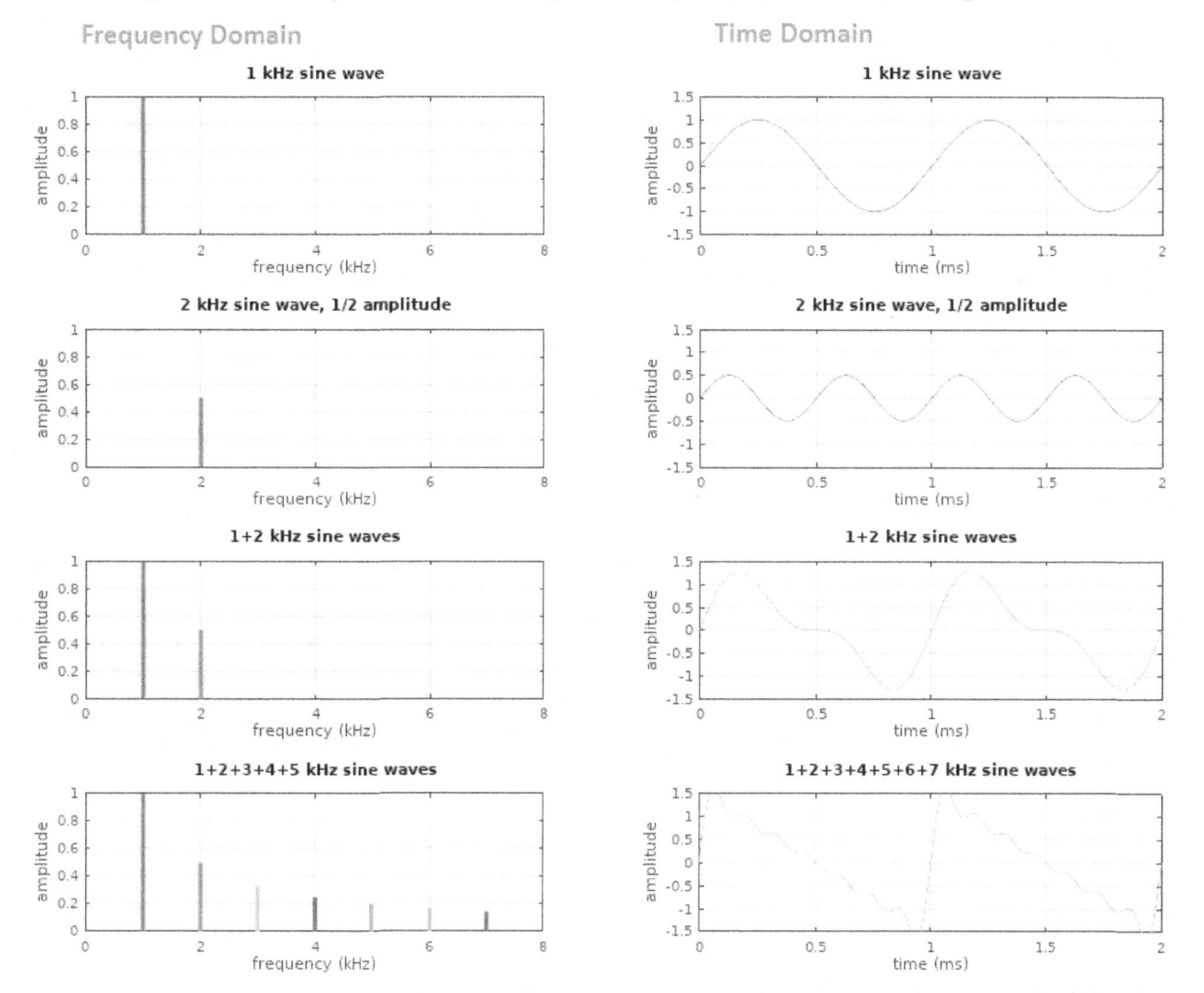

In the first row, we have a 1 kHz sine wave of amplitude 1. In the second row we have a 2 kHz sine wave of amplitude ½. In the third row we add the two together, resulting in a sawtooth-looking waveform. If we continue to add more sine waves at 3, 4, and 5 kHz at their fractional amplitudes accordingly, we get the fourth row, which looks remarkably like a sawtooth waveform in time domain.

We can thus perform **additive wave synthesis** to create more complicated waveforms from a series of simple sine waves. In 1807, Joseph Fourier made the bold claim that ***any* arbitrary function** (which, if you recall, represents relationship, and can be represented as waveforms) **can be represented by a Fourier series** (which is simply a collection of sine waves in various fractional amplitudes). The decomposition of a complex waveform into its sine wave components is known as **Fourier analysis**, and the building up of the complex waveform from simple sine wave components is known as **Fourier synthesis**.

4.3 Basic Periodic Waveforms

Here is a summary of the same strategy to additively synthesize what we call basic periodic waveforms:

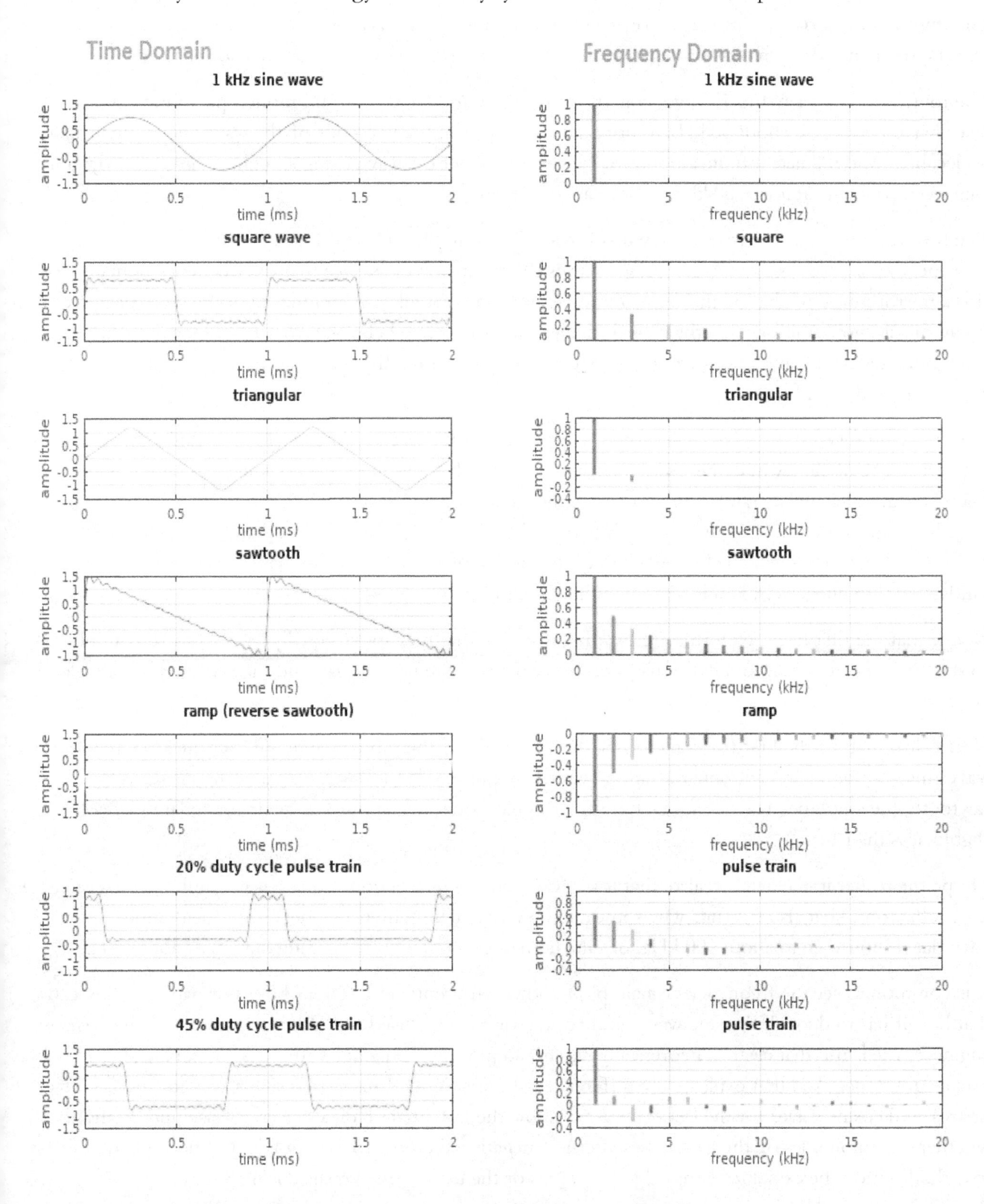

[Time Domain and Frequency Domain plots for: 1 kHz sine wave, square wave, triangular, sawtooth, ramp (reverse sawtooth), 20% duty cycle pulse train, 45% duty cycle pulse train]

The first trace, again, is the **sine wave**.

The second trace was constructed with 9 **odd** harmonics plus fundamental of 1 kHz (up to 20 kHz), and it begins to resemble a **square wave**. It takes harmonics up to about 100 kHz to not have the ripples in the wave becoming visually noticeable.

The third trace is a **triangular wave**, constructed with only 4 **odd** harmonics plus the fundamental. Notice that the frequency domain plot go negative alternatively with the harmonics, as we had seen earlier.

The fourth trace is a **sawtooth wave**, constructed with 20 (odd and even) harmonics plus fundamental. It takes harmonics up to about 50 kHz to not have the ripples at the two ends of the wave becoming visually noticeable. And it'll take infinite amount of harmonics to get the wave rising straight up immediately; it'll take again probably at least 100 kHz to have a sharper rise.

The fifth trace is a ramp or **reverse sawtooth wave**. It is simply a mirror-image of the sawtooth, mirrored over the x-axis. It looks like it is ramping up, thus the **ramp wave**. Notice that the frequency domain plot is also a mirror image of that of the sawtooth wave, meaning that all the amplitudes are shown as negative. Again this is another illustration that it is difficult to interpret time behavior from frequency domain plot - even though all wave amplitudes are negative in frequency domain, the resulting waveform still has positive peaks in time domain!

The sixth and seventh trace are 20% and 45% duty cycle **pulse trains** respectively. The duty cycle refers to the percentage of time where the wave is positive. Notice that although they look like a square wave (which is a 50% duty cycle pulse train) in time domain, the frequency response is completely different. The square wave has only odd harmonics response, while the other duty cycles uses both odd and even harmonics, and in different amplitudes, which alternates to negative in a sine-like cycle. Note that the amplitude of the fundamental frequency (1 kHz) is not 1 in either case (it is 0.988 for 45% duty cycle).

Notice that the first five waveforms are **bipolar**, meaning that they move equally away from zero amplitude, so that the average value would be zero. The last two waves are not bipolar, and the average is positive and not zero.

Notice also that for the first five waveforms, the amplitude of the fundamental is 1, but the amplifier of the waves may not be 1. In fact, only the sine wave is 1, because only the fundamental is involved. For sawtooth, the amplitude reaches about 1.5; for triangular wave, it is about 1.2; while for square wave, it is slightly less than 1.

I hope the reader immediately realize that it is *NOT* easy to create a good square wave, and to a lesser degree sawtooths too. That also explains why some audiophiles and amplifier designers want their amplifiers to have flat response up to about 100 kHz, despite human ears not able to hear anything past about 20 kHz.

It is common to see the frequency domain plots shown differently in different books or papers. How can that be? It has to do with the sine waves and cosine waves being used for additive synthesis, as the *language* engineers use is not that clear. Theoretically we should plot the frequency with phase in parallel, but that is usually not done. And that explains the differences. In this figure, I have used sine waves for the first five figures, and cosine waves for the last two, just to show the potential of differences. Please understand that whether the amplitudes of the harmonics alternate negative and positive or not is not as big a big deal as one may think - it can be explained simply by the phases or the use of sine versus cosine waves.

Additive synthesis had not been popular using analog hardware, as a lot of oscillators and mixers are necessary, each one adding to the expenses and the noises that may result.

4.4 RMS, Peak, Peak-to-Peak, Average, and Conversions

Many students don't want to tackle RMS (Root-Mean-Square) because it looks scary - or at least the *explanation* looks scary in many textbooks. We shall see that in reality the student had done something similar already…

Pythagoras Theorem

Given a right-angled triangle with a side of 3 m and 4 m, what is the hypotenuse? We first *square* the sides, $3^2 = 9$ and $4^2 = 16$. Then we *sum* them together: $9 + 16 = 25$. Then we *square-root* to get the answer of $\sqrt{25} = 5$ m. The operation can be called **root-sum-square** instead of Pythagoras… :)

Let us use the same two pieces of data: 3 m and 4 m to calculate RMS, peak, peak-to-peak, average…

Root-Mean-Square

To calculate root-mean-square, we simply substitute *sum* for *mean* (which mathematically means average). For 3 m and 4 m, we first *square* the sides, $3^2 = 9$ and $4^2 = 16$. Then we *average* them together: $(9 + 16) / 2 = 25/2 = 12.5$. (Note that there are 2 pieces of data and so we divide the sum by 2.) Then we *square-root* to get the answer of $\sqrt{12.5} = 3.54$ m. Notice this result is somewhere in between the two initial values of 3 and 4.

Peak

The peak of 3 m and 4 m is simply which appears taller (when placed vertically). Obviously the peak is 4 m.

Peak-to-Peak

The first "peak" in peak-to-peak is the peak we found previously. The second "peak" in peak-to-peak refers to the "bottom" peak (as in a monkey hanging in reverse). In this case the second peak is 3 m. So peak-to-peak is the difference between the two, or $4 - 3 = 1$ m.

Average

The average is simply $(3 + 4)/2 = 7/2 = 3.5$ m.

Examples

1. Calculate the RMS, peak, peak-to-peak, average for 3 V, 4 V, and 5 V.

We first *square* the data, $3^2 = 9$, $4^2 = 16$, $5^2 = 25$. Then we *average* them together: $(9 + 16 + 25) / 3 = 50/3 = 16.67$. (Note that there are 3 pieces of data and so we divide the sum by 3.) Then we *square-root* to get the answer of $\sqrt{16.67} = 4.08$ V. Notice again this result is somewhere in between the three initial values.

The peak is the highest number, or 5 V. The bottom peak is the lowest, or 3 V. Thus the peak-to-peak is $5 - 3 = 2$ V.

The average is $(3 + 4 + 5)/3 = 12/3 = 4$ V. Notice in this case the RMS is very close to the average.

2. Calculate the RMS, peak, peak-to-peak, average for 3 V, -3 V, 4 V, and -4 V.

We first *square* the data, $3^2 = 9$, $(-3)^2 = 9$, $4^2 = 16$, $(-4)^2 = 16$. Then we *average* them together: $(9 + 9 + 16 + 16) / 4 = 50/4 = 12.5$. Then we *square-root* to get the answer of $\sqrt{12.5} = 3.54$ V.

The peak is the highest number, or 4 V. The bottom peak is the lowest, or -4 V. Thus the peak-to-peak is 4 - (-4) = 8 V.

The average is (3 + (-3)+ 4 + (-4))/4 = 0/4 = 0 V. Notice in this case the RMS is very different from the average! The average is zero, showing the middle of the data, while the RMS is the middle of the peaks in absolute value, if you will.

RMS, peak, peak-to-peak, and average for waveforms

The same method applies when calculating the RMS and so forth for sine waves: conceptually we go through the same calculation for every data point in time. However, the calculation is tedious, but thankfully it has already been done for common waveforms of amplitude A, the result is as shown below:

Waveform with amplitude A	RMS	Peak	Peak-to-Peak	Average
Bipolar sine wave	$A/\sqrt{2}$	A	2A	0
Bipolar square wave	A	A	2A	0
Bipolar triangle wave	$A/\sqrt{3}$	A	2A	0
Bipolar sawtooth wave	$A/\sqrt{3}$	A	2A	0
Polar positive pulse train	$A/\sqrt{D}$	A	A	A * D

Note that for **bipolar** waves, they move equally away from zero, and are symmetrical about the x-axis, thus their average is 0. Since we define their peak to be A, and they are symmetrical about the x-axis, and thus their peak-to-peak is 2A.

For **polar** wave example of a positive cycle pulse train with duty cycle of D, the average is positive, depending on how often the pulse is turned on. D is between 0 and 1. When it is 0, it is never turned on, and so the average is A * 0 = 0. When it is 1, it is always turned on, and so the average is A * 1 = A, or the same as peak. The bottom peak is zero, and thus peak-to-peak is equal to peak, or A. The RMS is shown mathematically to depend on the square root of D.

Please be careful when comparing this chart with the waveform chart earlier. This chart uses amplitude A for the *composite* waveform, while the earlier chart uses amplitude A for the *fundamental* sine wave.

Examples

1. For a 120 V mains, calculate all the respective values.

First note that the mains specify RMS. In other words, RMS is 120 V. Using the table above, the peak would be RMS*$\sqrt{2}$ = 120 * $\sqrt{2}$ = 169.7 V. The peak-to-peak would be twice as much, or 2*RMS*$\sqrt{2}$ = 339.4 V. The average is zero, which means the DC component is zero.

2. If the amplitude of a bipolar sine wave is 10 V, calculate all the respective values.

First note A = 10 V. Then RMS = $A/\sqrt{2}$ = 7.07 V. Peak-to-peak = 2 * A = 20 V. Average is 0 V for a bipolar wave.

4.5 COMPARING WAVES: MAGNITUDE & PHASE

When we have two or more waveforms, we can compare their **magnitudes**, using RMS, peak, peak-to-peak, and average, concepts we have introduced earlier.

We can also compare their **phases**. Note that phase is relative - while magnitude is determinable with only *one* waveform, phase must be determined between *more than one* items.

The unit for phase is either **degrees** or **radians**. The conversion factor is simply visualized by a circle, which has 360 degrees or 2π radians. In other words, **2π radians = 360 degrees**.

Examples

1. Convert -90 degrees to radian.

    Simply multiply by the conversion factor 2π radian/360° :

    $$-90° * 2\pi \text{ radian}/360° = -\pi/2 \text{ radian}$$

    Notice I deliberately put in the unit in the conversion factor so that we can ascertain that the unit of degree is cancelled and we get the radian at the end. This is a very good double check that we apply the right conversion factor.

2. Convert $\pi/4$ radians to degrees.

    Simply multiply by the conversion factor 360°/2π radian :

    $$\pi/4 \text{ radian} * 360°/2\pi \text{ radian} = 45°$$

Phase

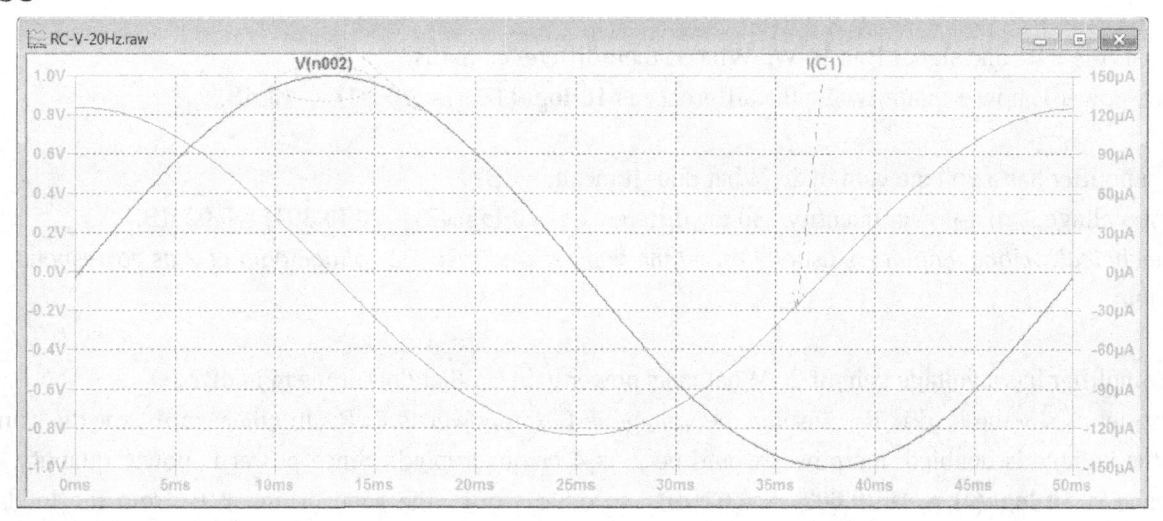

In the chart above, we have two waves: V(n002) and I(C1) - they don't even need to be the same type (in this case one is voltage and the other is current). They are both sinusoidal, and V(n002) crosses zero from positive to negative at 25 ms, while I(C1) crosses zero at 12.5 ms. In other words, the current leads voltage by 12.5 ms. The period is 50 ms for both waveforms. We say 50 ms represents 360°, so current leads voltage by 12.5 ms * 360°/50 ms = 90°. The phase shift between voltage and current for this circuit is thus +90°, with current leading voltage.

4.6 BASICS OF SOUND, RATIO, dB, AND CONVERSIONS

When we think about sound, we often think of this sound being twice as loud as the other sound. We are essentially thinking in terms of **ratios**. Engineers think of using the *shortcut* of **decibel** (notated as **dB**) to help with their calculations and understanding. Students usually think of learning *one more thing* a bad idea, but let us *first* learn about ratios and dB and their conversion, and then see how that can be useful…

Converting ratios to dB

First we need to classify the ratio as whether it is ratio of **power quantity** or not. If it is power or acoustic or luminous intensity, it is power quantity. The other category is **root-power quantity**, previous called **field quantity**. Examples are current and voltage, which are really related to square roots of power (hence the name).

The reference is power quantity, and the formula for **change of power is L = 10·log$_{10}$(y/x) dB**.

In order for dB to mean the same for root-power quantities, **change of power is L = 20·log$_{10}$(y/x) dB**.

The factor of two compensates for the differences in power or root-power so that a 3 dB has the same effect, whether for power or current or voltage.

Examples

1. Signal A is 10 V$_{RMS}$, and signal B is 1 V$_{RMS}$. What is their difference in dB?
   First, RMS voltage is root-power quantity. So the difference is **20·log$_{10}$(10/1) = 20·1 = 20 dB**.

2. Signal A is 1 A, and signal B is 10 A. What is their difference in dB?
   First, current is root-power quantity. So the difference is **20·log$_{10}$(1/10) = 20·(-1) = -20 dB**.
   Notice that ratio 10 yields 20 dB, while ratio 0.1 yields -20 dB. Depending on which quantity is numerator or denominator in the ratio, the result is + or - 20 dB, or one is higher or lower.

3. Signal A is 1 W, and signal B is 10 W. What is their difference in dB?
   First, power is power quantity. So the difference is **10·log$_{10}$(1/10) = 10·(-1) = -10 dB**.

4. An amplifier has a voltage gain of 2. What does it mean in dB?
   First, voltage is root-power quantity. So the difference is **20·log$_{10}$(2) = 20·(0.301) = 6.02 dB**.
   In technical writing, engineers usually round the number and take the voltage gain of 2 as corresponding to 6 dB of gain.

5. An amplifier has a voltage gain of 2. What is its power gain? What does it mean in dB?
   First, let us say the load is R. Assume the voltage is E, the power is E$_2$/R. In other words, for the same load, when the voltage is doubled, the power would be 2$_2$ = 4 or quadrupled. Since power is power quantity. So the difference is **10·log$_{10}$(4) = 10·(0.602) = 6.02 dB**. In other words, the acoustic intensity from the loudspeaker would be increased by 6 dB, just as it was indicated earlier in example 4.

This last example shows why we have to understand whether a ratio is of power or root-power quantity. By compensating between the two, we maintain the same dB gain for an amplifier without having to specify whether it is a 6 dB for power or 6 dB for voltage or current.

Chapter 4 – Waves

Converting dB to ratios

First we need to classify the ratio as whether it is of **power quantity** or **root-power quantity**, as before.
If one is comfortable with the math, one can use the same formula as before, and apply antilog arithmetic.
The reference is power quantity, and the formula for **change of power is L = 10·log$_{10}$(P/P$_0$) dB**.
In order for dB to mean the same for root-power quantities, **change of power is L = 20·log$_{10}$(V/V$_0$)) dB**.

But for those not comfortable with the math, it is probably easiest to just memorize another set of formulas:
For power quantity, **P = 10$_{(L/10\ dB)}$ · P$_0$**
For root-power quantity like voltage, **V = 10$_{(L/20\ dB)}$ · V$_0$**

Examples

1. Calculate the voltage gain corresponding to +4 dB.
   First this is a root-power quantity.   **V/ V$_0$ = 10$_{(+4\ dB/20\ dB)}$ = 10$_{(0.2)}$ = 1.585**

2. Calculate the power gain corresponding to +1 dB.
   First this is a power quantity.   **P/ P$_0$ = 10$_{(+1\ dB/10\ dB)}$ = 10$_{(0.1)}$ = 1.259**

dBV

The suffix V after dB is a shorthand for dB with a reference of **1 V**. Note that dBV is *not* recognized by international standards organizations ISO or IEC, although still commonly used.

dBm

The suffix m after dB is a shorthand for dB with a reference of **1 mW**.

dB VU

The **VU meter** was developed in 1939 for audio metering, where 0 dB VU was generally defined as the "turning point" of a piece of audio gear beyond which distortion would start to increase due to clipping and other phenomena. Above 0 dB VU is shown in red, but the meter continues forward past +3 dB VU, with most modern pro audio equipment having a **headroom** of 15 dB or more, meaning it continues to function well without much distortion until +15 dB VU or so. Note that the mid-point of a conventional VU meter is typically at -4 dB VU, so psychologically it was designed to make audio engineers feel like the average of the music should be at around -4 dB VU point, and as long as it peaks not past the red area in the meter, it should be fine.
For consumer audio, the nominal line value of 0 dB VU is -10 dBV.

dBu

The suffix u refers to *unloaded*.   During the early telephone era, signal levels were written in dBm referenced to a standard telephone line impedance of 600 Ω. It is a *loaded* dB value. When we *unload* that standard impedance, that is without reference of the standard impedance, but keeping its voltage value, we call it dBu.
For pro audio, the nominal line value of 0 dB VU is +4 dBu.

dB FS

The *de facto* reference for analog audio has been 0 dB VU. However, for digital audio, it is 0 dB FS, with a very different usage. FS means **full scale**, beyond which the digital system cannot handle any voltage larger. Another way of saying it is that it clips digitally. So for a digital system we *never* wants to reach 0 dBFS, but instead we try to always get the whole music program just a little bit lower, so that at maximum peak it is still a tiny bit lower than 0 dBFS. **There is no headroom past 0 dBFS**.

In contrast, recall that in pro audio we usually have 20 dB or more headroom past 0 dB VU. Some analog systems try to incorporate dBFS into their labeling, which of course occasionally means that one may go above 0 dBFS. In digital systems, one never goes above 0 dBFS; dBFS is a negative number.

dB SPL

The suffix SPL refers to **Sound Pressure Level**, with the reference being "**threshold of hearing**" which is usually given as **20 micropascals** at 1 kHz.

dBA or dB(A)

The suffix A or C denote the use of different **weighting filters** when used to measure dB SPL. A-weighting is commonly used to emphasize around 3-6 kHz to which the ear is most sensitive, and is used to measure environmental noise. B- and C- weighting are common for measuring louder sounds.

Examples

1. Calculate the voltage corresponding to 0 dBu.

The RMS voltage is defined for 1 mW at 600 Ω. Applying Ohm's Law and power formula (or just remembering the power wheel), we know $P = E^2/R$, or $E = \sqrt{(P \cdot R)} = \sqrt{(0.001 \cdot 600)} = \sqrt{(0.6)} = 0.775$ V.

2. Calculate the nominal line voltage for pro audio corresponding to +4 dBu.

We have calculated earlier that +4 dB voltage gain is 1.585 x. We just calculated the RMS voltage for 0 dBu to be 0.775 V. So +4 dBu corresponds to 1.585 x 0.775 V = 1.228 V RMS.

3. Calculate the nominal line voltage for consumer audio corresponding to -10 dBV.

Recall that dBV is referenced to 1V. So we directly apply the dB formula: $V = 10^{(-10 dB/20 dB)} \cdot 1$ V = 0.316 V RMS.

4. How much is pro audio voltage higher than consumer audio?

This is a little tricky. If we calculate the ratio between the above two numbers, we get 1.228 V / 0.316 V = 3.883 or nearly four times higher. In terms of dB, it is $L = 20 \cdot \log_{10}(3.883) = 11.78$ dB or nearly 12 dB.

However, it should be noted that consumer audio is unbalanced, while pro audio is balanced with differential signaling, the latter means that two lines are used in complementary fashion, so the voltage swing per line is actually *half* the value calculated, or nearly two times and 6 dB.

5. If the maximum defined voltage on consumer unbalanced cable is 2 V RMS, what is its maximum *possible* headroom?

The difference between the 0.316 V calculated earlier for -10 dBV and the 2V provided would be the headroom. We just need to convert that ratio into dB: $L = 20 \cdot \log_{10}(2 V/0.316 V) = 16.02$ dB.

Consumer equipment seldom takes advantage of the maximum possible, so usually they have far less headroom above 0 dBVU than the maximum of 16 dB allowed under the interface standard.

In contrast, there was *no* interface standard defined for pro audio headroom; yet 20-24 dB of headroom is common for high end pro audio.

4.7 REVIEW EXERCISES

Chapter 4 Review Guide

Theory
- **concepts** of waves, time and frequency domains, basic periodic waveforms
- **differentiation** time and frequency domains and their implications
- **concepts** of RMS, peak, peak-to-peak, average, phase
- **concepts** of dB, ratio, dB SPL, dB FS
- **avoidance of wrong concepts**

Practice
- ability to **convert** between wave parameters
- ability to **convert** between RMS, peak, peak-to-peak, average
- ability to **convert** between dB and ratio

Exercise
Define these terms:
1. *Wavelength*, **period**, *frequency*. *(Illustrate with diagrams, show relationship, symbols, units)*

2. **Phase** *(define, illustrate using diagrams, symbol, unit)*

3. **RMS**. *(Full name, alternate name, how is it defined formally, how is it useful & defined informally)*

4. Diagram and label **five** *basic repetitive* **waveshapes** covered in class (with peak amplitude of A):

	Name	Time Domain Response
1		
2		
3		
4		
5		

Chapter 4 – Waves

5. If the frequency is *1.360 kHz* and the speed of sound is *340 m/s*, what is the **wavelength** in meters?

6. If the wavelength is **two meters** and the speed of sound is *340 m/s*, what is the **frequency**? (*unit*)

7. If the frequency is *500 Hz*, what is the **period**?

8. If the period is *0.002 s*, what is the **frequency**?

9. Provide the missing conversion value in table below and provide the steps in your calculation. Round your answer in dB to whole numbers, and your answer in ratio to 3 significant figures.

Question	Ratio	dB	Calculation
31	Unity (1)		
32	Power ratio of 2		
33	Power ratio of _____	10	
34	Voltage ratio of 1/2		

4-19

35	Voltage ratio of _____	+4 dBu	
36	Voltage ratio of _____	-10 dBV	

10. Convert voltage slope of *20 dB/decade* into *dB/octave*, using the following table to show the steps.

Step title	Numerator	Denominator

Answer: 20 dB/decade means _____ dB/octave

11. If the typical 'AC Mains' voltage in America is 120 V at 60 Hz, calculate the following voltages (or state which is the mains voltage):

Vavg

VRMS

Vdc

Vpeak

Vpp

12. Convert *phase* to respective units in table below. (*show steps/reasoning, write basic conversion factor*)
Basic conversion factor:

Question	In degrees	In radian
44		π
45	$-90°$	

13. In *consumer audio*, the reference of 0 dBV is defined as _____ V RMS regardless of impedance load. The nominal line level is _____ dBV. Calculate the *effective voltage* at nominal.

14. In *pro audio*, the reference of 0 dBu is defined as _____ VU and 1mW into 600 ohms load. Calculate the *effective voltage* across the load. (Hint: *you may have to apply two laws in sequence.*)

15. In pro audio, the nominal line level is _____ dBu. Use result in above question to calculate the effective line voltage at nominal *in dBu*.

16. Calculate the nominal line voltage for ***pro audio*** (obtained earlier) in table. (*show steps*)

peak voltage	peak-to-peak voltage	in dBV

17. What is the difference, *in dB*, of nominal line levels between consumer and pro audio? (*hint: use dbV values calculated earlier*)

What is the *ratio* of nominal line level effective voltages of consumer and pro audio?

Supplement T - Project Notebook & Report

A generic project report outline follows:

1) **Executive Summary** or *Abstract* (summarize what the project is about, and what you had accomplished, and what is novel, if any - for this course, usually one paragraph to one page). You should pay special attention to this section, as executives often only read this section (thus the name executive summary).

2) **Theory of Operations** or *Background* (explain as if to a client or boss what the circuit does, and why, and how it is to be used. Show schematics and equations and explain how it is expected to work. section by section. For this course, this should be 1-3 pages.)

3) **Materials** (summarize what you used, where you got them; show plenty of pictures, assuming someone will only read the report and not see the real project; plus, any equipment you used, listing specific test equipment you used and what for. Explain the choice you made concerning the enclosure.)

4) **Method** (summarize your project diary; specifically include the order of soldering - which sets of parts go first, and why; followed by what kind of testing - that shows the method of your madness, or lack thereof, plus any planning which is a plus. For this course, this should be 1-2 pages.)

5) **Results** (summarize results of your troubleshooting and testing - both intermediate and final; use tables and graphs as appropriate.)

6) **Discussions** (mention anything you learned, anything worth noting because of your project. Does the project work as expected? Why or why not? Does your enclosure fit? What will you do differently in the future? Anything novel that you did?)

Notice this is essentially the same outline for writing a paper to submit to journals or research conferences. You should be concise and follow the guidelines published by the sponsors, especially in terms of length of the report.

The Project Notebook

It is usual for companies to require employees to use a project notebook, where all the above materials can be written as evidence when copyrights and patents are in question. The materials should be regularly witnessed and dated by people who understands the inventions in question. When a project notebook is used regularly, it becomes easy to write a project report - just copy and paste and edit.

For legal reasons, the project notebook must be a bound notebook that cannot be added to or deleted from without being noticed. One can add printed pages to it (for example, computer printouts) by simply gluing the printout on top of a project notebook page, and certifying that it is a glued page on the regular page, with witnesses and certification on both pages. Thus, if one wants the benefits of computer automation, the project notes *can* be kept on computer, but printed out regularly onto the project notebook as stated above.

A common weakness of many engineers is poor communication and writing skills. That needs to be developed early in life. Otherwise, one will have to have good enough engineering skills that the company will assign a technical writer to help with documenting your work; essentially the technical writer will be taking money away from your salary, as far as the company is concerned. Thus, this chapter is born to help you get started to improve your writing; send me whatever money you would otherwise divert to technical writers in the future in appreciation of this advice… :)

The opinions expressed above should not be considered legal advice; consult a legal counsel whenever necessary.

Chapter 5 AC ANALYSIS

> "AC Analysis" is the second foray into the world of analytical circuit analysis. By extending DC analysis through newly defined impedances and executing the Bode Plot, one performs basic AC analysis for linear systems, applicable especially for audio filters and audio amplifier frequency responses.
>
> *Expected Learning Outcome*: The student will be able to read and write Bode Plots and frequency responses for various audio filters. The student will also be able to tackle the majority of circuits encountered in audio electronics without using complicated math by extending DC analysis to obtain AC transfer functions and draw Bode Plots to obtain frequency responses.

CONTENTS

Virtual Lab: Series RC with AC	5-3
Complex Numbers for AC	5-12
Introducing Impedance & Reactance for AC	5-13
AC Analysis for Series RC	5-16
Virtual lab: Series L/R with AC	5-21
AC Analysis for Shelving Lowpass Filter	5-24
Time Constants: RC, L/R, $\sqrt{LC}$	5-28
Review Exercises	5-29

These are the steps we'll follow to perform *detailed* AC Analysis:

Steps:	Example:
1) Calculate **impedances** of components in circuit	$Z_R = R$; $Z_L = sL$; $Z_C = 1/sC$
2) Calculate **Transfer Function** of circuit (We use *same* analysis method as in DC analysis)	For circuit with two passive components, it is simply Z_1/Z_1+Z_2

3) Calculate key **parameters** of transfer function:
   a) the **order**
   b) when $f = 0$ (DC)
   c) when $f \to \infty$ (HF)
   d) **corner frequencies** or **time constant**
	For RC LPF:
	   a) order = 1
	   b) $H = 0$ when $f = 0$
	   c) $H = 1$ when $f \to \infty$
	   d) $\tau = RC$

4) Draw **Bode Plot** from 3)	(Bode plot showing LPF response: A vs Log F) LPF
5) Make any final magnitude observations	Use pythagoras theorem

Before we go through these steps, we'll first wander through our virtual lab to see the time and frequency domain responses, to bridge the learning from previous chapter about waveforms and dB. We'll realize why using dB is beneficial, as it is the *critical success factor* for Bode Plot.

After that, we'll review complex numbers, and then define impedances and move on with the steps.

Supplement V shows **how to visually identify common passive filters quickly**. In contrast to the above approach, it can be considered a *simplified* AC Analysis. Once we are familiar with the approach, we can evaluate the circuit mostly in our head. In essence, it executes *only* steps 3) and 4) above, skipping the math-intensive steps of 1) and 2) as well as 5), especially skipping complex numbers about which students often complain.

5.1 VIRTUAL LAB: SERIES RC WITH AC

Welcome to our Virtual Lab! The engineer has set up a simple circuit consisting of a resistor and a capacitor in series with a variable frequency sine wave generator and is about to probe the circuit....

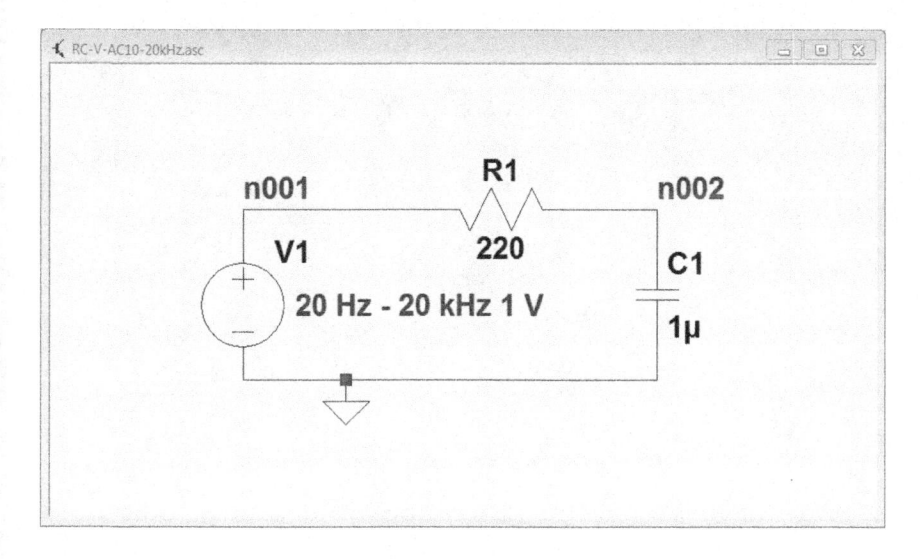

At Low Frequency

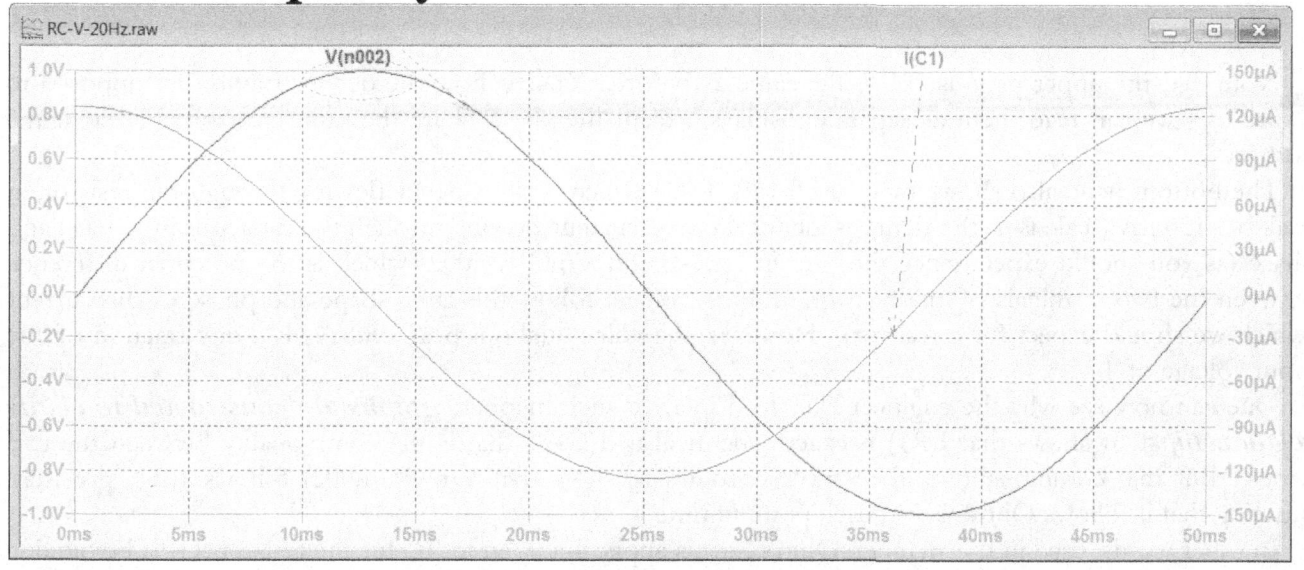

Do we need to peek to find out what was set on the signal generator? Not really - the engineer is probing to find out! (After all, care means the engineer will typically *not* trust the settings but will instead verify by testing.) Notice that the period is 50 ms, meaning that the frequency is 1/50m = 20 Hz. (*Make sure you know how to observe and make such calculations.*) The peak is 1.0 V for V(n002), as one can see from the waveform, and one could have probed V(n001) to be sure, but the engineer can already infer from reading the circuit diagram that the signal generator was set for 1.0 V peak at 20 Hz. (That is the approach we'll learn...)

The capture shows two waveforms, V(n002) which is the potential difference of the capacitor relative to ground, with scale on the left; and I(C1), pointed to by an arrow and slightly lighter in color in the chart, with scale to the right, which is the current flowing through the capacitor. The current is leading the voltage by 90 degrees, because the zero-transition points were at 12.5 ms and 25 ms respectively, making the lag 12.5 ms, or

¼ of the period. (*Make sure you know how to do this. If not, review Chapter 4 again.*) "**ELI the ICE-man**" can be used to remember this, where ICE means I (current) leads E (voltage) for C (capacitors).

However, engineers typically like to squeeze more captures into a limited display window. One needs to expect to see such complications, and just deconstruct the panes of the display separately to understand them individually, and then relate them together so that one can understand *why* the engineer likes to display them in such a manner. Here is the next capture…

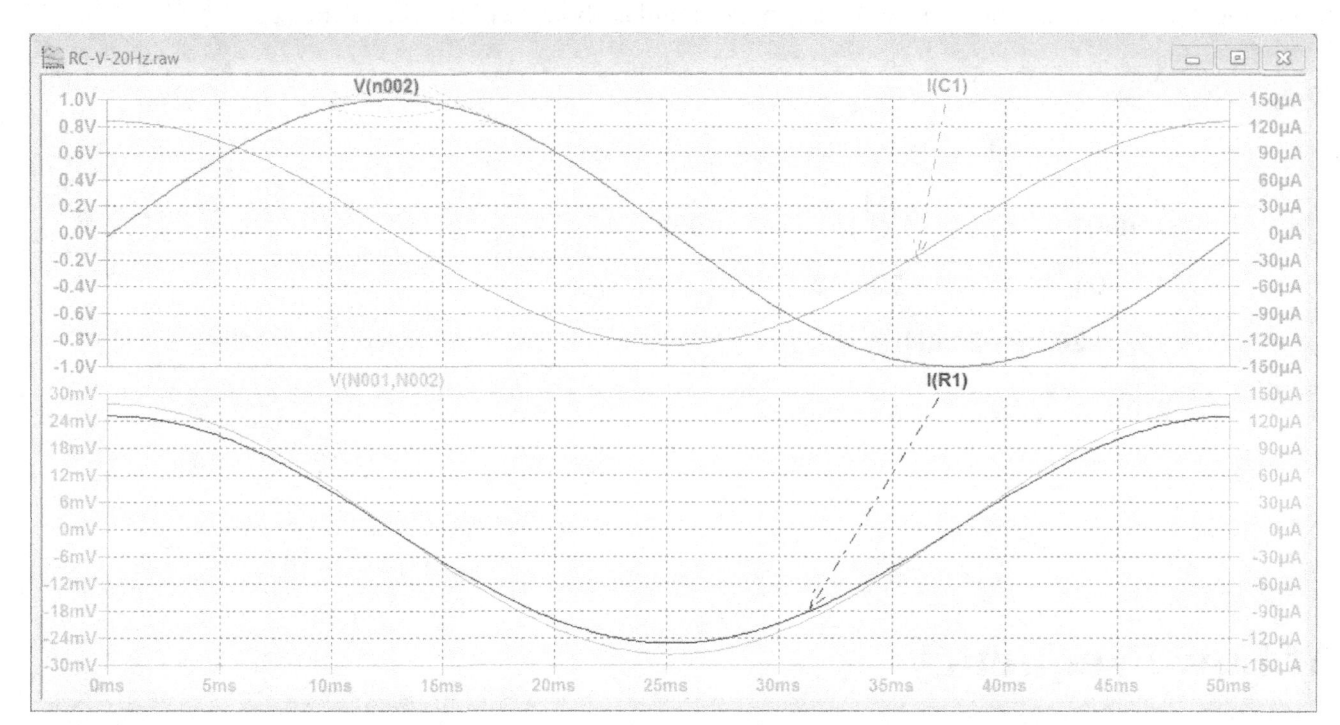

Note that the upper pane is *exactly* the same as before. That is how the deconstruction is supposed to assure us - we can read them as separate displays, except they both share the same horizontal scale at the bottom.

The bottom pane also shows two waveforms, I(R1) which is the current flowing through the resistor, at peak of 120 µA (scale is to the right), is identical to the current flowing through the capacitor in voltage and phase (as you should expect since they are in series); and V(n001, n002) which is the potential difference between the two terminals of the resistor, and the voltage follow the same shape and phase as the current (which we should expect for a resistor). However, the voltage (about peak 25 mV) is a tiny fraction of the input voltage of 1 V.

One can now see why the engineer likes to display in such manner: **graphs are constructed to clarify relationships**! It shows that I(R1) is exactly identical to I(C1) - that is the commonality between the two panes. But that commonality is then turned around to show that V(n001, n002) follows I(R1) precisely, meaning that it follows Ohm's law at each point in time.

In other words, used in this manner, Ohm's Law is applicable even in AC for the resistor R1. The old dog can still play new tricks in AC as well as DC: haha AC/DC…

At High Frequency

The engineer proceeded on, changing the frequency of the source from the audio low of 20 Hz to the audio high of 20 kHz. This is called (low and high) **boundary conditions** testing.

Again, you should know by now that a period of 50 μs means that the frequency is $1/50\mu = 20$ kHz.

Also, note that engineers minimize changes of the setup. Since we are already used to the double pane display approach, it will be shown that way from now on, all the probes being unchanged. Onlookers coming in at this point will most likely be confused. But if you were with us from the beginning, you should have no trouble reading the displays at this point.

You will notice similarities, and differences - and that is the point. Similarity first…

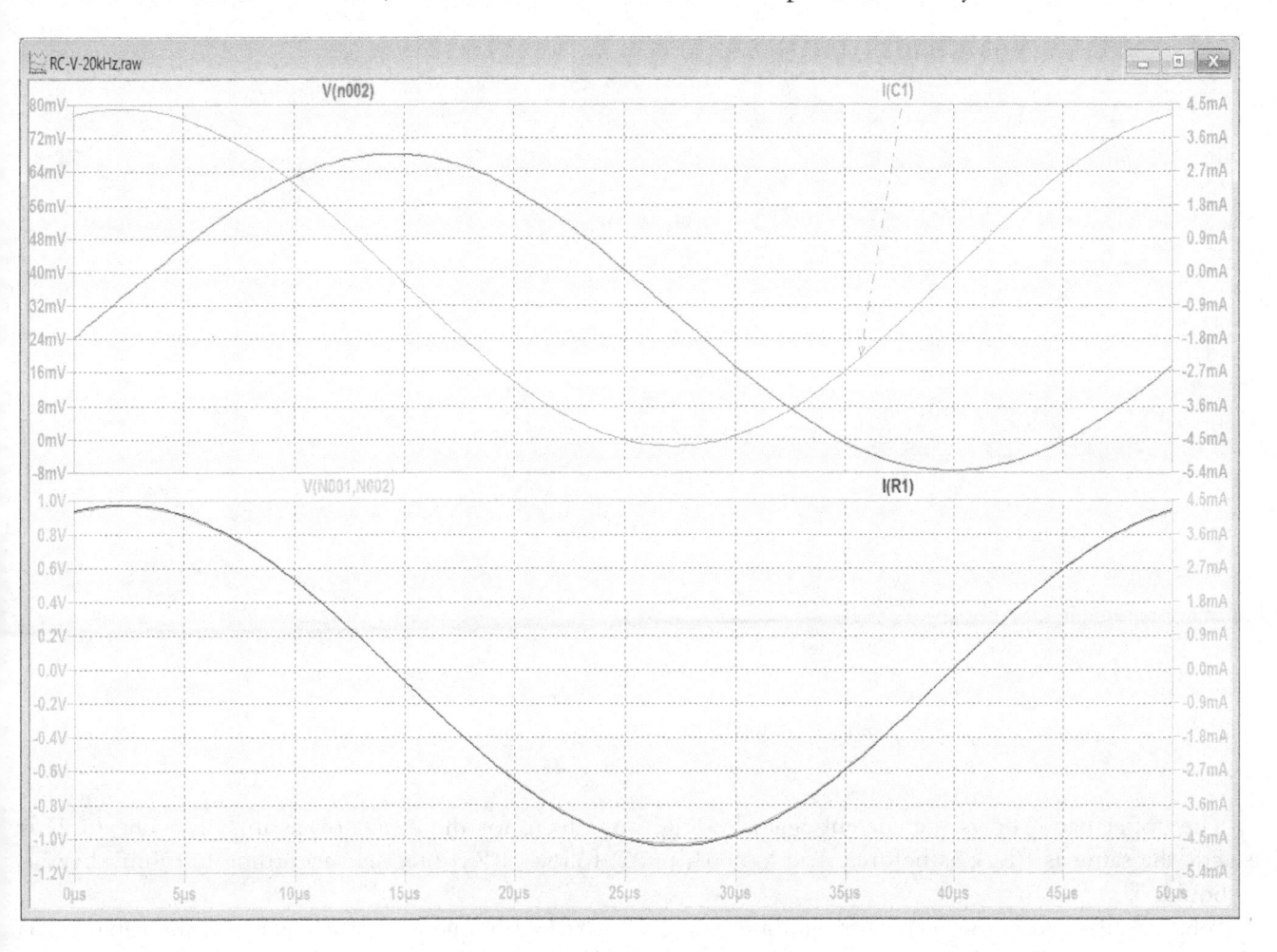

The upper pane shows the current again leading the voltage for the capacitor, as in ICE. The I(C1) is exactly the same as I(R1) in the lower pane, as before. And V(n001, n002) follows I(R1) precisely according to Ohm's Law, as it should. Make sure you are comfortable reading all these from the chart.

What is different is the relative magnitudes involved. At low frequencies, the voltage at the capacitor is high (near 1 V peak) and the voltage at the resistor is low (near 25 mV peak). At high frequencies, it is the opposite: the voltage at the capacitor is low (about 68 mV peak) - less than a tenth of the former, while the voltage at the resistor is high (about 1 V peak) - about the same as before.

At Mid Frequency

The engineer proceeded on, changing the frequency of the source to 1 kHz, which is considered the "middle" of the audio frequencies. This is commonly referred to as **nominal** testing (rather than *boundary* testing). Verify that you see a period of 1 ms and that means 1/1m = 1 kHz.

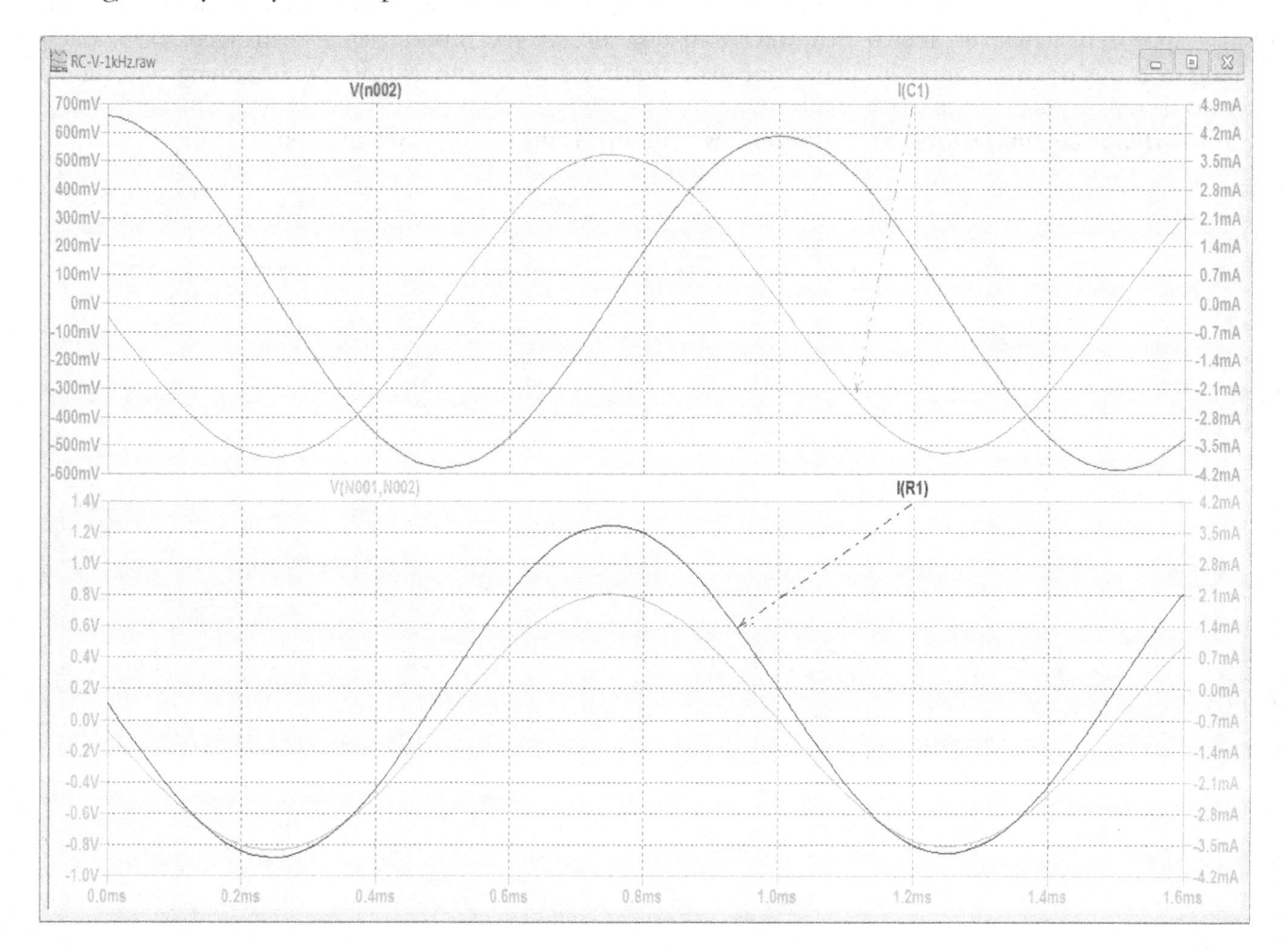

The upper pane shows the current again leading the voltage for the capacitor, as in ICE. The I(C1) is exactly the same as I(R1), as before. And V(n001, n002) follows I(R1) precisely according to Ohm's Law, as it should.

What is different is the relative magnitudes involved. At low frequencies, the voltage at the capacitor is high (near 1 V peak) and the voltage at the resistor is low (near 25 mV peak). At high frequencies, it is the opposite: the voltage at the capacitor is low (about 650 mV peak) while the voltage at the resistor is high (about 0.8 V peak). At mid frequencies, it is in the middle: the voltage at the capacitor is medium (about 600 mV or 0.6 V peak - you should read it from the middle of the plot rather than from the beginning which may have experimental trigger-induced errors - just to warn you about what happens in real life) while the voltage at the resistor is also medium (about 0.8 V or 800 mV peak). One should expect that at a certain frequency, the peak of the two voltages should be exactly the same. Can we find that frequency without trial and error? The plot thickens...

Frequency Domain Plot

Do you think the engineer will continue to make more plots at more frequencies? It is certainty possible, but that means a lot of effort. Remember the engineer prefers shortcuts? Now that one has three points of data, for low, mid, and high frequencies, one can really use a shortcut to see the rest. Here comes the **frequency domain plot**. (Previously we were doing **time domain plots**.) The student unfamiliar with the process may be confused at first, but once it is understood that it is a shortcut for efficiency's sake, then everything will make more sense. (That also explains why we preceded this chapter with a discussion of time and frequency domains.)

Since this is the first time the engineer attempting frequency plot, one would expect the start-small approach from a "sane" person, looking at *only* one signal first...

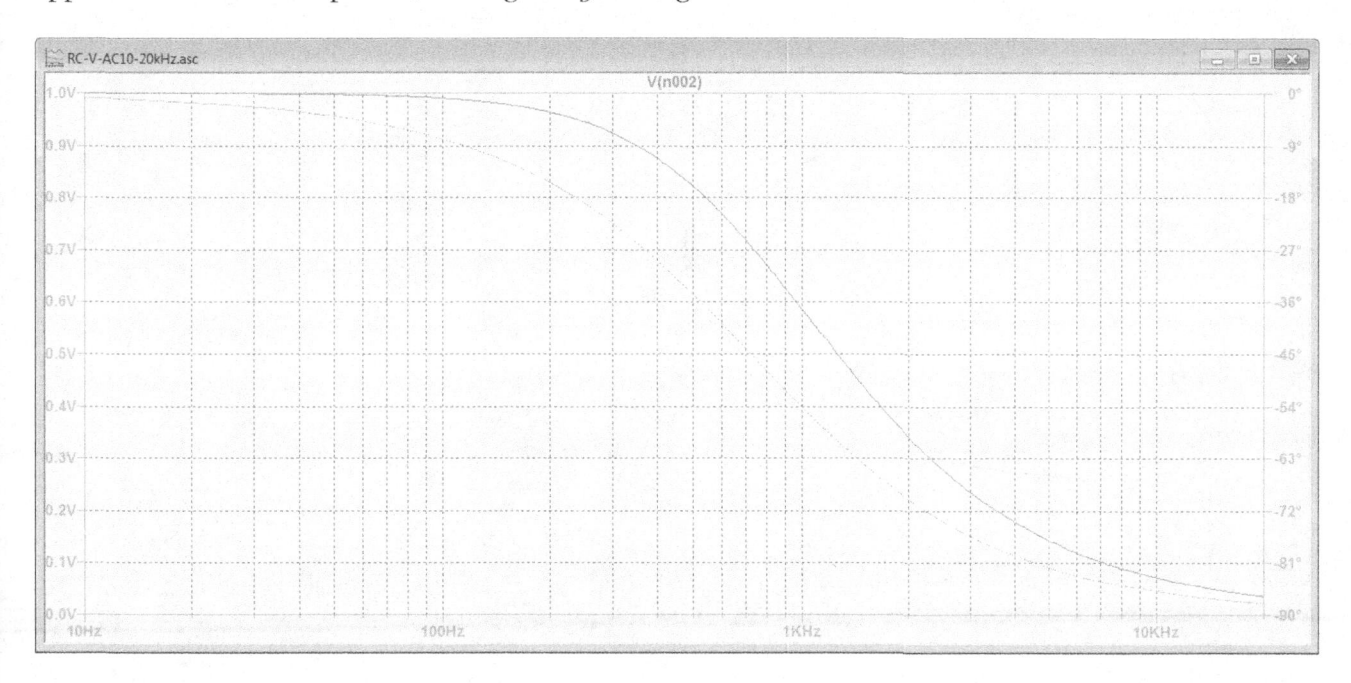

There are conventions to everything, and frequency domain plot is no exception. Some gave it a special name, the **bode plot**, but there are of course doubts if this would qualify (we'll soon see) Note that there are two curves, with one shown dotted. The convention (which means you won't be told otherwise) is that the dotted curve represents the phase, which means that its scale would be in degrees, which is shown on the right side in this plot. The solid main curve uses the scale on the left, which in this case is from 0 to 1 V, and it is presented linearly, meaning distance between 0 to 0.1 and 0.9 to 1 V are equal. **It is very important to note the scales in plots, especially to note if they are linear or logarithmic**. In the case of the horizontal scale, one can see that it is nonlinear, with 10 to 20 Hz much wider than 90 to 100 Hz, and the same width as 100 to 200 Hz. One should remember this characteristic - it is a **logarithmic scale**. Audio frequencies are often graphed using logarithmic scales, in this case specifically it is in **decade scale**, meaning that orders of ten are shown in prominence.

This plot shows exactly what we had known so far: at low frequency the voltage at the capacitor is at the full 1 V; at high frequency it is near zero (near 25 mV); at medium frequency it is 0.6 V. Refer to the past graphs and convince yourself that is the case.

But it also shows us the ***phase relationships***: at low frequency it is about -1°, at high frequency it is about -88°, at mid frequency it is at about -45°. (By the way, the fact that they are *not* exactly 0° and -90° explains why our time-domain plots earlier were slightly off in zero-crossing in terms of timing.)

Once the first plot (some may call it a prototype) is understood, then one would expect that the same multiple pane approach be used for observing the same four signals we had seen in the past...

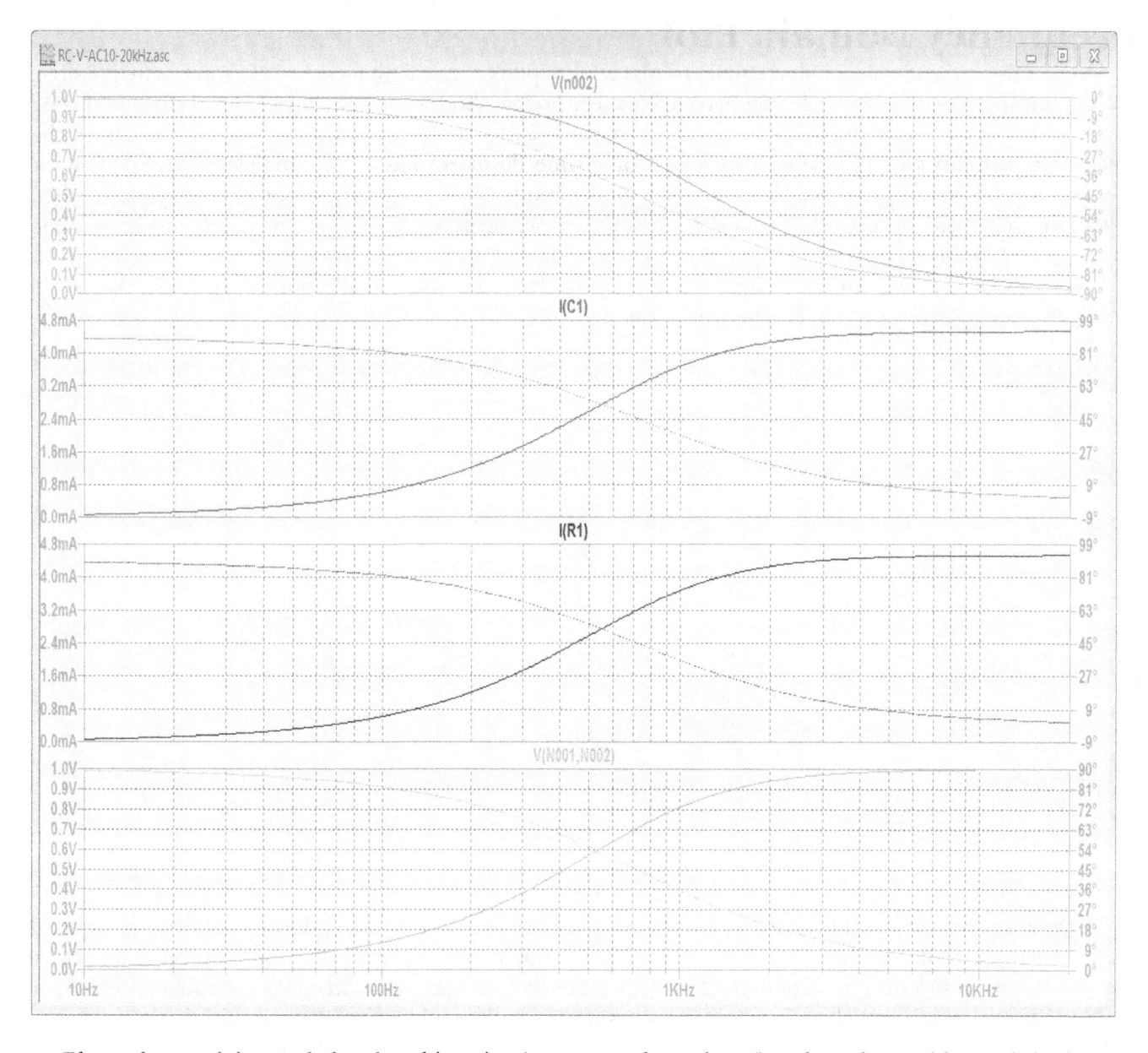

Please do not *feel* overwhelmed at this point (we are *not* through yet!) - please bear with us - it is showing essentially the same thing we had observed thus far:
1) voltage in capacitor changes from 1 V to half volt to nearly zero when frequency goes from low to mid to high (seen on first pane)
2) current in capacitor changes from zero to medium to max of about 4.4 mA from low to mid to high frequencies; current in capacitor and resistor is identical in magnitude and phase, as expected
3) voltage in resistor follows its current precisely as predicted by Ohm's Law

Please spend time to ponder and understand the diagram. Most students will skip and thus never understand the significance. It takes time to realize why this is so important a diagram.

Bode Plot

We've gone through a lot in the our virtual lab tour. Is that it? Well, another surprise is in store for us! There is a reason why we learn dB. In fact, engineers prefer to work with dB. Proceeding cautiously, the engineer displays one signal in dB first…

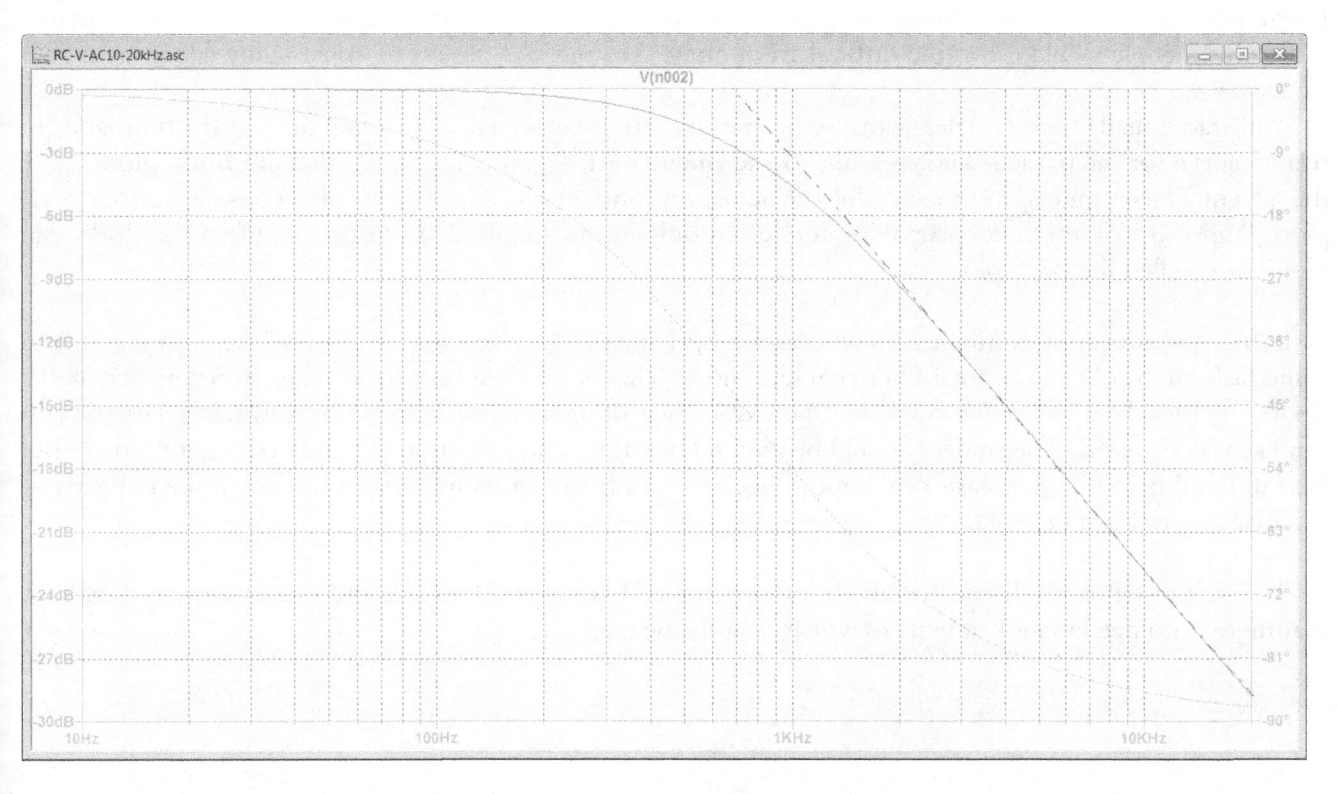

Compare it with the plot for the *same* signal V(n002) earlier, repeated below:

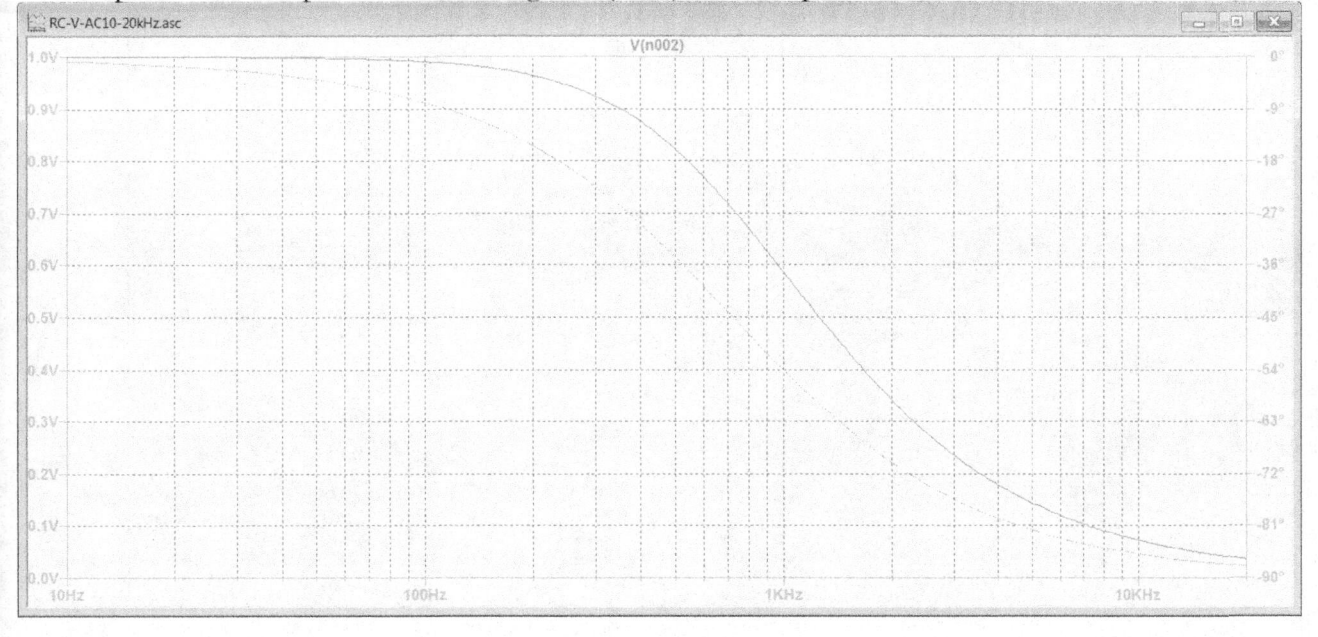

The important thing is to notice is that the **same** voltage relationship displayed linearly and logarithmically has **different shapes**! The linear scale shows a typical S-curve, while the logarithmic scale shows a typical slope, known as **roll-off**, annotated by a dotted line above. That explains one important reason why dB is used - by plotting this way, one can easily estimate the value at different frequencies, since they fall on a straight slope. See for yourself from the plot that the slope is **-20 dB per decade** for this simple circuit. (In general, the roll-off for first order filter is + or -20 dB/decade.) Note again one does *not* see a roll-off in linear scale, but only in *log-log scale*, and it shows more clearly that it goes asymptotically to zero. (Yes, asymptote is a math jargon you should know by now.)

The point where the roll-off crosses the horizontal asymptote line (at 1 V in this case) is called the Bode Pole (some called both lines the Bode Pole), and the frequency corresponding to that point the **corner frequency**, f_c. For filters, sometimes people would call f_c the **cutoff frequency**. For other circuits, sometimes people would call f_c the **critical frequency**. For me, it is easiest to consistently call it the corner frequency.

And traditionally the two lines with the proper axis are together known as the **Bode gain plot**, and the related curve for the phase is known as the **Bode phase plot**; together they are called the **Bode plot**. With the advent of computers, we usually plot the actual response curve, which some also considers as the Bode plot. And most of the time, people are interested only in the amplitude response, so often the Bode gain plot is also called the Bode plot.

Before the advent of computers, engineers use Bode plots of component filters, add them up together in some fashion, to obtain the total filter characteristics. That is possible because of the characteristics of the Bode gain plot, being a log-log curve, can be *added* when we expect responses to be *multiplied*. The roll-offs can likewise be *added*. The student should be grateful that this is no longer taught and no longer needed. But bear in mind because of its historical importance, it is still the main communications tool in the industry, so it is essential to learn this well.

As you can tell from the next chart, both currents and voltages are in dB, and one immediate benefit is that there is no need to look at units of voltages and amperages.

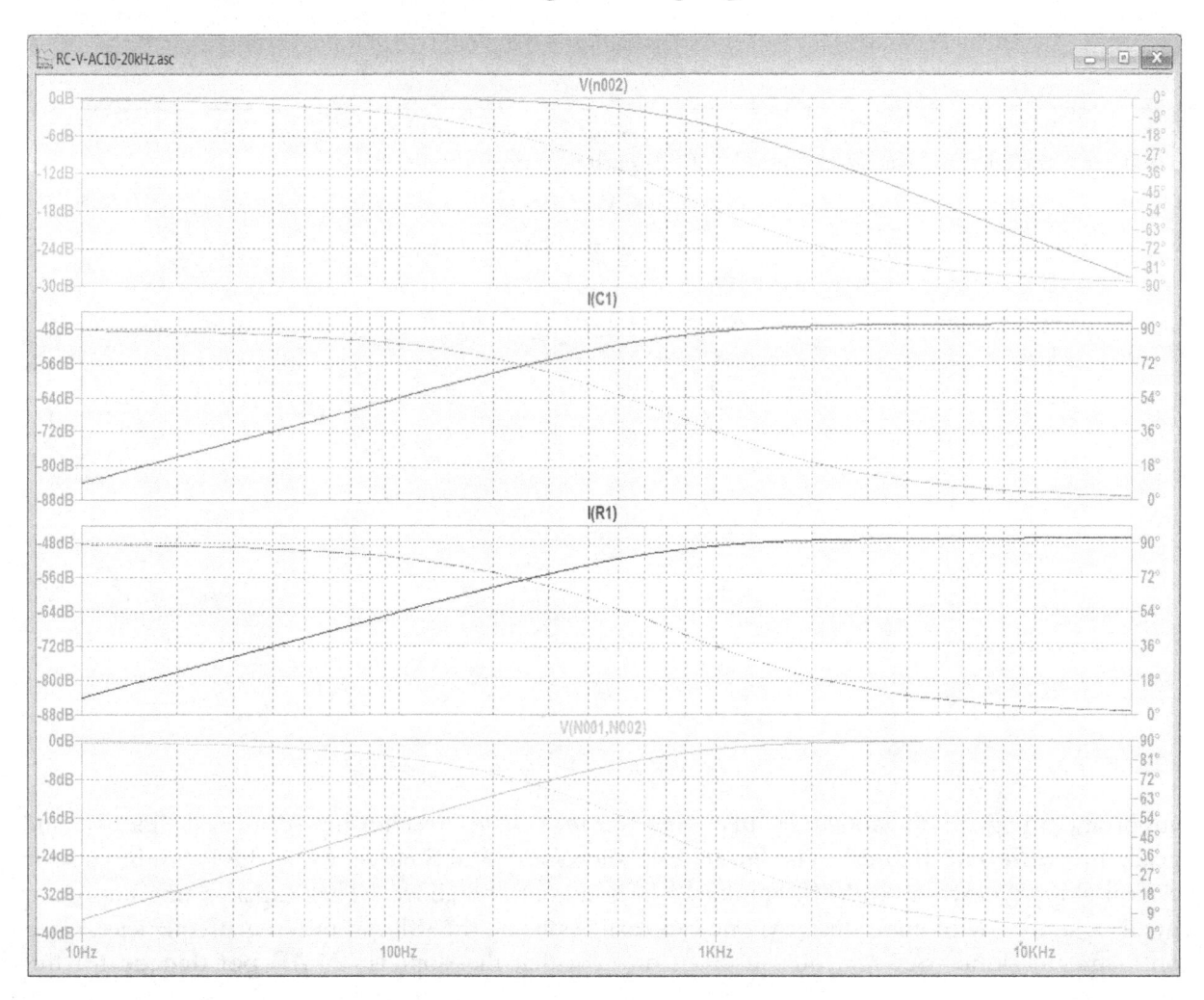

Playing Favorites

Sometimes after all these the engineer may focus on and select *only* two specific waveforms for comparison. Here are the two for comparison. Notice they are like mirror images.

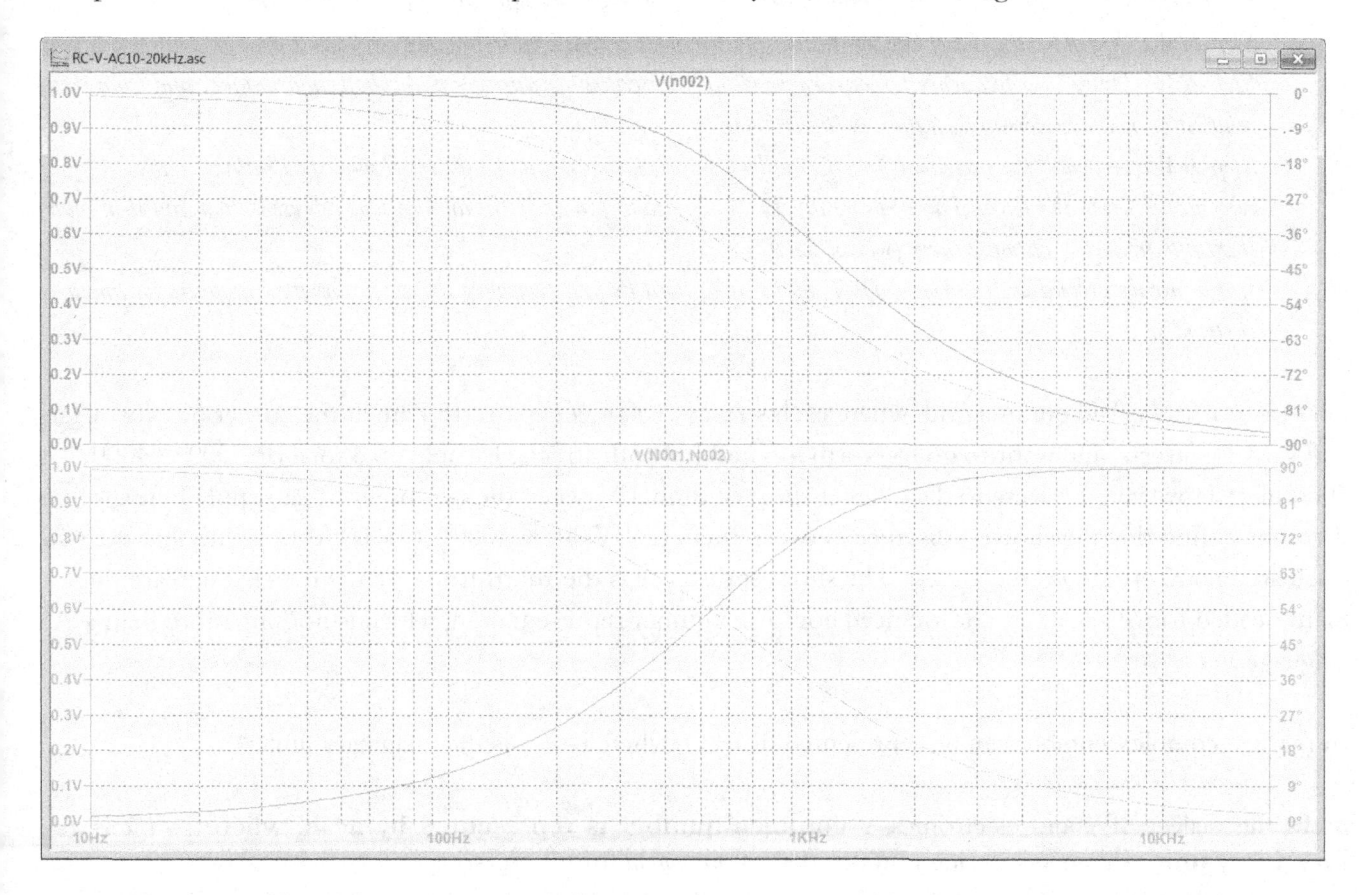

Your turn: please explain the similarities and differences between there, and why the engineer chose these for the final report.

5.2 Complex Numbers for AC

How do we simplify AC analysis? Through the "re-use" of skills we have already learned so far!
Step 1: We "extend" the concept of resistance into the more generalized impedance (sections 5.3).
Step 2: We "mine" the frequency response curve visually for critical frequencies and rolloff information, and from there relate it to its algebraic representation (sections 5.4).
Step 3: We "extend" the same tools in DC analysis, substituting impedance for resistance, specifically visually recognizing series and parallel and voltage divider configurations, to arrive at an algebraic representation and then map them into frequency response curve (section 5.4.).
But before we proceed with learning step 1, let us ease into it by first reviewing the use of complex numbers for audio electronics.

Let's say our city is laid out in a grid, where each street is 1000 feet apart. If I am at the corner of North 1st and East 1st streets, and want to go to North 4th and East 5th streets, I would go a total of 3000 + 4000 = 7000 feet. However, if I were to dig a direct shortest tunnel between the two points, I can apply Pythagorean theorem to find that the shortest distance between them is $\sqrt{(3000^2 + 4000^2)}$ = 5000 feet. (Note that the two calculated numbers are *not* the same!) The shortest distance is the **magnitude** of the travel. Such are the math needed for AC analysis, and we need complex number representation, which is nothing more than a language...

In math, a complex number can be represented by a+bi, where i=$\sqrt{-1}$ is the imaginary unit.

In the dialect of audio electronics, **a complex number is represented by a+jb, where j=$\sqrt{-1}$ is the imaginary unit**. Engineers decided on this dialect *convention* for two reasons:
1) Frequently i is used to represent current; to reduce confusion, a different character j is chosen.
2) Placing j at the front of a potentially complex algebraic expression is far easier to read than at the end, and emphasises its complex nature when it is put first, right after the plus sign rather than at the end.

Other than for this distinctive dialect modification, all the math remains the same.

> A complex number is represented by a+jb, where j=$\sqrt{(-1)}$ is the imaginary unit, a is known as the real part, and b is the imaginary part.
> The magnitude of the complex number is represented as:
> |a+jb| = |$\sqrt{(a^2 + b^2)}$| (Magnitude of Complex Number)
> The angle (or phase) of a complex number is represented as:
>   ∠(a+jb) = arctan (b/a)   (Phase Angle of Complex Number)

Given any two of the four numbers: real part, imaginary part, magnitude, angle, we should be able to find the other two numbers. *Make sure you know how to do that.* Similarly, you should be able to do simple mathematical operations with two complex numbers.

But first the good news: engineers like to avoid complex numbers as much as students. In the next section we shall learn step 1, and then in following sections we shall see how complex numbers are often *not* required, before we start using them *sparingly* in sections further down.

Chapter 5 – AC Analysis

5.3 Introducing Impedance & Reactance for AC

Impedance sounds similar to resistance. In fact it is the generalized version of resistance. Both impedance and resistance are the measure of the opposition of the flow of current through a circuit when a voltage is applied. Specifically, impedance is the ratio of voltage to current in an alternating current (AC) circuit in complex number form. (See supplement X "Why Different Formulas for Impedance & Reactance".)

> For zero initial conditions:
> $Z_R = R$ (Impedance for Ideal Resistor)
> $Z_L = sL$ (Impedance for Ideal Inductor)
> $Z_C = 1/sC$ (Impedance for Ideal Capacitor)
> Imagine s as a complex frequency variable.
>   (Actually $s = \sigma + j\omega$, where $\omega = 2\pi f$, and $\sigma = 0$.)

How to calculate the **magnitude** will be next.

Talk About Complex

We have already hinted that impedance is a complex number. We may as well tell you how much you need to know about complex numbers - actually not much other than what is shown next...

> Impedance Z is a generalization of resistance.
> $Z = R + jX$ (Impedance Equation)
>   where R is resistance (Real part of Impedance)
>   X is reactance (Imaginary part of Impedance)
>   and j is imaginary unit.
> Often when people say impedance, they refer to its *magnitude*.
> $|Z| = |\sqrt{(R^2 + X^2)}|$ (Magnitude of Impedance)
>   where | | is the absolute value (throw away the sign).
> $\angle Z = \arctan(X/R)$ (Phase of Impedance)

In other words, you apply the Pythagorean Theorem to get the magnitude of impedance. That is the primary thing you need to remember. You are seldom asked to *calculate* the phase.

For the *ideal* capacitor and inductor, $R = 0$.

Example: $|Z| = |\sqrt{(R^2 + X^2)}|$. *If R is 3 Ω and X is 4 Ω, what is $|Z|$?*
Many students give up when they see square roots and squares, or when they see an unfamiliar symbol like $|Z|$. What is shown here is simply the Pythagoras theorem. $|Z|$ simply means dropping the sign for Z. It's not as bad as it seems...
Substituting with R and X, we get $|Z| = |\sqrt{(3^2 + 4^2)}| = |\sqrt{(9+16)}| = |\sqrt{25}| = 5\ \Omega$

Example: $|Z| = |\sqrt{(R^2 + X^2)}|$. *If R is zero and X is 4 Ω, what is $|Z|$?*
Many students give up when they see square roots and squares, or when they see an unfamiliar symbol like $|Z|$. What is shown here is simply the Pythagoras theorem. $|Z|$ simply means dropping the sign for Z. It's not as bad as it seems...
Substituting with R and X, we get $|Z| = |\sqrt{(0^2 + 4^2)}| = |\sqrt{(0+16)}| = |\sqrt{16}| = 4\ \Omega$

In other words, when **R is zero, $|X| = |Z|$**. This is a very important shortcut to know and remember.

Example

Fill in the table below using engineering units (with metric prefix) for your calculated answers.

	Ideal **Resistor** (value R)	Ideal **Inductor** (value L)	Ideal **Capacitor** (value C)
Impedance formula	$Z_R = R$	$Z_L = sL$	$Z_C = 1/sC$
Resistance value	R	0	0
Reactance value (substitute $s = 2\pi f$)	0	$2\pi fL$	$1/(2\pi fC)$ (or properly, $-1/2\pi fC$)
Magnitude of Impedance	R	$2\pi fL$	$1/(2\pi fC)$
Phase Angle (arctan (X/R))	0	+90°	-90°
	Resistor **1 kΩ**	Inductor **1 mH**	Capacitor **1 μF**
Magnitude of Impedance at 1 Hz	1 kΩ	2π·1·1m = 6.28 mΩ	1/(2π·1·1μ) = 159 kΩ
Magnitude of Impedance at 1 kHz	1 kΩ	2π·1k·1m = 6.28 Ω	1/(2π·1k·1μ) = 159 Ω
Magnitude of Impedance at 1 MHz	1 kΩ	2π·1M·1m = 6.28 kΩ	1/(2π·1M·1μ) = 0.159 Ω

Notice how we use engineering notation to simplify and re-use the calculations.

Notice also how inductors **increase** in impedance as frequency increases, while capacitors **decrease** in impedance as frequency increases, and resistors **maintain** its impedance as frequency changes (up to a point - remember Ohm's law is empirical).

Notice the phase angle: it is zero for resistors, +90° for inductors (by convention we put in the plus sign), and -90° for capacitors. "ELI the ICE man" can help remember: E leads I for inductors represented by L, and I leads E for capacitors represented by C, and thus the phase angle is negative.

Chapter 5 – AC Analysis

Admittance

The reciprocal of impedance is admittance, abbreviated as Y. Acting as a generalized version of conductance, its unit is also siemens, abbreviated as S (in older books it is denoted as *mho*, in other words, *ohm* spelled backwards). We shall *not* use it in practice, but only as an aid or shortcut to help with memorization of formulas.

$$Y = 1/Z$$
$$Z = 1/Y$$
(Admittance)

The Take-Away

This chapter deals with the concept and implication of impedances. Several things you need to know:
1) Impedance is simply a **generalization of the concept of resistance**. Impedance impedes the flow of current when a voltage is applied.
2) The concept is simple, but the bad news: impedance is a *complex* subject, and its implementation requires then, obviously, **complex numbers**. Specifically, impedance $Z = R + jX$, where R is resistance, X is reactance (which is similar to resistance, except it works in the imaginary plane), and j is the imaginary unit usually denoted by i, but since i is used quick often in audio electronics for current, the convention is to use j to represent it.
3) In the case of a simple (ideal) conductor or resistor, reactance X is zero, thus impedance degenerates to resistance (in other words, $Z = R$ for resistors and conductors).
4) In the case of a simple (ideal) capacitor and inductor, the resistance is zero, thus reactance is simply the imaginary part of impedance. (Or impedance degenerates to the reactance multiplied by the imaginary unit.)
5) The good news: engineers don't like to communicate with complex numbers! Thus often when you hear them say impedance, they were really giving you the *magnitude* of impedance. The **magnitude of impedance is the root-addition-square of resistance and reactance**, calculated using Pythagorean Theorem. (That explains why when X is zero, $|Z| = R$. Likewise, when R is zero, $|Z| = |X|$. The latter simplifies our life when we deal with ideal capacitors and inductors, which we shall visit in the next sections.)
6) In audio electronics, you are seldom called upon to calculate phase angles. But since you may hear about them, you need to at least understand what it is. (Hint: we won't talk about phase angles for speakers in the upcoming section.) Plus you may need to at least have a general understanding of whether it is capacitive or inductive (the equivalent of you need to know whether you are going northbound or southbound), so remember ELI the ICEman!

Historically speaking, impedance was coined by Oliver Heaviside in 1886. Arthur Kennelly first represented it in complex number form in 1893.

The Magic of It All

Since we use s in the formulas, the approach is known as the s-domain approach. **The magic of the s-domain approach is that you *already* know how to execute the front end of AC analysis - it is the *same* as DC analysis!** You just need to add Bode Plot analysis to finish the job - we shall explain that in following sections.

S-domain analysis uses DC analysis as a front end. That's why we need to know DC analysis well. Since $s = \sigma + j\omega$, when we set $s = 0$, we obtain DC analysis! And when we set $s = j\omega$, we get AC analysis results. How cool! Really *one ring to rule them all*...

5.4 AC Analysis for Series RC

The above lab tour is to help you gain insight into the engineer's' workflow, and in particular to get you acclimated to reading graphs. That is probably the most difficult part. It is also another review of chapter 4; if you haven't noticed yet, you cannot fall behind in that chapter, as everything from now on depends on it.

Let us now resume the AC analysis for the RC circuit we had witnessed in the lab…

Step 1: Calculate the Impedances

The circuit is repeated here:

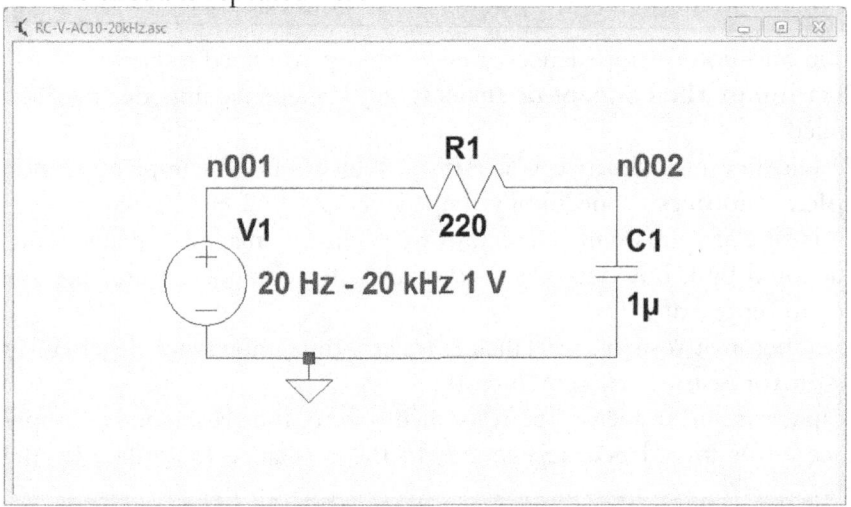

We see that there is a voltage source (V1), a resistor (R1) and a capacitor (C1). Let us calculate the impedances for the components:

	Resistor R1	Capacitor C1
Impedance Formula	$Z_R = R$	$Z_C = 1/sC$ $(s = j2\pi f)$ $Z_C = 1/sC = 1/j2\pi fC = -j(1/2\pi fC)$
Impedance at 20 Hz		$Z_C = -j(1/(2\pi \cdot 20 \cdot 1\mu))$ $= -7.96 \text{ k}\Omega$
Impedance at 1 kHz	$Z_R = R = 220 \text{ }\Omega$	$Z_C = -j(1/(2\pi \cdot 1k \cdot 1\mu))$ $= -159 \text{ }\Omega$
Impedance at 20k Hz		$Z_C = -j(1/(2\pi \cdot 20k \cdot 1\mu))$ $= -7.96 \text{ }\Omega$

Again, please learn to take advantage of the unit prefixes. Once you find the value for 20 Hz, the value at 20 kHz should be obvious… You can use the calculator to double check, of course.

Also, this is written out just as an example. In reality, all we need to do is complete the first row, and that is enough to go to step 2… :)

Step 2: Calculate the Transfer Function
The circuit is a voltage divider. The transfer function is thus $H(s) = Z_C / (Z_R + Z_C)$
If we continue the calculation in s domain, we get $H(s) = Z_C / (Z_R + Z_C) = (1/sC) / (R + (1/sC))$
Multiplying both numerators and denominators by sC, we get:
$$H(s) = 1 / (sCR + 1)$$

Alternatively, we can calculate in f domain. Substituting what we got in step 1: $Z_R = R$, and $Z_C = 1/sC = 1/j2\pi fC = -j(1/2\pi fC)$, we get:
$$H(f) = Z_C / (Z_R + Z_C) = -j(1/2\pi fC) / (R + -j(1/2\pi fC))$$
$$= -j / (2\pi fCR - j)$$, and multiplying both numerator and denominator by j,
$$= 1 / (j2\pi fCR + 1)$$

As one can see, it is much easier to keep the calculation in the s domain, i.e., $H(s) = 1 / (sCR + 1)$. We can get the same result by simply substituting $s = j2\pi f$. But we just did "double checking" to make sure we have got the right answer either way...

Step 3: Calculate Key Parameters
a) The "order" of the transfer function is 1. This is formally defined as the highest power of s in the transfer function. And for $H(s) = 1 / (sCR + 1)$ that means an order of 1.

b) At DC, that is, when f = 0, the transfer function $H(0) = 1 / (0 + 1) = 1$

c) At infinitely high frequency, when $f \to \infty$, the transfer function $H(\infty) = 1 / (\infty + 1) = 0$

d) The roots of the denominator of the transfer function tells us the corner frequencies.
For this simple case, there is only one "pole", and the answer can be obtained by simply setting the denominator of the transfer function to zero.
$H(s) = 1 / (sCR + 1)$. The denominator is $(sCR + 1)$. Setting the denominator to zero,
$sCR + 1 = 0$, and simplifying to obtain s,
$s = -1/RC$

The value "RC" occurs so frequently that we often refer to it as the **RC constant**. The corner frequency of this circuit can be derived from the reciprocal of the RC constant.
One may "imagine" that $s = j2\pi f$ and take the magnitude of the equation $s = j2\pi f = -1/RC$ and thus simplifying, $f = 1/(2\pi RC)$. The proper way to come up with the answer is shown in step 5.
In this case, substituting R = 220Ω and C = 1µF, we get $f = 1/(2\pi \cdot 220 \cdot 1\mu) = 723$ Hz

Step 4: Draw the Bode Plot
We got the key parameters in step 3, enabling us to draw the Bode Plot. Remember to first draw the log-log axes.
a) At DC, that is, when f = 0, the transfer function H(0) = 1, meaning the gain is 20 log (1) = 0 dB. We draw a horizontal line at 0 dB.

b) We stop that line at the corner frequency, 723 Hz.

c) We start a new line segment at the corner frequency, sloping down at 20 dB/decade, because the order of the transfer function is first order. (First order systems have one 20 dB/decade rolloff. Second order systems either have *one* 40 dB/decade rolloff or *two* 20 dB/decade rolloff. And so forth.)

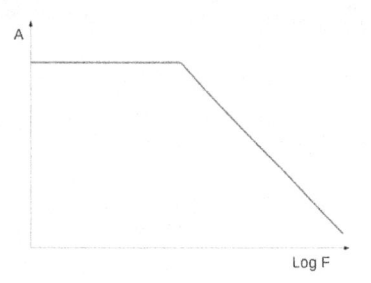

Step 5: Other Requested Calculations

We got the key parameters already, now comes the (relatively) easy part...

$$\omega = 2\pi f \quad \text{(Definition of angular frequency)}$$
$$X_c = -1/\omega C \quad \text{(Capacitive Reactance)}$$
$$X_c = -1/2\pi f C \quad \text{(combining the two)}$$
$$|X_c| = 1/\omega C \quad \text{(Magnitude of Reactance)}$$

Again the difficult part has to do with convention and language - historically people play loose on the word "reactance" when they often really meant "*magnitude* of reactance". You'll just have to learn to accept such looseness of language (and not complain to your boss or colleagues). In other words, those who treat reactance as magnitude of reactance will remove the negative sign from the value. In that case, **you will have to learn to add the negative sign to the impedance on your own.**

$$Z_c = jX_c \quad \text{(Impedance for Capacitance)}$$
$$Z_c = -j|X_c| \quad \text{(Impedance for Capacitance when using magnitude)}$$

In audio electronics, j is used instead of i to represent the imaginary number. For this book, you don't need to know much more about complex numbers beyond this.

Again, by convention and language, often the *magnitude* of impedance is used in conversation. For example, when we say the impedance of a speaker is 8 ohms, we are really referring to the *magnitude* of the impedance of the speaker being 8 ohms.

In the case of a perfect capacitor, the magnitude of the impedance is simply the magnitude of the reactance.

Voltage Divider Calculation

The circuit can be seen as a voltage divider configuration. We can thus calculate the voltage levels using the magnitude of the reactance and impedance.

First, let us apply the definitions of reactance and impedance we learned so far and calculate their values for the capacitor we had in the circuit at the low, medium, and high frequencies...

Chapter 5 – AC Analysis

Formula	f = 20 Hz	f = 1 kHz	f = 20 kHz
$X_c = -1/2\pi fC$	$X_c = -1/(2\pi \cdot 20 \cdot 1\mu)$ = -7.96 kΩ	$X_c = -1/(2\pi \cdot 1k \cdot 1\mu)$ = -159 Ω	$X_c = -1/(2\pi \cdot 20k \cdot 1\mu)$ = -7.96 Ω
Magnitude of X_c	7.96 kΩ	159 Ω	7.96 Ω
$Z_c = jX_c$	$Z_c = -j \cdot 7.96$ kΩ	$Z_c = -j \cdot 159$ Ω	$Z_c = -j \cdot 7.96$ Ω
Magnitude of Z_c	7.96 kΩ	159 Ω	7.96 Ω

Once the table is read, it is obvious why the third row of Z_c is seldom communicated. That also explains why the magnitude of reactance and impedance is often used in communications.

Notice again I used the SI unit prefixes in the calculations. That is because once you understand them, it is very quick to do calculations with them without having to convert to the scientific notation for calculations, and then back into the prefixes again. Notice also that one does not need to use the calculator to obtain the values for 20 kHz once the values for 20 Hz is known.

Now that we have found the magnitudes of the impedances, we can substitute into the voltage divider formula to obtain the values of V(n002). But that requires complex number calculation (which is equivalent to the use of Pythagoras formula), so I'll just do the calculation below:

Formula	f = 20 Hz	f = 1 kHz	f = 20 kHz
$\|V(n002)\| = \|Z_c / (220 + Z_c)\| \cdot 1$ V	V(n002) = 7960 / $\sqrt{220^2 + 7960^2}$ · 1 V = 0.999 V	V(n002) = 159 / $\sqrt{220^2 + 159^2}$ · 1 V = 0.586 V	V(n002) = 7.96 / $\sqrt{220^2 + 7.96^2}$ · 1 V = 0.036 V

Compare the calculated results with the graphs above, and you can see that it matches. This demonstrates the use of reactances and impedances.

A Final Quest

All good things must come to an end (and bad things too). Let us watch how the engineer find the corner frequency of this circuit.

The **corner frequency is 3 dB below the Bode pole**. So we can simply look up the Bode plot and find that frequency to be at around 700 Hz.

It turns out that the corner frequency occurs when the magnitude of the impedances for the two components match. Recall that the magnitude of the impedance for the resistor is simply 220Ω. The magnitude of the impedance for the capacitor is $1/2\pi fC$. So the corner frequency occurs at $1/2\pi fC = 220$ or $f = 1/(2\pi \cdot 220 \cdot 1\mu) = 723$ Hz.

> To calculate the corner frequency f_c for a basic RC circuit, simply *equate the magnitudes of the two impedances*. In other words:
> $1/2\pi f_c C = R$ (Approach: equate the two impedances)
> Thus, $f_c = 1/2\pi RC$ (Resulting formula)

The Take-Away

The Bode plot or frequency response chart shows the variability of the capacitor and the basic RC circuit responding at different frequencies. In fact, they are the main communication tools for *any* audio circuit. One needs to learn how to read and write them as that is a fundamental part of the language of audio electronics.

Visualizing Frequency Response for ANY circuit

If what we see so far is overwhelming, consider that the audio engineer has to be able to *visualize* the frequency response for **any** circuit one comes into contact. There has to be a shortcut; thankfully there is.

First you need to be able to do DC analysis on the circuit, for example, eventually seeing that it is in voltage divider configuration, as in the case here.

Second, you need to be able to do boundary condition analysis, as the engineer sort of did during probing. It goes like this:

a) At DC, the impedance of the capacitor is infinite (i.e., it blocks DC), thus it is **effectively an open circuit**. Thus no current flows through the capacitor, which means all current flows through the resistor and into the load. Assuming the load has infinite impedance, then the voltage out would be essentially the voltage in at DC.

b) At high frequencies, the impedance of the capacitor is very low, thus it is **effectively a short circuit**. Current appears as if it flows through the resistor to ground. Thus the load appears to be shorted to ground as well, receiving little output voltage.

> The capacitor can be <u>visualized</u> as an **open circuit at DC** and a **short circuit at high frequencies**.

Supplement V carries on this approach to show how we can visually identify common passive filters quickly.

The RC Constant

Multiple the target resistance by the capacitance: this is known as the RC constant. The corner frequency is simply $1/2\pi$ times the reciprocal of the RC constant. The RC constant plays an important role in designing and analyzing filters and transmission lines, and thus a special term is given. Often only an approximation is needed, and so the corner frequency is simply 1/6 the reciprocal of the RC constant.

> RC constant = R · C   (*doh!*)
> Corner frequency $f_c = 1/2\pi \cdot 1/RC$
>   or approximately 1/6 of 1/RC

If there is only one corner frequency (in the case of the series RC circuit above), mark that corner frequency with a dotted line, and mark that dotted line intersection with the voltage output value that is horizontal with an X mark. Identify the rolloff, then you can draw the Bode plot. The frequency response is just asymptotic to the Bode plot, deviating from it by at most 3 dB which is at the corner frequency.

Remember all these take-aways, as we shall soon utilize them in analyzing a real-life audio shelving filter.

5.5 Virtual Lab: Series L/R for AC

Welcome again to our Virtual Lab, where an audio engineer is probing this series L/R circuit:

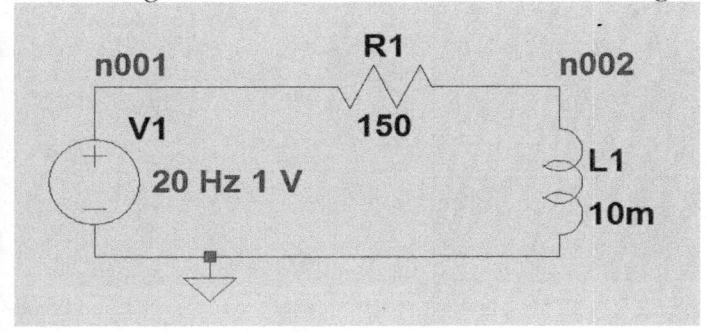

At High Frequency

Here is the time-domain response chart gathered at 20 kHz...

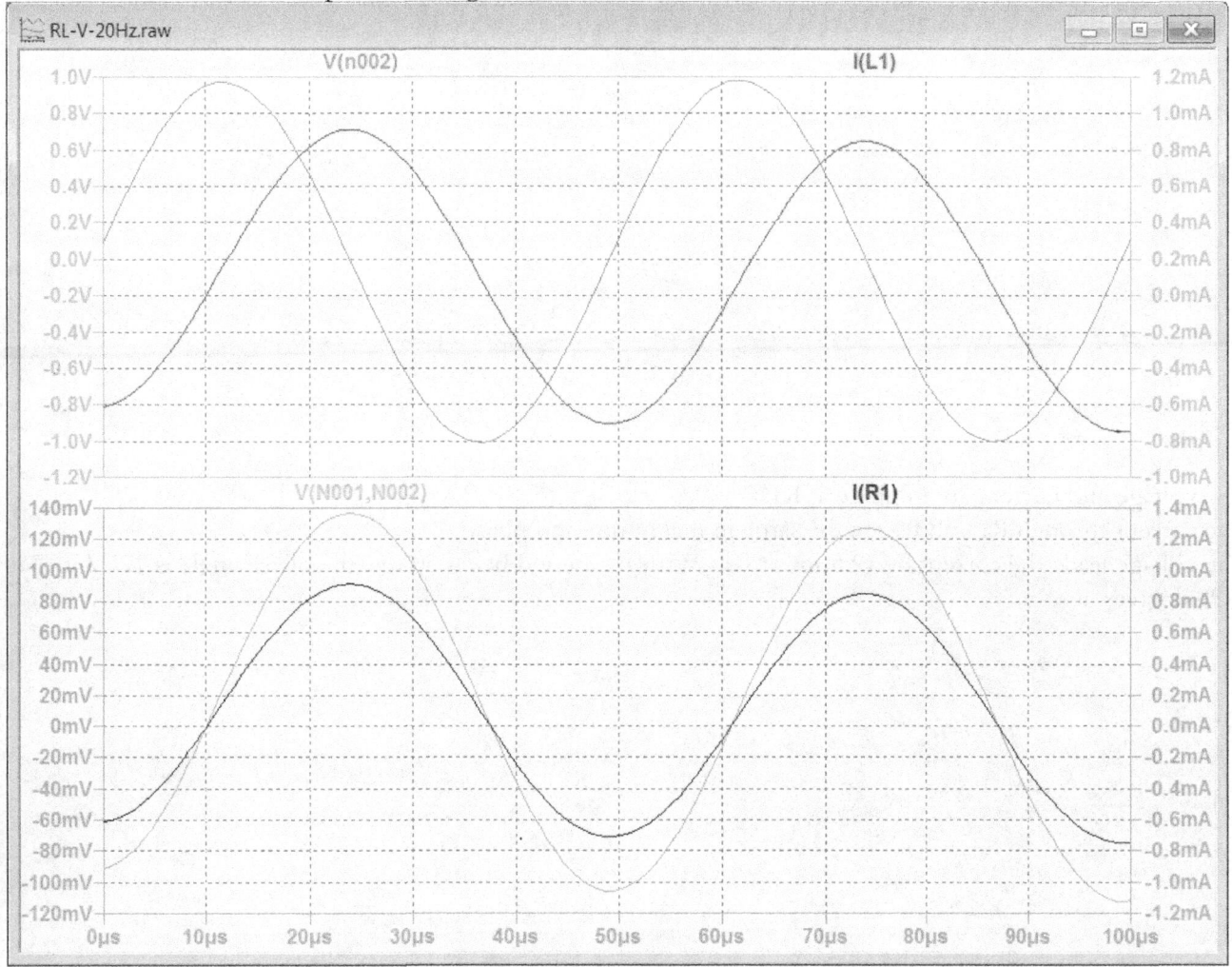

The voltage and current are in phase at R1.
The current through R1 and L1 are the same in magnitude and phase.
The voltage leads the current by 12.5 μs at L1. With the period being 50 μs, the phase angle is 12.5/50*360 = 90 degrees.

At Low Frequency

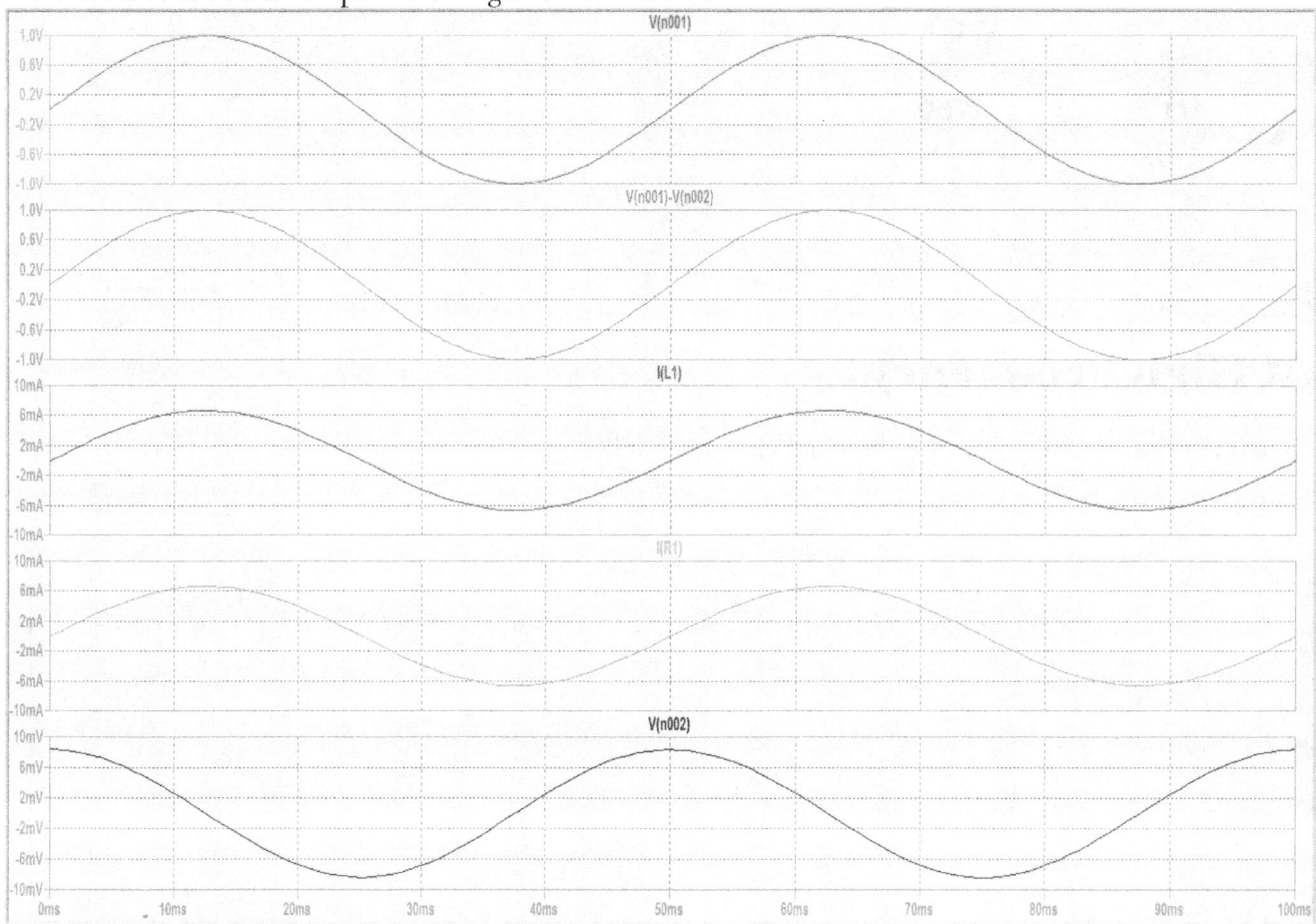

Here is the time-domain response chart gathered at 20 Hz...

The voltage and current are in phase at R1.
The current through R1 and L1 are the same in magnitude and phase.
The voltage leads the current by 12.5 ms at L1. With the period being 50 ms, the phase angle is 12.5/50*360 = 90 degrees.

Frequency Sweep

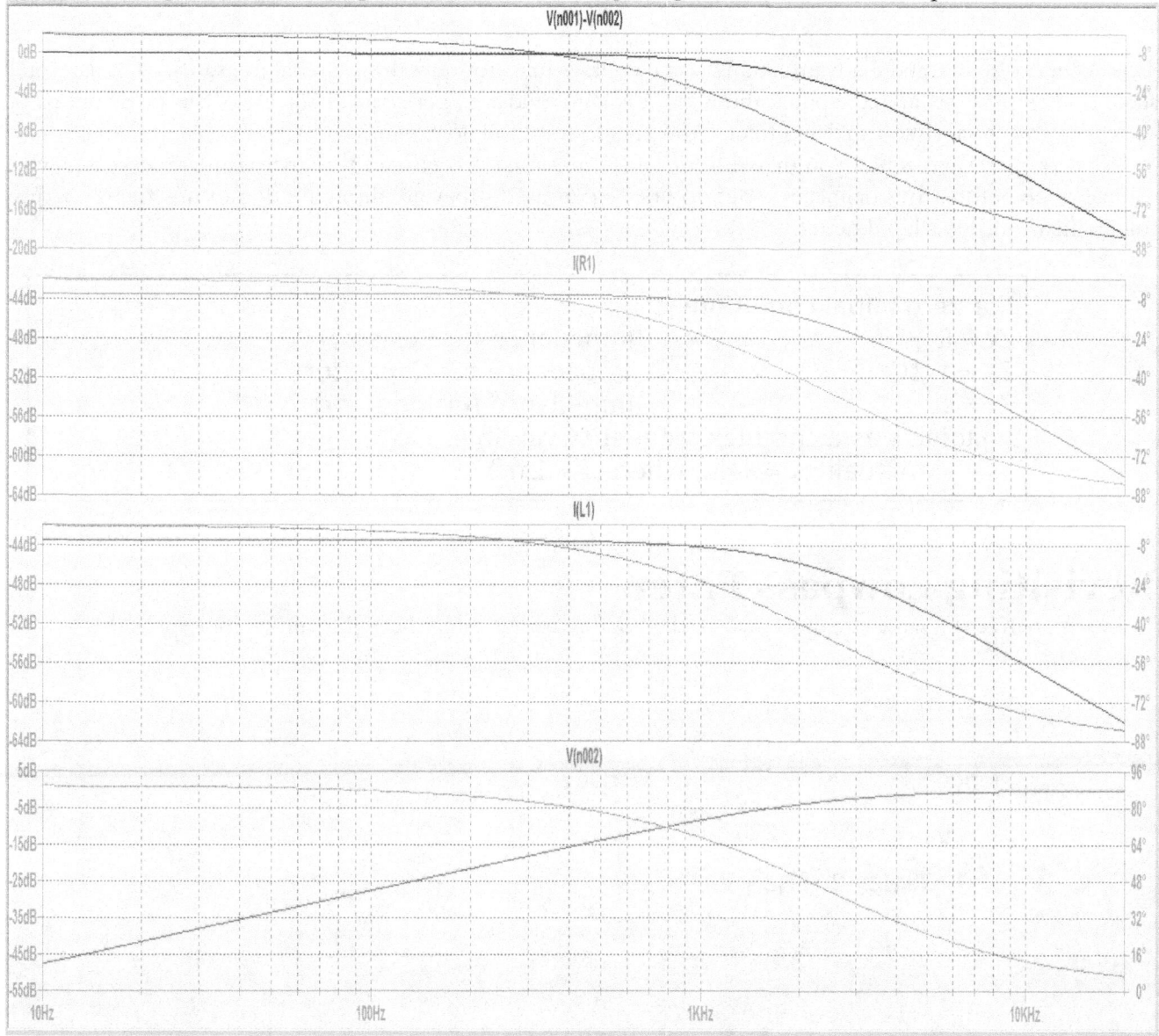

Here is the frequency-domain response chart when the input signal V1 is 0 dB at all frequencies...

The voltage and current are in phase at R1 at all frequencies.
The current through R1 and L1 are the same in magnitude and phase at all frequencies.
The voltage leads the current by 90 degrees at L1 at all frequencies.
Te voltage across L1 are complementary to the voltage across R1, in other words, the two added together equals V1 at all frequencies, as expected. Notice that this is a little bit difficult to see because of the use of dB as unit.at the y-axis.

> To calculate the corner frequency f_c for a basic RL circuit, simply *equate the magnitudes of the two impedances*. In other words:
> $2\pi f_c L = R$ (Approach: equate the two impedances)
> Thus, $f_c = R / 2\pi L$ (Resulting formula)

5.6 AC ANALYSIS FOR LOW-PASS & SHELVING FILTERS

This section is a little more advanced, and some courses may not quiz this material in exams. Because you'll see S-domain analysis in many publications, so it is a good idea to get acquainted. Also, this turns out to be the easiest way to analyse a shelving filter. And you do want to know, right?

What is interesting is that you may discover a lot of similarity between this analysis method versus what we had done before. It is deliberate. But do not be confused - they are different; do not mix them up. Just follow the procedures listed here.

> For zero initial conditions:
> $Z_R = R$ (Impedance for Resistor)
> $Z_C = 1/sC$ (Impedance for Capacitor)
> $Z_L = sL$ (Impedance for Inductor)
> Imagine s as a complex frequency variable.
>   (Actually $s = \sigma+j\omega$, where $\omega = 2\pi f$)

Revisiting Lowpass Filter

Let us analyze the same circuit we had studied before earlier using this s-domain method instead:

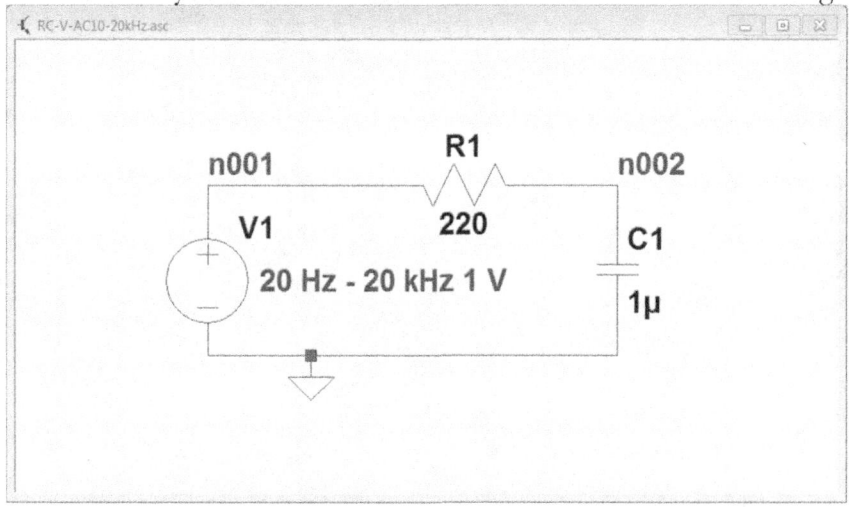

[Circuit diagram: RC-V-AC10-20kHz.asc — V1 (20 Hz - 20 kHz, 1 V) source at node n001, R1 = 220 between n001 and n002, C1 = 1μ from n002 to ground]

This is still a voltage divider circuit, so we can apply the same approach:
$V_{out}/V_{in} = Z_C / (Z_C + Z_R) = (1/sC) / (1/sC + R) = 1 / (1 + sCR)$

In other words, we just observed the **transfer function** $H(s) = V_{out}/V_{in} = 1 / (1 + sCR)$

> First step in s-domain analysis is to express the transfer function in terms of poles and zeroes, i.e., in this form:
> $H(s) = a \cdot (s-Z_1)\cdot(s-Z_2)\cdots(s-Z_m)/(s-P_1)\cdot(s-P_2)\cdots(s-P_n)$
> $$H(s) = a \cdot \frac{(s-Z1)\cdot(s-Z2)\cdots(s-Zm)}{(s-P1)\cdot(s-P2)\cdots(s-Pn)}$$

So we divide both numerator and denominator by the RC constant to obtain:
$H(s) = V_{out}/V_{in} = (1/RC) / (1/RC + s) = (1/RC) / (s + 1/RC)$
It is now in the correct form, with $a = 1/RC$, $P_1 = 1/RC$.

> Second step in S-domain analysis is to construct the Bode Plot, as follows:
> Each pole has a horizontal asymptote, changing to a -20 dB/decade rolloff·
> Each zero has a horizontal asymptote, changing to a +20 dB/decade rolloff
> Add all the individual Bode Plots to form the composite response.

Our transfer function has only one pole, which is $1/RC$, and so our corner angular frequency $\omega_c = 1/RC$. Applying $\omega = 2\pi f$, we get the same answer as before: $f_c = 1/2\pi RC$. Because it is a pole, we have a -20 dB/decade rolloff, predicting exactly its behavior.

Analyze Shelving Filter

Let us analyze the shelving filter circuit shown below using this s-domain method:

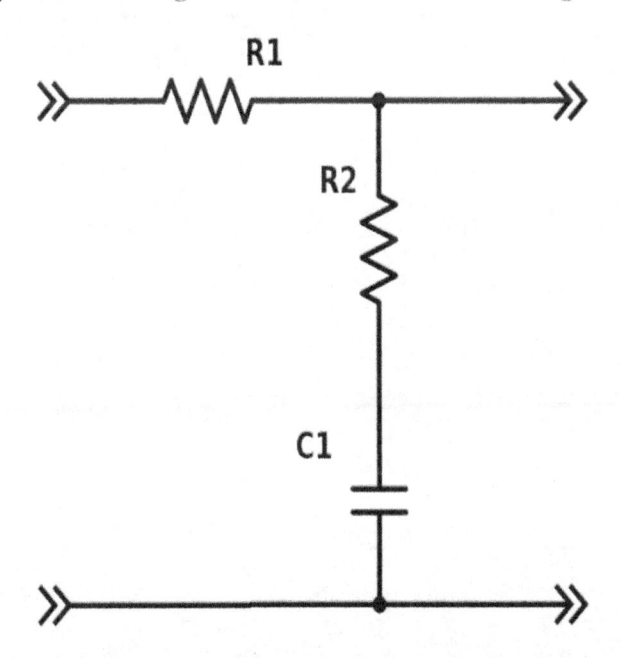

This is also a voltage divider circuit; the transfer function is:
$V_{out}/V_{in} = (R_2 + 1/sC) / (R_1 + R_2 + 1/sC) = (1+sCR_2) / (1+sC(R_1+R_2))$
Rearranging to the standard form by multiplying numerator and denominator by $1/C(R_1+R_2)(1/CR_2)$:
$V_{out}/V_{in} = (1/C(R_1+R_2)) (s + 1/CR_2) / (1/CR_2) (s + (1/C(R_1+R_2)))$
$= R_2 (s + 1/CR_2) / (R_1+R_2) (s + (1/C(R_1+R_2)))$
It is now in the correct form, with $P_1 = 1/C(R_1+R_2)$, $Z_1 = 1/CR_2$, and $a = R_2/(R_1+R_2)$

In other words, we have a pole at $1/C(R_1+R_2)$ and a zero at $1/CR_2$.
Applying $\omega = 2\pi f$, we get from the poles and zeroes:
$f_P = 1/2\pi C(R_1+R_2)$
$f_Z = 1/2\pi CR_2$
If you compare a with f_P and f_Z, you'll notice that $f_P/f_Z = (1/2\pi C(R_1+R_2)) / (1/2\pi CR_2) = R_2/(R_1+R_2) = a$.
Let us use a table to help us determine how to add the roll-off's together.

	Low frequencies	Mid frequencies @f_P	High frequencies @f_Z
Pole Bode Plot	Pole horizontal	Pole rolloff -20 dB/decade	Pole continues rolloff -20 dB/decade
Zero Bode Plot	Zero horizontal	Zero horizontal	Zero rolloff +20 dB/decade
Combined Bode Plot	horizontal asymptote	rolloff -20 dB/decade @f_P	horizontal asymptote @f_Z

Notice that when we add a -20 and +20 dB/decade rolloff, we get a horizontal asymptote at high frequencies. This is the shelving part of the filter.

Thus we can graph the Bode Plot. Of course we can also go to our virtual lab to obtain the same result, where R1 = 220 Ω, R2 = 110 Ω, C1 = 1 μF:

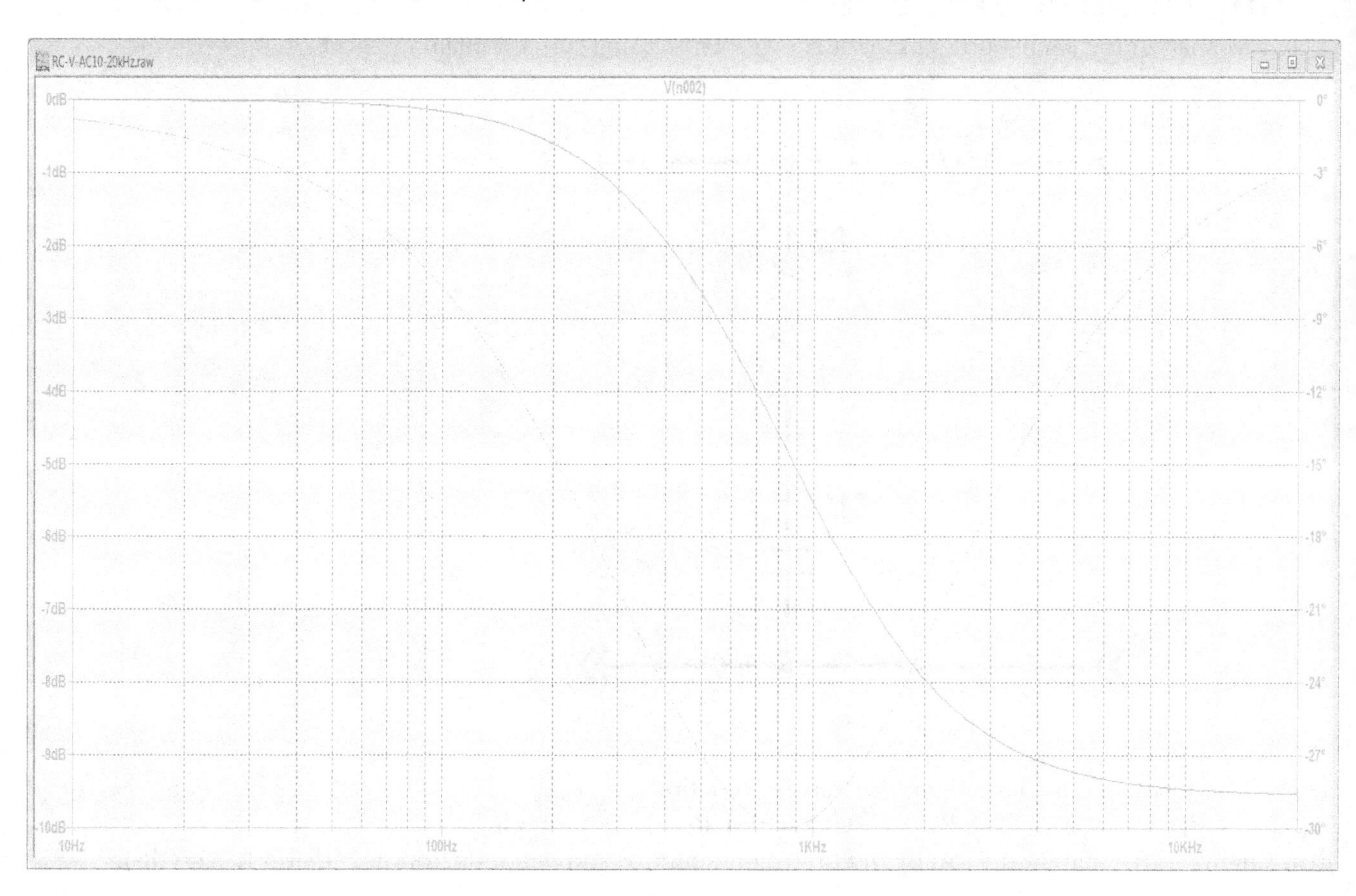

Substituting values, we get a = 110/(110+220) = 1/3. Converting to dB, we get a = 20 log (1/3) = -9.54 dB.
$f_Z = 1/2\pi CR_2 = 1/2\pi(1\ \mu)(110) = 1447$ Hz
$f_P = 1/2\pi C(R_1+R_2) = 1/2\pi(1\ \mu)(110+220) = 482$ Hz

The calculated values match the results from our virtual lab.

Historical Note & Big Picture

We have to thank Pierre Laplace for this shortcut method which does not require us to use differential or integral calculus. However, to prove that this works requires us to use differentials and integrals and more advanced math. That is why we left out all the details of the proof and proceeded directly to the method.

We have covered time and frequency domains, and now s-domain. Can we convert from s-domain to time domain? Yes! But it requires advanced math, which is outside the prerequisite of this course.

However, the fact that we can use the same algebraic tools as we had done with DC analysis for time-varying signals is a big deal. Please treat it as such. Remember, simplicity is often built upon extreme complexity, which is true in this case, thus we skipped all the foundational complexities to arrive at this simple shortcut.

More importantly, most publications use s-domain analysis. If you want to find a particular filter or circuit, most likely it'll have s-domain transfer function information. That is why you need to know how to make use of such information, at least in the most basic way, which is what was presented here.

Order of a Filter

The filters so far are known as **first order filters**. There are many ways to find out the order of a filter, and you'll need to know *all* of them. Notice on the Bode Plot the highest rolloff possible is either +20 or -20 dB/decade (or correspondingly + or - 6 dB/octave) - that is the behavior of a first order filter.

A second order filter has + or - 40 dB/decade maximum rolloff. An nth order filter has + or - 20·n/decade maximum rolloff.

It is usually true also that the number of reactive elements (i.e. inductor or capacitor) in a passive filter circuit determines the order of the filter. So far the filters use one reactive element, and so we got first order filters.

The order came from the math of the transfer function. Specifically, highest order of the numerator or denominator polynomial is the order of the filter for an s-domain analysis. Likewise for z-domain analysis.

When we use delay elements to implement active, the maximum delay in samples is the filter order. The largest order in the difference equation is the order.

If we want to have steeper rolloffs, we can **cascade** filters together. Cascading two first order filters will give us a second order filter with + or - 40 dB/decade maximum rolloff. However, cascading properly is tricky as we also need to fulfil impedance bridging requirements (that we'll discuss later). Thus usually an op-amp is used, either as a buffer between the cascaded filters, or as an active filter. (This will also be discussed later.)

> Order of a filter is the highest order of the polynomial describing the filter.
> It usually reflects the number of reactive elements in the circuit.
> An nth order filter has + or - n·20/decade (6 dB/octave) maximum rolloff.

5.7 Time Constants: RC, L/R, √LC

Capacitors and inductors attempt to hold respectively voltages and currents constant. So voltages and currents decay exponentially. The time it takes to decay to 63% of its initial value is known as the **time constant**. The good news is that time constant is easy to remember: for series RC circuits the time constant is RC. And for series L/R circuits the time constant is L/R. In fact, *now* one can see why we call it series L/R circuit and not RL or LR, as the term "series L/R" helps us remember that when inductor and resistor are in series, the time constant is L/R!

The unit for time constant is - obviously - second (s). For audio frequencies, μs is most convenient. Its symbol is the Greek letter *tau*, τ.

Supplement X shows an easy method to visually identify common passive filters quickly. It also indicates how to calculate the corner frequencies of the filters using a simple and consistent concept of time constants. Here is a quick summary…

> $f_c = 1/(2\pi\tau)$ where τ is the time constant, and f_c is the corner frequency
> For RC circuits, τ = RC
> For RL circuits, τ = L/R
> For LC circuits, τ = √LC

The key is that regardless whether the filter is lowpass or highpass or notch, the corner frequency is identified by their time constants stated. This is far easier to memorize than any other formulas. As a result, time constants are used even in defining industry standards, like in the case for RIAA (Recording Industry Association of America).

For those interested in what the decay actually looks like mathematically, the universal time constant formula is:

$$\text{change} = (\text{final} - \text{start})\left(1 - (1/e^{(t/\tau)})\right)$$

Remember the formula is applicable to what is being held as constant as possible, which is voltage for capacitors and current for inductors.

Examples

1. What is the time constant for a series RC circuit with 75k and 0.001μF?
    τ = RC = 75k * 0.001μ = 75 μs

2. What is its corner frequency?
    $f_c = 1/(2\pi\tau) = 1/(2\pi \cdot 75\,\mu) = 2122$ Hz

5.8 Review Exercises

Chapter 5 Review Guide

Theory
- **concepts** of impedance, reactance
- **concepts** of time constant, order, Bode plot
- **avoidance of wrong concepts**

Practice
- ability to calculate **impedance** and **reactance**
- ability to perform **AC analysis** given schematics
- ability to produce **Bode Plot** from basic parameters
- ability to **calculate time constant** and **corner frequencies**

Exercise A

1 What load does the amplifier see when two 8Ω speakers were connected in series and in parallel?
In series:

In parallel:

2. Fill in the table, using engineering units (with metric prefix) for your calculated answers. (*Hint: once you get some answers, others may be derived quickly.*)

	Ideal **Resistor** (value R)	Ideal **Inductor** (value L)	Ideal **Capacitor** (value C)
Impedance formula	$Z_R =$	$Z_L =$	$Z_C =$
Resistance value			
Reactance value			
Magnitude of Impedance			
Phase Angle (*hint: ICE/ELI*)			
	Resistor **10 kΩ**	Inductor **10 mH**	Capacitor **10 nF**
Magnitude of Impedance at 20 Hz			
Magnitude of Impedance at 1 kHz			
Magnitude of Impedance at 20 kHz			

Exercise B

Fill in the blanks for the circuits supplied - you should first discover the order (*hint: count the number of capacitors and inductors*), then calculate the voltage transfer response ratios at DC and high frequency indicated (*hint: use voltage divider formula and treat capacitors and inductors as open or short circuits as appropriate*), then derive the characteristic curve, and then say what type of filter it is (*not by reversing the steps and avoiding simple calculations that can even be done in the head*). Remember always to label the axis of your plots, as the shape shifts depending on the axis.

Circuit	Order	DC response	high freq resp.	characteristic curve	filter type name
R1 series, L1 shunt					
C1 series, R1 shunt					
C1 series, L1 shunt					
L1 series, C1 shunt					
R1 series, L1 shunt, C1 shunt					
L1 series, R1 shunt					
R1 series, R2 shunt, C1 shunt					

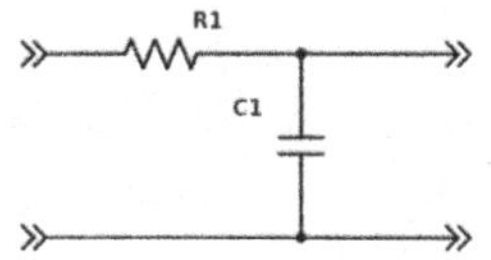 R1 C1					
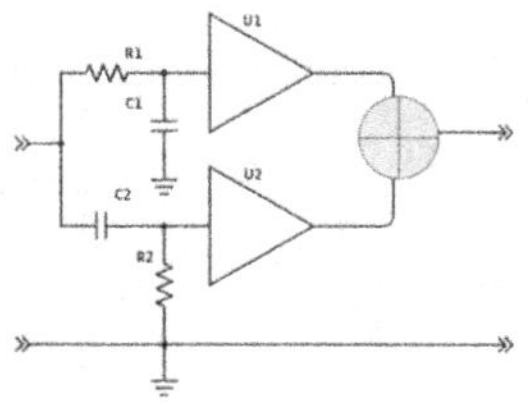					
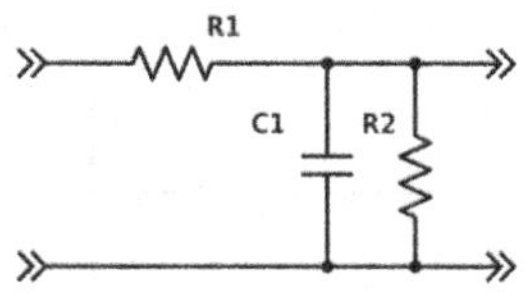 R1 C1 R2					
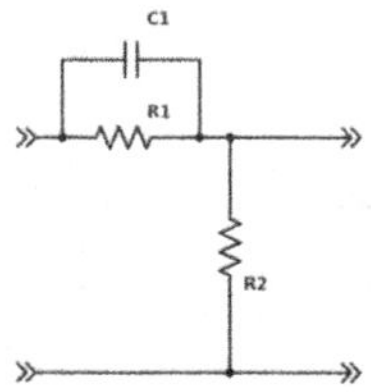 C1 R1 R2					
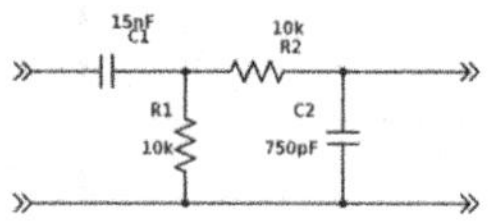 15nF C1 10k R2 R1 10k C2 750pF					
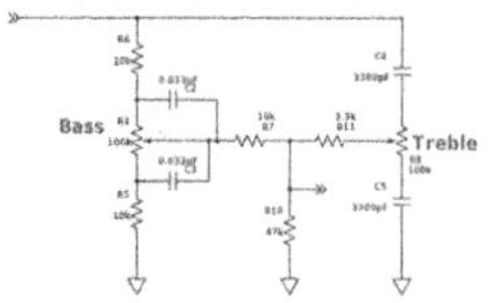 Bass Treble					
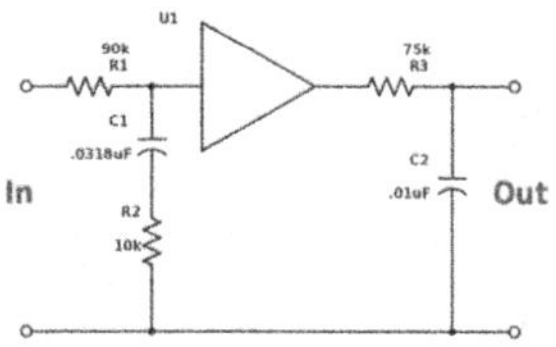 U1 90k R1 C1 .0318uF In R2 10k 75k R3 C2 .01uF Out					
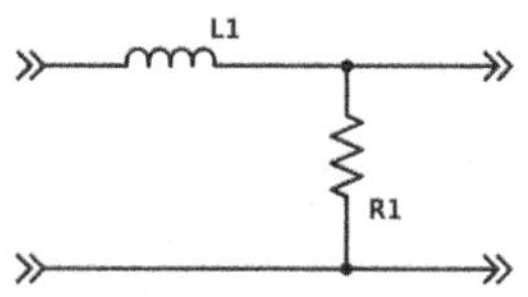 L1 R1					

SUPPLEMENT U - R WITH RC AND LOUDSPEAKERS IN SERIES & PARALLEL

Normal real-life waveforms contain **both** DC and AC components. Impedance contains **both** resistance and reactance. So how do we handle AC with resistors? And how do we handle loudspeakers which have reactances when we connect them together?

R with AC

In a nutshell, in simple situations with resistances only, treat impedance just like resistance. Use the same formulas as we had used for DC analysis. That's it! Do not be afraid…

Examples

1. Two 1 kΩ resistors are in series. What impedance is the equivalent circuit at 1 kHz? How about DC? Applying the series formula: 1k + 1k = 2 kΩ. Its impedance is 2 kΩ at *all* audio frequencies, including DC .

2. Two 1 kΩ resistors are in parallel. What impedance is the equivalent circuit?
Applying the parallel formula: 1/(1/1k + 1/1k) = 1/(2/1k) = 500 Ω.

♫ Speakers in Series & Parallel

With resistors out of the way, let's talk about speakers in series and parallel. In a nutshell, engineers goes out of their way to make things simple, even arbitrarily; in simple situations like that, treat impedance just like resistance. Engineers often use "nominal values" in these situations. Use the same formulas as we had used for DC analysis. That's it! Do not be afraid…

In reality, the impedance for speakers are complex, and varying over frequency as well as loudness and other factors. But for practicality, only a nominal value which is *not* complex is given. Consumer speakers are often rated at 4-8 ohms; professional speakers 8-16 ohms; and headphones 16-32 ohms, all nominally.

Examples

1. Two 8Ω speakers are in series. What impedance does the amplifier see?
   Applying the series formula: 8 + 8 = 16Ω

2. Two 8Ω speakers are in parallel. What impedance does the amplifier see?
   Applying the parallel formula: 1/(1/8 + 1/8) = 1/(1/4) = 4Ω

3. Is it a good idea to put four 8Ω speakers are in parallel?
   Applying the parallel formula: 1/(1/8 + 1/8 + 1/8 + 1/8) = 1/(1/2) = 2Ω
   This is lower than the usual rated impedance that many amplifiers will support, so it is *not* a good idea.

Supplement V - How to Visually Identify Common Passive Filters Quickly

When given a schematic, how do we quickly identify what kinds of passive filters are involved?

Basic Concept

The *capacitor* behaves as an *open circuit* at DC, and a *short circuit* at high frequencies.
The *inductor* behaves as a *short circuit* at DC, and an *open circuit* at high frequencies.
The *resistor* behaves the *same* at all (audio) frequencies.
The *order* of a simple passive filter is the total number of *capacitor* and *inductor*.

Using that basic concept, we can obtain two separate equivalent circuits at DC and high frequencies - it can be just in our heads...

Example

Original Circuit	Equivalent DC circuit	Equivalent High Freq Circuit
R1 in series, C1 shunt to ground	R1 only (C1 open)	R1 with shunt short (C1 short)

We can see from visual inspection that the DC transfer function response must be 1 (because of the open) and the high frequency response must be 0 (because of the short). Specifically, at DC, if we apply 1 V in, we get 1 V out, so the transfer function is $H_{DC} = 1/1 = 1$. For high frequency, it is 0 V out for 1 V in, so the transfer function is $H_{HF} = 0/1 = 0$. The transfer function is usually between 0 to 1, because for *passive* filters, no external power is available. Plus, we have one capacitor, so the order of the filter is one. From those three pieces of information and the concept of Bode Plot (see Chapter 5), we can draw the characteristic curve or frequency response immediately.

When the circuit is more complicated, we may need to use the voltage divider formula to calculate the transfer function. Again, the formula is:

$$H = V_{out} / V_{in} = Z_2/(Z_1+Z_2)$$

Know the First Order 2-component Filters

It is best to understand how to derive the characteristics curves for the basic RC, CR, RL, LR filters, as shown in Chapter 5, and then take advantage of the Bode Plot to give a simplified answer.

The *alternative* simplified approach is to derive the result in our head using the basic concept outlined above. For example, for the RC filter, we apply the voltage divider formula in our head, and we can see that the voltage transfer function is $Z / (R + Z)$ where Z is the impedance of the capacitor, which is open at DC, meaning very large; and short at high frequency, meaning very small. Thus the voltage transfer function is $H_{DC} = 1$ at DC, and $H_{HF} = 0$ at high frequency. (Substitute values into the formula to convince yourself that it is the case.) With that, and knowing Bode Plot, we know the characteristics curve is that of a low-pass filter.

Similarly, we can quickly figure out the responses of the four basic filters:

Circuit	Order	H_{DC}	H_{HF}	Corner Freq (ω_c)	Characteristics Curve
R-C	1	1	0	$1/(RC)$	LPF
C-R	1	0	1	$1/(RC)$	HPF
R-L	1	0	1	R/L	HPF
L-R	1	1	0	R/L	LPF

As a reminder, as shown in chapter 5, the order of a passive filter can be easily determined by counting the number of capacitors and inductors.

Notice that the axis of a plot *must* be labeled (if it is *not* log frequency as the x-axis, the curve will look different!).

Notice also that duality applies here, and using that concept means BOGO, or you have to remember only half to get all... :) (See supplement W "Duality".)

The corner frequency is easiest memorized without the 2π, in the form of **angular frequency** ω_c, as $1/(RC)$ for RC circuits and R/L for RL circuits, or as the time constant τ, RC and L/R which are their reciprocals. In the form of f_c, we have to include the 2π, with $f_c = 1/(2\pi RC)$ or $R/(2\pi L)$. Notice that the corner frequency is the *same* whether it is high-pass or low-pass. Please convince yourself that they are really the *same* circuit with a different *output port*, and thus having the same corner frequency should not be surprising at all...

Know How Adding a Resistor Changes Characteristics

The idea is that adding one resistor to the basic circuits above changes only one thing. Take example of the first circuit below. Remove R1 and it is a basic series RC circuit with HPF characteristics as seen in row 2 of above table. Now calculate the DC response. It is no longer 0, but rather R2/(R1+R2). At high frequencies, its response is still 1 as the capacitor acts as a short to R1. Thus we simply add one more horizontal line to the Bode Plot, as shown. It is known as a **shelving filter**, as now the rolloff is stopped by the additional shelf.

Circuit	Order	H_{DC}	H_{HF}	Corner Freq (ω_c)	Characteristics Curve
C||R-R	1	R2/(R1+R2)	1	$\omega_z = 1/R_1 C_1$ $\omega_p = 1/(R_1\|\|R_2)C_1$	Shelving HP
R-C+R	1	1	R2/(R1+R2)	$\omega_z = 1/R_2 C_1$ $\omega_p = 1/(R_1+R_2)C_1$	Shelving LP (20 dB/dec rolloff)
R-C||R	1	R2/(R1+R2)	0		LPF

Note for the second example, it does not matter whether R2 is above or below C1 in the schematic. Series is series. We can again start with the basic RC filter and then add the extra horizontal line due to R2.

Notice that adding a resistor does not necessarily make a shelving filter. In the third example, all it does is to reduce the output at DC from 1 to R2/(R1+R2). It is still a LPF in that case. Likewise, convince yourself that a voltage divider followed by a capacitor is still a HPF, although again with output reduced by the voltage divider.

There are two corner frequencies in the first two cases. One of which is similar to the basic RC filter. The other has to do with the two resistors in series or in parallel. Remember $R_1\|\|R_2$ means the two resistors in parallel, or $R_1 R_2/(R_1+R_2)$

Know the Second Order 2-component Filters

The second order filters can also be set up for lowpass and highpass; but in addition, they can function as bandpass and bandstop as well.

The corner frequency is easiest memorized without the 2π, in the form of ω_c, as $1/\sqrt{(LC)}$ for series LC circuits, or reciprocal of the square root of the LC constants. (The LC constant is square-rooted because this is a second-order filter; the RC and R/L constants are not square-rooted because they are first order filters.) In the form of f_c, we have to include the 2π, with $f_c = 1/(2\pi\sqrt{(LC)})$. Notice that the corner frequency is the *same* whether it is high-pass or low-pass. Please convince yourself that they are really the *same* circuit with a different *output port*, and thus having the same corner frequency should not be surprising.

Circuit	Order	H_{DC}	H_{HF}	Corner Freq (ω_c)	Characteristics Curve
L-C	2	1	0	$1/\sqrt{(LC)}$	LPF, 40 dB/dec rolloff
C-L	2	0	1	$1/\sqrt{(LC)}$	HPF, 40 dB/dec rolloff
R-C||R	2	0	0	$1/\sqrt{(LC)}$	BPF

V-4

Know How Adding a Capacitor or Inductor Changes Characteristics

Adding one capacitor or inductor to the basic circuits above changes the characteristics further. It may be best to calculate the response using the method described in chapter 5. However, for a simple R-LC circuit shown below, we can use the basic concept first listed to take a stab for the response…

The voltage transfer function at DC is 1 because of the capacitor, and at high frequency it is also 1 because of the inductor, since they behave like open circuit in both situations. So we can deduce that at an intermediate frequency, the energy stored in the electric field of the capacitor will transfer to the energy stored in the magnetic field, and vice versa, creating a dip in the characteristic curve at that point. We call that a notch filter.

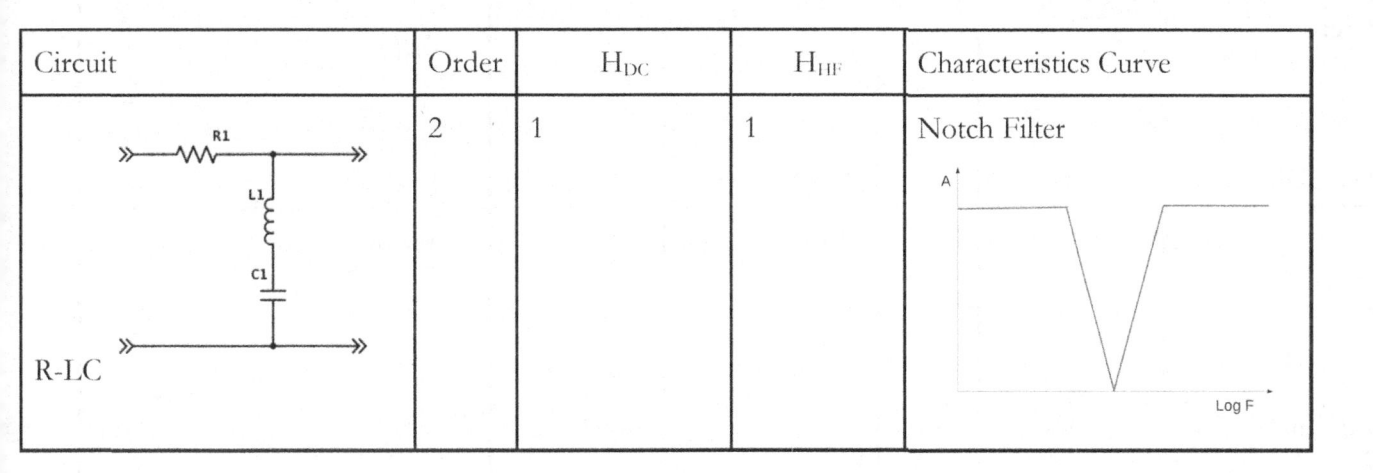

Circuit	Order	H_{DC}	H_{HF}	Characteristics Curve
R-LC	2	1	1	Notch Filter

Remember the Time Constants

> Recall the corner frequency for 1st order RC filter is $\omega_c = 1/RC$
> and $f_c = 1/2\pi RC$, thus period $T = 1/f_c = 2\pi RC$
> We define time constant $\tau = T/2\pi = 1/\omega_c$ and thus for 1st order RC filter, $\tau = RC$

There is a reason why engineers keep on using the terms RC constant, L/R constant, and $\sqrt{(LC)}$ constant. Those are the ones often used in the literature. They can be used to find the corner frequencies for typical passive filter circuits, as long as we remember the use of reciprocal and inclusion of the 2π.

One should also notice that the last two examples do not follow Bode Plot principles. That happens when inductors and capacitors are intertwined, with energy transferred between their respective electric and magnetic fields. Despite that, the corner frequency is still $1/\sqrt{(LC)}$.

Know How Basic Filters can be Connected in Series and in Parallel

More complicated filters can be created by combining simpler circuits in series or in parallel. Often because of impedance bridging requirements, an active buffer is needed to separate the circuits. However, occasionally, the circuits can be combined in series without buffers. An example is shown below.

Circuit	Order	Filter 1	Filter 2	Characteristics Curve
Series	2	HPF	LPF	Bandpass
Parallel	2	LPF	HPF	Bandstop

Please note that the language of electronics is not completely standardized. Some people equate bandstop filter with notch filter, as their characteristic curves look similar. Others separate them by saying bandstop filter have more gentle rolloffs, like the above, while the notch filter has steep rolloffs, like the example earlier, generated through resonance phenomenon.

Take-away

One can figure out the frequency response of typical passive filters without doing a lot of calculations. In fact, with sufficient practice, one can do most of that determination in our head. Please practice with the above examples, and create your own, so that you can identify passive filters by just looking at the schematics.

Application Example: RIAA Equalization

The Recording Industry Association of America (RIAA) established a standard for recording and playback of phonograph records. During recording, "pre-emphasis" was applied. During playback, "de-emphasis" is to be applied to neutralize the pre-emphasis. Together, the two should yield a flat response.

Why go through all that trouble? By emphasizing the high frequencies during recording, and reducing it during playback, we attenuates high frequency noise such as hiss and clicks common in phonograph playback, thus improving perceived sound quality.

RIAA specified the time constants for the transition points as 75 µs, 318 µs and 3180 µs, which correspond roughly to 2122 Hz, 500 Hz and 50 Hz respectively.

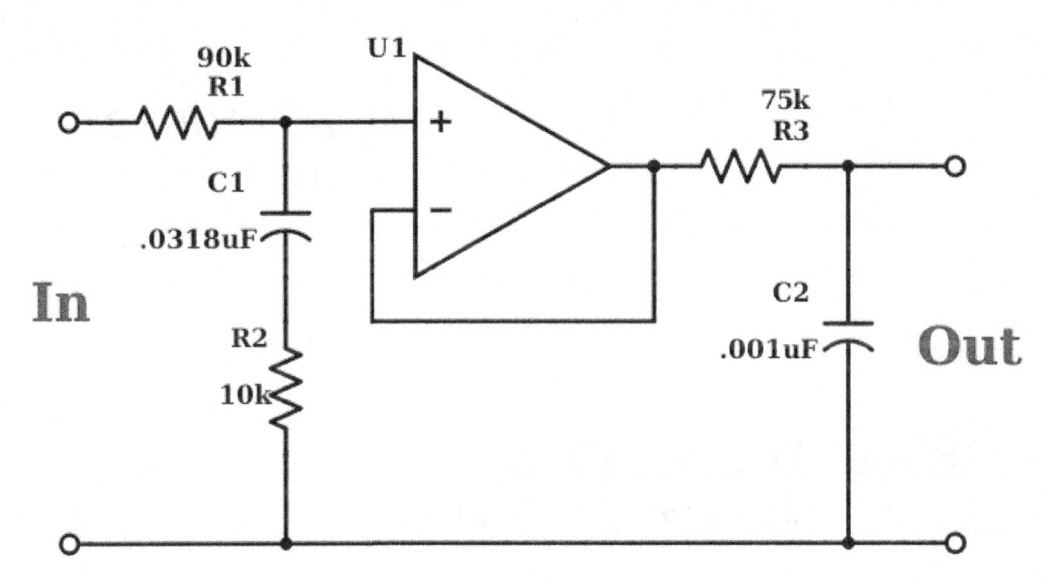

In this circuit, we can see the amplifier U1 is a non-inverting buffer between the two circuits in series. The first circuit to the left of U1 is a shelving lowpass filter as seen from the table earlier, while the second circuit to the right of U1 is a basic lowpass filter. Since the circuits are cascaded together in series, we can simply add their responses to obtain the final response.

The two RC constants for the first circuit (can be seen from the table earlier) are

$R_2 C_1 = 10k * 0.03181\mu = 318$ µs

$(R_1+R_2)C_1 = 100k * 0.03181\mu = 3180$ µs

The RC constant for the second circuit is

$R_3 C_2 = 75k * 0.001\mu = 75$ µs

The RC constants are *exactly* as specified by RIAA as mentioned earlier (provided we can find the precise resistors and capacitors). This also explains why we use angular frequency and not apply the 2π until we really need the physical frequency. It is because the standards organization and many engineers prefer to work with RC constants directly, and the unit for those constants are simply µs.

The Bode Plot would then look like this:

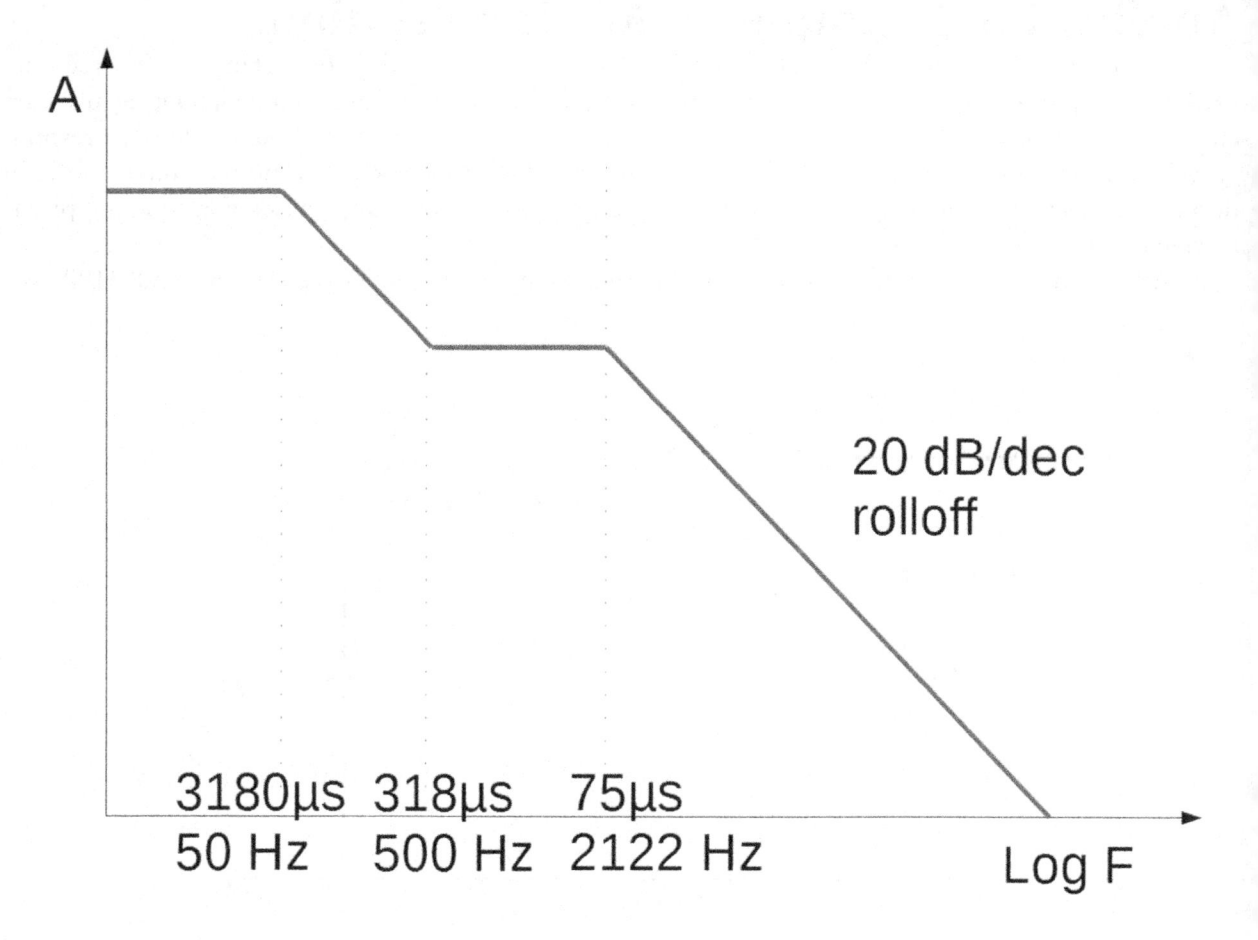

What is the amplitude of the shelf in the Bode Plot? It is simply R2/(R1+R2) as shown in the table earlier, which can be calculated using the voltage divider formed with R1 and R2. Thus the shelf is at R2/(R1+R2) = 10k/(10k+90k) = 10k/100k = 0.1

Converting to dB, the result is 20 log (0.1) = -20 dB.

Thus the shelf occurs at -20 dB. The two rolloffs are -20 dB/decade as they are first order filters.

The implementation above is known as a **split RIAA filter** implementation, because we split into two simple filters for the implementation, one being a shelving lowpass filter, another a simple lowpass. A single passive filter implementation is possible, while most implementations now use active filtering.

Supplement W - Duality Principle in Electronics

If I tell you there is a *BOGO* (Buy One, Get One free) sale, and all you need is to know a coupon, will you take advantage of it? Well, here is your coupon for *BOGO* in electronics that saves memorizing half of the laws or formulas. Are you interested?

Examine these pairs of formulas; can you guess how the pairs hint at duality?

Series and Parallel Circuits
The formula for series circuit is: $Z = Z_1 + Z_2$
The formula for parallel circuit is: $Y = Y_1 + Y_2$

> Remember $Y = 1/Z$, or $Z = 1/Y$
> Thus $Y = Y_1 + Y_2$ means $1/Z = 1/Z_1 + 1/Z_2$
> or, $Z = 1/(1/Z_1 + 1/Z_2)$, which is our usual formula

Voltage and Current Dividers
The formula for voltage divider is: $V_o/V_i = Z_1 / (Z_1 + Z_2)$
The formula for current divider is: $I_o/I_i = Y_1 / (Y_1 + Y_2)$

Kirchhoff's Laws
1st Law (KCL):   Sum of currents at node = zero
2nd Law (KVL):   Sum of voltages at loop = zero

Capacitor and Inductor
The formula for inductor's impedance is: $Z = sL$
The formula for capacitor's admittance is: $Y = sC$

> Remember $Y = 1/Z$, or $Z = 1/Y$
> Thus $Y = sC$ means $1/Z = sC$ or, $Z = 1/sC$, which is our usual formula

Key Capacitor and Inductor Behavior for Quick Circuit Analysis
At DC, the capacitor behaves as open circuit, the inductor behaves as short circuit.
At very high frequencies, the capacitor behaves as short circuit, the inductor behaves as open circuit.

> The *capacitor* behaves as an *open circuit* at DC, and a *short circuit* at high frequencies.
> The *inductor* behaves as a *short circuit* at DC, and an *open circuit* at high frequencies.
> The *resistor* behaves the *same* at all (audio) frequencies.
> The *order* of a simple passive filter is the total number of *capacitor* and *inductor*.

Circuit	Order	H_{DC}	H_{HF}	Corner Freq (ω_c)	Characteristics Curve
C-R	1	0	1	1/(RC)	HPF
L-R	1	1	0	R/L	LPF

Capacitors are reactive to try to hold *voltage* constant.
Inductors are reactive to try to hold *current* constant.

Duality Summary

Current I	Voltage V
Node	Loop
parallel	series
Admittance Y	Impedance Z
Resistance R	Conductance G
Inductor L	Capacitor C
short circuit	open circuit
Norton's Theorem	Thevenin's Theorem
KCL	KVL

Take-away

One should always be in the lookout for BOGO... :) Remember, the engineer's approach to simplifying life is *not* in "forced over-simplification" (leading to incorrect answers; KISS is for the "Stupid"), but rather in *managed complexity* - learning new concepts and theorems are the best ways to simplify life!

Sometimes that means one needs to be proficient in reciprocals - that is the weakest link many incoming students need to overcome... :(

Duality was first proposed by Alexander Russell in 1904.

Supplement X - Why Different Formulas for Impedance and Reactance

If there are still any doubts left about audio electronics being an issue of language or dialect, this chapter, especially the topic of impedance, will hopefully settle it.

Mirror, Mirror on the Wall, Who is the Fairest of Them All?

I can think of at least the following different formulas given by various people for the impedance of a capacitor. What's wrong with this picture?

Abby	Brad	Charlie	Denis	Evan
$1/2\pi fC$	$1/sC$	$1/j\omega C$	$-j/\omega C$	$1/\omega C$
Different ways to compute impedance for capacitor					

Music student Francis encountered Abby, Brad, Charlie, Denis, and Evan, all audio engineers, all giving a different way to compute impedance for capacitor (in fact, some of them call it condensor, another dialect variation). What should Francis do?

Most people would think if *one* if right, then *others* with *different* answers must be wrong. (That is called linear logic.) But such is *not* the case. They are all valid under *some* circumstances or interpretations. Whether one is applicable depend on the specific dialog and circumstance where it may be used.

If that is the case, is there *one* way which is better than all others? Or, who is the fairest of them all?

One Ring to Rule Them All?

Can we just learn *one*? Let us come up with a set of criteria. That is usually how an engineer solves problems.
1) Needs the fewest brain cells to remember.
2) Can use that formula under the widest circumstances. Last the longest in a professional career. No need to keep changing and learning a new formula every year.
3) Insiders use it. Experts in audio electronics routinely use it to write papers and publish reports. That is typically used in a reference of audio filters, for example.
4) The formulas does not have imaginary units or complex numbers in it (that complicates memorization as well).

For the first criterium, Brad and Evan won. In fact, they look ridiculously the same, other than s versus ω.

For the second criterium, Brad won. If students encounter a new formula every year, Brad's formula is usually the *last* one to encounter. And then no more changes. Longevity. It last the rest of the professional career of an audio engineer. (Until someone invents something newer, of course.)

For the third criterium, Brad won hands down. Most professional publications use Brad's formula.

For the fourth criterium, Brad, Abby, and Evan met it.

So it looks like Brad's formula won! If so, why would students often learn Abby's formula one year, Evan's another year, and then Denis, Charlie, and Brad?

It is because textbooks often follow the physicist's outlook. The physicist has to explain everything with math. Where the math gets tough, that's when corners are cut and simplification happens. That way everything can be understood clearly from first principles. The math for Abby's is the easiest, using only real number, and so is Evan's, even though it is obviously designed so that students can transition easier to the next (complex) forms that Denis and Charlie use. And Brad's? It requires the most complex math of all. That obviously makes sense, as we already heard that insiders use it.

But can we use it? If we follow the engineer's outlook, then yes we can. We want shortcuts, and that is the simplest, quickest way to compute impedance for capacitance.

But not only that, if Francis wants to talk with Abby, that is no big deal. Just substitute $s = 2\pi f$.
If Francis wants to chat with Evan, it's even easier, just substitute $s = \omega$.
If Francis wants to discuss with Charlie, just remember to substitute $s = j\omega$, where j is the engineer's imaginary unit.
And if Francis wants to communicate with Denis, one just needs to know $j^2 = -1$, which is the definition of the imaginary unit. Thus by multiplying both numerator and denominator of Denis' verse, we get Charlie's. In other words, they are *the same* mathematically speaking. One emphasizes the j being at the right mathematical place, which is at the numerator (Charlie), the other emphasizes the connection j has with ω, as $s = j\omega$ and always go together.

But the inquirer wants to know, *why some of the formulas have imaginary unit, and others don't?* The answer may be surprising: *engineers use the term impedance loosely* - sometimes they reply with the *value* or *magnitude of impedance* when impedance is requested. It has to do with audio electronics as a language. Different dialects come into being as shortcuts for communication. When only the magnitude is needed, often the reply is simply the magnitude. That is Abby's and Evan's formula - only the magnitude. Charlie's and Denis' formula both have the imaginary number, so they are a complete formula, not just the magnitude. Brad's formula is also a complete formula, as it is a complex number. (For inquiring minds, s is defined as $s = \sigma + j\omega$, a true complex number, and what makes it work so well as a shortcut. And when one sets σ to be zero, we get the other representations! But *you don't need to know how it works to use the shortcut.*)

So Brad's is the fairest of them all, and we shall use its representation primarily. In fact it is not difficult to remember the formula for inductance or resistance either:

> For zero initial conditions:
> $Z_R = R$ (Impedance for Ideal Resistor)
> $Z_L = sL$ (Impedance for Ideal Inductor)
> $Z_C = 1/sC$ (Impedance for Ideal Capacitor)
> Imagine s as a complex frequency variable.
>   (Actually $s = \sigma + j\omega$, where $\omega = 2\pi f$, and $\sigma = 0$.)

However, bear in mind that audio electronics is a language. As much as it is impolite to tell a Brit that he spelled "colour" wrong, and we do not want to tell Abby, Charlie, Denis, or Evan, that they were wrong either. So **we need to know how to *read* all those dialects, but concentrate to learn to *write* using *only* Brad's dialect.**

Reactance Formulas

Using the above shortcut, and since for ideal capacitance the resistance is zero, we have:

	Abby	Brad	Charlie	Denis	Evan
Impedance	$1/2\pi fC$	$1/sC$	$1/j\omega C$	$-j/\omega C$	$1/\omega C$
Reactance	$1/2\pi fC$	*	$-1/\omega C$	$-1/\omega C$	$1/\omega C$

Different ways to compute impedance & reactance for capacitor

Engineers actually *seldom* use and reference reactance. It was only *taught* first in order to introduce impedance in a step-by-step pedagogical fashion. But once impedance is known, that is the one used most of the time. Likewise, since we use Brad's formula, we shall not dwell over reactance, other than noting that for ideal capacitors, reactance is the same for Abby and Evan (as $\omega = 2\pi f$), but has a different sign for Denis and Charlie (and especially confusing for Charlie converting from impedance). Just remember that Abby and Evan were referring to the *magnitude* of the reactance. * Brad does not bother to remember the reactance; when needed, the reactance is calculated using the $Z = R + jX$ formula, where the reactance is simply the imaginary part of the impedance.

> Resistance (R) is the *real* part of Impedance.
> Reactance (X) is the *imaginary* part of Impedance.
> That's what $Z = R + jX$ meant.

(this page deliberately left blank)

Chapter 6 AMPLIFIERS

> Active components like BJT, FET, and vacuum tubes are first covered, then configured for single stage amplification, then multiple configurations are cascaded or cascoded together, coupled and biased, leading to amplifier classes, and the final packaged amplifiers can be further paralleled or bridged for more amplification.
>
> *Expected Learning Outcome*: The student will be able to visually recognize active components, configurations, and figure out inverting or non-inverting, biasing, and voltage amplification ratios. In addition, the student will be able to understand terms commonly used in the industry, like push-pull, Darlington pairs, amplifier bridging, classes, and how/when to apply those principles.

CONTENTS

Bipolar Junction Transistors (BJT)	6-3
Common Emitter Amplifier	6-4
Other BJT Amplifier Configurations	6-6
FET and Vacuum Tubes	6-8
Common Source and Common Cathode Amplifiers	6-10
Cascade and Cascode	6-12
Amplifier Coupling	6-14
Amplifier Classes and Biasing	6-16
Amplifiers Bridged and Paralleled	6-21
Amplifier Distortions	6-22
Review Exercises	6-24

There are too many amplifier configurations that can be studied in one semester; the approach taken here is to show the key features and approaches, so that one can understand a new amplifier encountered by **analogy** with one that had previously been learned.

Here is a summary of the nine possible configurations for BJT, FET, and vacuum tubes. Even though we only give five examples, the nine possible configurations can be deduced as indicated.

(for BJT)	Common Emitter	Common Collector (Emitter Follower)	Common Base
(for FET)	Common Source	Common Drain	Common Gate
(for triode)	Common Cathode	Common Anode	Common Grid
Voltage Gain	can be high (R3/R4)	nearly (slightly less than) 1	High
Current Gain	High	High	somewhat less than 1
Power Gain	High	moderate	moderate
Phase Inversion	Inverting	Non-inverting	Non-inverting
Input Impedance	moderate	exemplary high	very low
Output Impedance	moderate	exemplary low	very high

Add the concepts of cascade and cascode, push-pull, Darlington pairs, paralleled and bridged, plus different amplifier classes and coupling methods, and hopefully one can see why we have so many variety of amplifiers today...

6.1 Bipolar Junction Transistors (BJT)

Bipolar junction transistors (BJT) are three-terminal active devices. The three terminals are base, emitter, and collector. The emitter is shown with an arrow, either going in or out. The base is shown in between the other two terminals. The transistor can be shown with or without a circle; the usual connotation is that the circle represents the can or the package, thus with the circle drawn it indicates a discrete transistor. Without the circle it is unknown; integrated circuits are usually drawn with transistors shown without a can, as all the transistors are packaged together in one unit.

NPN and PNP

The bipolar junction transistor can be NPN or PNP, as indicated below, with or without the circle. (The P and N in Q4 and Q3 are *not* part of the symbol but only there for explanation below.)

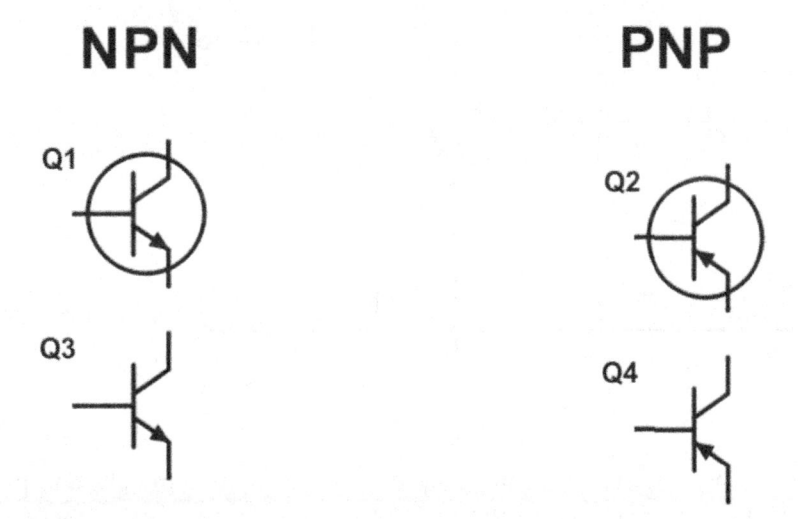

It is important to learn how to read the symbol - just look at the arrow. Imagine it flows from P to N according to the arrow. Once you write P and N on those two terminals, the third terminal is the letter which alternates. Thus Q2 below is an NPN, and Q3 is a PNP.

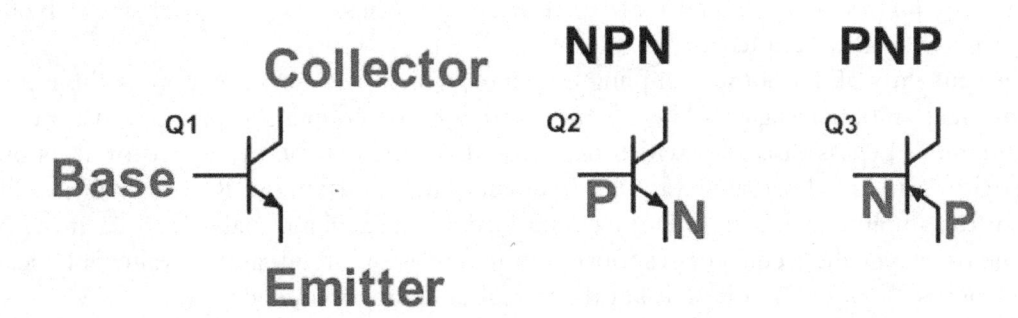

In these drawings, the **base** is the terminal that stands alone on one side, the **emitter** is where the arrow is drawn, and the **collector** is the third terminal. Conventionally circuit schematics are typically drawn with inputs on the left and outputs to the right, and the most common use of bipolar junction transistors are with inputs going to the base. Thus although the transistor can be drawn in any orientation, the base is usually shown on the left.

6.2 Common Emitter Amplifier

Here is a typical NPN common emitter amplifier, which is the most common of all bipolar junction transistor amplifiers, with full complement of coupling capacitors. "Common emitter" refers to the emitter being connected as "common" or ground to both inputs and outputs, while the other two terminals (base and collector) are used for the other terminals for input and output respectively.

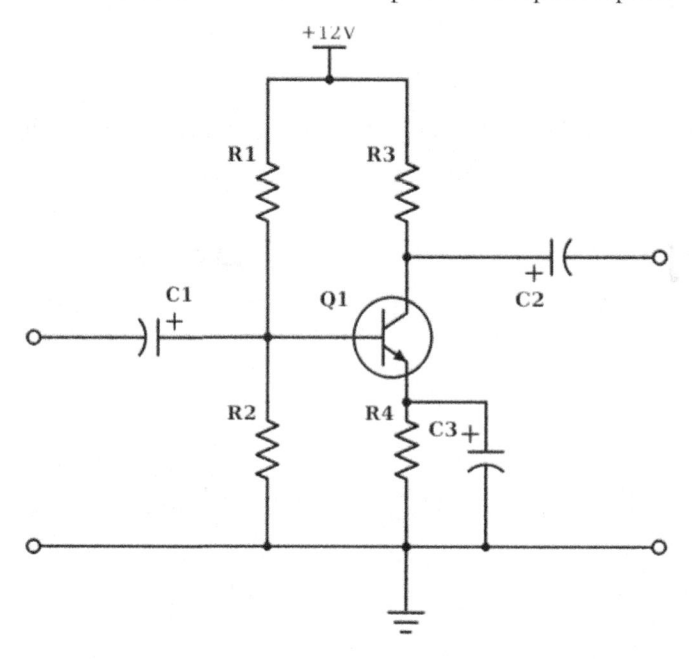

The input is applied from the left terminals; the output is at the right terminals, and power is supplied at Vaa, more commonly labeled as Vcc and a common ground at the bottom of the schematic. C1 and C2 are often polarized capacitors as they have relatively high values, and they act as blocking capacitors or coupling capacitors. Blocking refers to their action of blocking DC, while coupling refers to how they are used to couple input circuits into this circuit and out to the next circuit. Essentially they separate the AC signals from the influence of the DC biasing voltage applied through the voltage divider implemented by R1 and R2. Thus the bias condition is steady and not affected by the AC voltages, and is given simply by the voltage divider ratio, or Vb = Vaa [R2/(R1+R2)].

The output is sent through an output coupling capacitor, which *could* also function as the input coupling capacitor for the next amplifier stage. (There is no need for two coupling capacitors in series with each other.) C3 is an emitter bypass capacitor, which does not affect the bias of the transistor at its output side. This bypass capacitor acts as a short circuit at high frequency, thus shorting out R4 essentially, and letting the transistor drive at full amplification with only the circuit load and a small internal resistance at the output. In other words, it helps shapes the frequency response of the transistor circuit, and its value is typically chosen so that its impedance is about one-tenth of R4 at the lowest operating frequency.

The voltage gain of this circuit is defined as $A_v = V_{out}/V_{in}$ and is roughly

$$A_v = -R3/R4$$

Notice the gain is negative, meaning this is an inverting amplifier. Notice the gain is depending only on the load resistor R3 and the emitter resistor R4 (plus the internal emitter resistance of the transistor which is much smaller), and not on the current gain or beta of the transistor, or the input bias circuit, at least not when everything is operating properly. Not depending on the intrinsic and unpredictable behavior of the transistor improves stability and reduces distortion, at the expense of using up the current gain of the transistor itself.

The +12V shown in the schematic could be any suitable positive voltage suitable for the active device.

Here is a typical PNP common emitter amplifier. Notice that it is essentially the same as the NPN version, other than that anything associated with polarity are reversed: notice the polarized capacitors switched polarity, and so does the power supply, plus of course the transistor also switched polarity. Again the negative power supply does not have to be -12 V, but rather what is suitable for the transistor. (Note by convention the negative power supply is usually put at the bottom; here it is put at the same location as before to avoid having to mirror everything.)

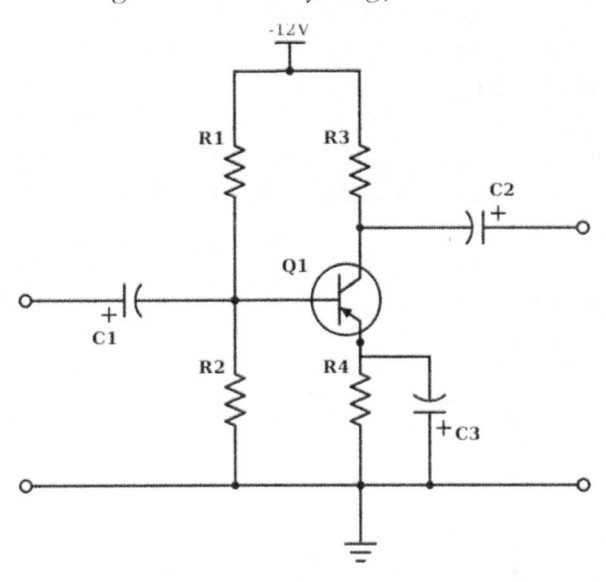

Example
Analyze the following circuit:

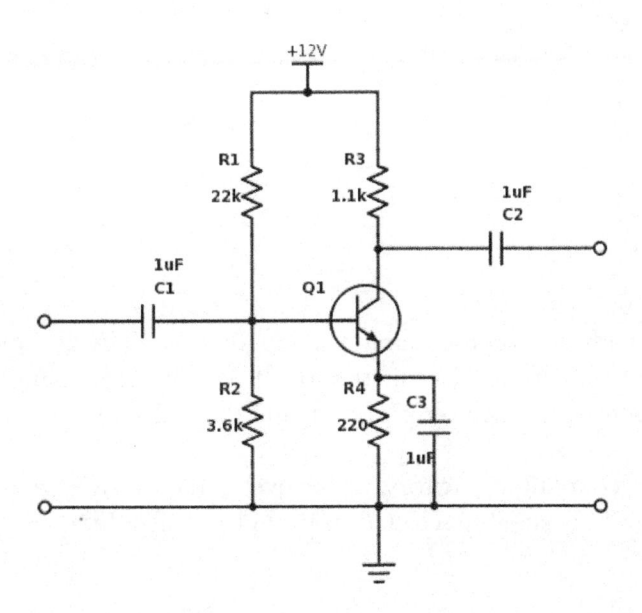

This is an NPN common emitter amplifier.
Its voltage gain is roughly - 1.1k/220 = -5, determined by R3 and R4, the negative indicates it is inverting.
C1 and C2 are coupling capacitors at input and output.
R1 and R2 is the bias network, biasing the base terminal of the transistor at 3.6k/(3.6k+22k) = 1.7 V.
C3 is the emitter bypass capacitor, increasing the gain at high frequency to compensate and obtain a flatter frequency response.
The output impedance of this circuit is roughly R3, or 1.1 kΩ.

6.3 Other BJT Amplifier Configurations

We have covered common emitter, the other choices would be common collector and common base. Again, there are three terminals, one input, one output, and the third is the "common"...

Common Collector (Emitter-Follower)

The next most common amplifier type is the emitter follower. If we use the same terminology as before, we'd call it a common collector amplifier. However, emitter follower is a more common term.

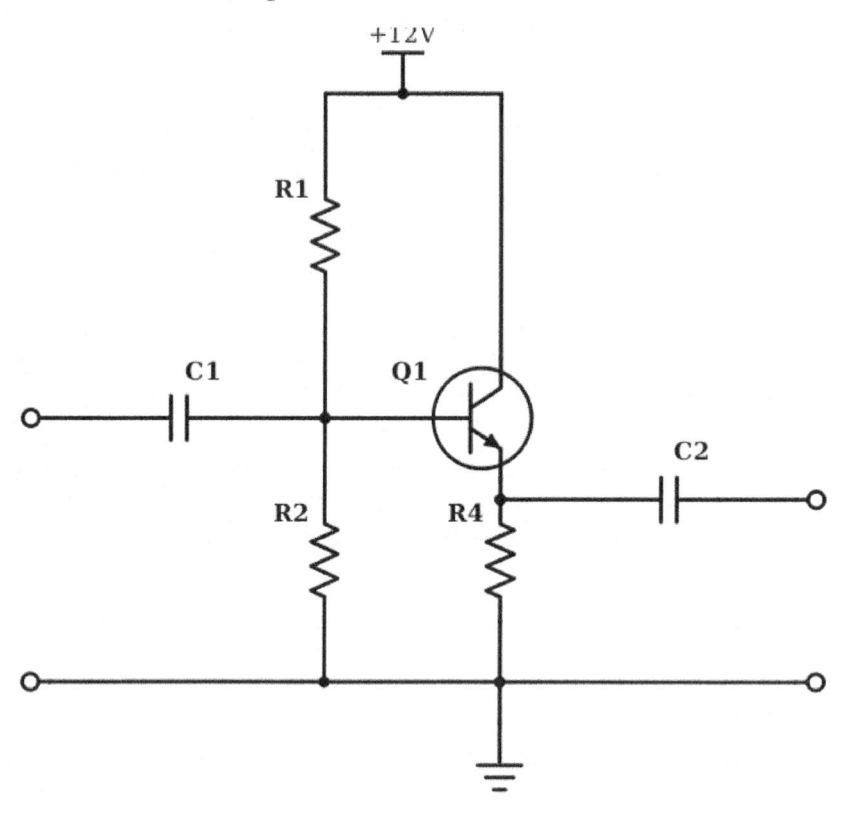

One may easily notice the similarity of the circuit with the common emitter configuration. The difference would then be easy to see: a) the output comes from the emitter; b) the previous R3 is shorted, giving the common collector designation (the base is input, as before, but emitter is output, and the collector is thus the common, common to both inputs and outputs).

The C1 and C2 likewise are the input and output coupling capacitors, respectively. R1 and R2 provides the DC bias network. (R2 is often omitted as long as R1 is suitably selected to provide suitable bias with the input resistance of the transistor.)

The **voltage gain** of the emitter follower circuit is roughly 1 (actually a minuscule amount smaller than 1). In other words, it is non-inverting, and its output *follows* the input (hence the common name of emitter follower), other than for an addition of ~0.7 V of V_{BE} diode drop.

If the voltage gain is unity, then what is significant about this circuit is that the **current gain** is large. This configuration is especially valuable as output stage of a class B or class AB amplifier, as we shall study later, when we desire large current gain.

Chapter 6 – Amplifiers

Darlington Pair

The following circuit is identical to the early common collector amplifier, other than that the active element is a Darlington pair of transistors: in this particular example the two transistors are cased in a single can or package, as indicated by the symbol of the circle enclosing both transistors. In other cases the Darlington pair may also consists of discrete transistors (or in other words, two cans or two circles in the schematic).

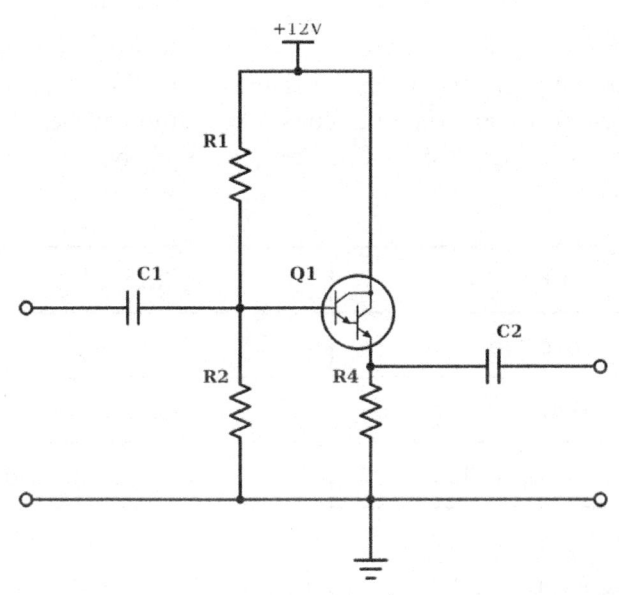

The Darlington pair is a favorite for power amplifiers when we want as much current gain as possible. The current gain is roughly the product of the individual transistor's open-collector current gains.

Common Base

Common base is least common, and most easily confused, as it can be drawn many ways. Remember the principle: identify the input and output, the third terminal is the common.

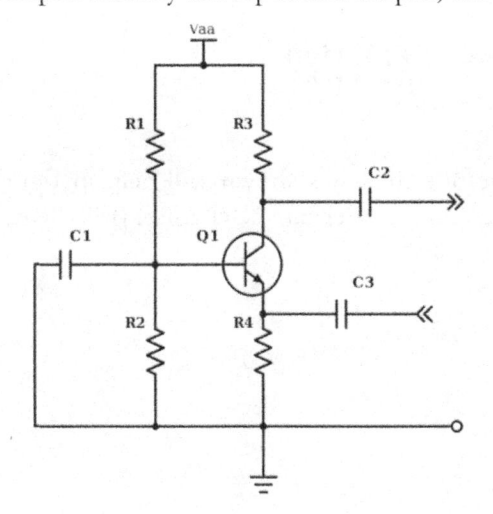

Notice that in the case the input is connected to C3, *very unusual* as usually we draw so that signals flow from left to right. In this case the signal comes from the right, and also goes to the right at output of C2.

For common base, the **current gain** is less than 1, and it is non-inverting. Instead, it has a large voltage gain.

The common base amplifier has better frequency response compared with the other two configurations.

6.4 FET AND VACUUM TUBES

Here are the other common active devices used in electronics for use in amplifiers.

FET

Field Effect Transistors (FET) can be junction type (JFET) or Insulated Gate (IGFET) which is often called Metal Oxide Silicon (MOSFET) in the world of integrated circuits. The latter can be divided further into depletion type or enhancement type, both of which can be P-channel or N-channel. The former can only be depletion type, and can be P- or N- channel.

	JFET	MOSFET
Input impedance	relatively low	very high
Device Type	Bipolar	Unipolar
Gate	Current controlled	Voltage controlled

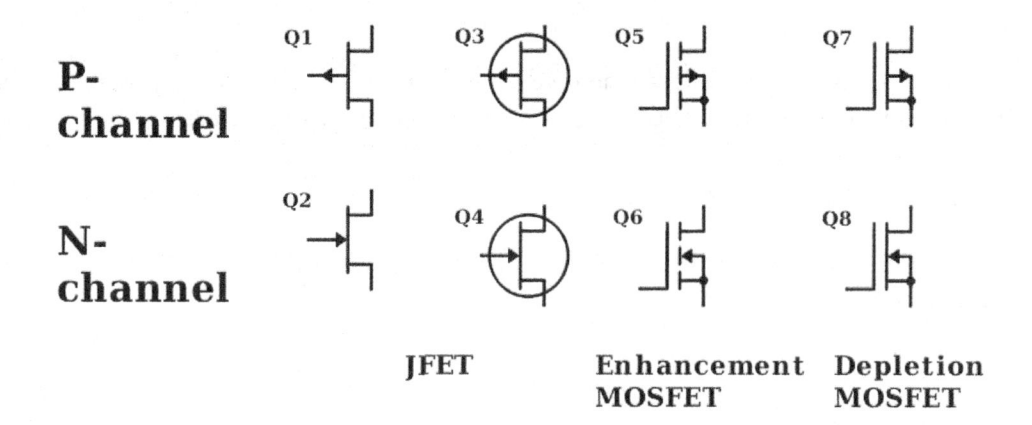

P-channel / N-channel symbols for JFET, Enhancement MOSFET, Depletion MOSFET.

There are many variations of schematics for MOSFETs; hopefully the ones shown will help in figuring out which is which when a different schematic symbol is encountered. An alternate N-channel JFET symbol is shown below.

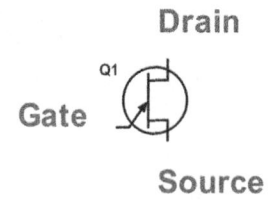

Drain
Gate
Source

The **gate** is the terminal on the left. The "**source**" is the terminal at the bottom for an N-channel FET. Often the gate is drawn nearer the source to make it clear, as shown in the example above. The "**drain**" is the terminal at the top. The substrate is the terminal that is shown connected to the source in MOSFETs in the earlier picture..

Vacuum Tubes (Valves)

Vacuum tubes are called valves in the other side of the pond. It is often shortened to just "tubes".

Triode   **Tetrode**   **Pentode**

V1   V2   V3

V4   V5   V6

Twin Triode   **Triode Pentode**   **Twin Pentode**

The number of active terminals (i.e., not including filaments, which is usually not shown connected anyway) determine how they are named: triodes have three active terminals, tetrode four, pentodes five. Twin triodes contains two triodes with a shared filament, twin pentodes have two pentodes, and triode pentode has a triode and a pentode in one tube. Other combinations and arrangements are also possible. Essentially just study the diagrams to deduce the exact configurations. (Diodes obviously contains two active terminals.)

We'll only study the triode: the control **grid** is the terminal connected to the dotted line (one can visualize the mesh of the grid as controlling the flow of charge carriers), the **cathode** is at the bottom, and the **plate** or **anode** is at the top (visualize it as catching the charge carriers emitted below), plus the optional (and if present, is usually shown unconnected) filament at the very bottom.

There are many variations of schematics for vacuum tubes; hopefully the ones shown will help in figuring out which is which when a different schematic symbol is encountered.

6.5 Common Source and Common Cathode Amplifiers

Common Source JFET Amplifier

The most common JFET amplifier looks very similar to the common emitter amplifier we studied earlier. In a sense we just substitute the N-channel JFET in place of the NPN transistor...

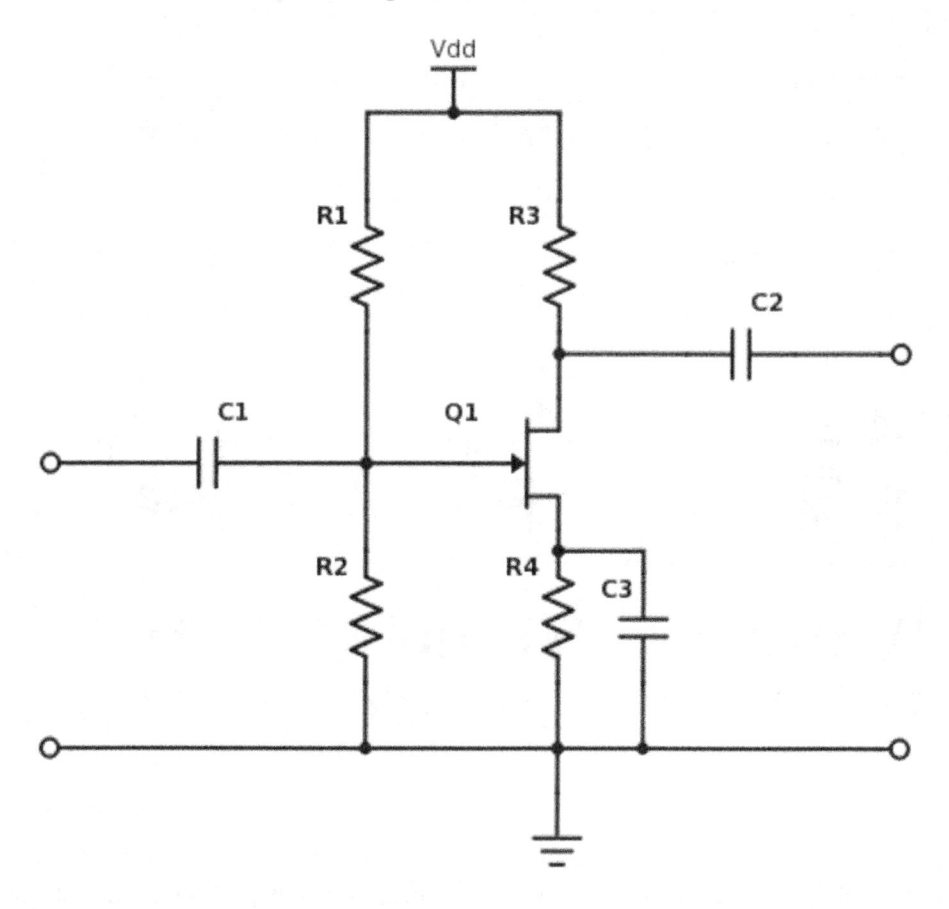

At the macroscopic level, the operation of the surrounding circuitry are the same: R1 and R2 set the bias, C1 and C2 are the coupling capacitors, and R3 is the load resistor.

The values are usually larger: R1 and R2 typically above megohms, bypass capacitor around 22 to 100 μF, and coupling capacitors around 10 μF. R4 could be quite small, and may even be a preset potentiometer to make it more convenient to adjust.

At the microscopic level, of course the N-channel JFET transistor works differently from the NPN transistor. The junction FET takes very little input gate current, behaving with a very high input impedance. Biasing is required, primarily to put the JFET at the ohmic region. In contrast, R4 is used to set the quiescent operating point of the JFET amplifier, putting the gate-source junction reverse-biased.

R1 can often be removed (making it open circuit).

As you may expect, the common drain and common gate configurations are similar to the corresponding ones for BJT and not further elaborated here.

Common Cathode Amplifier

The most common vacuum tube amplifier looks very similar to the common emitter amplifier we studied earlier. In a sense we just substitute the triode in place of the NPN transistor...

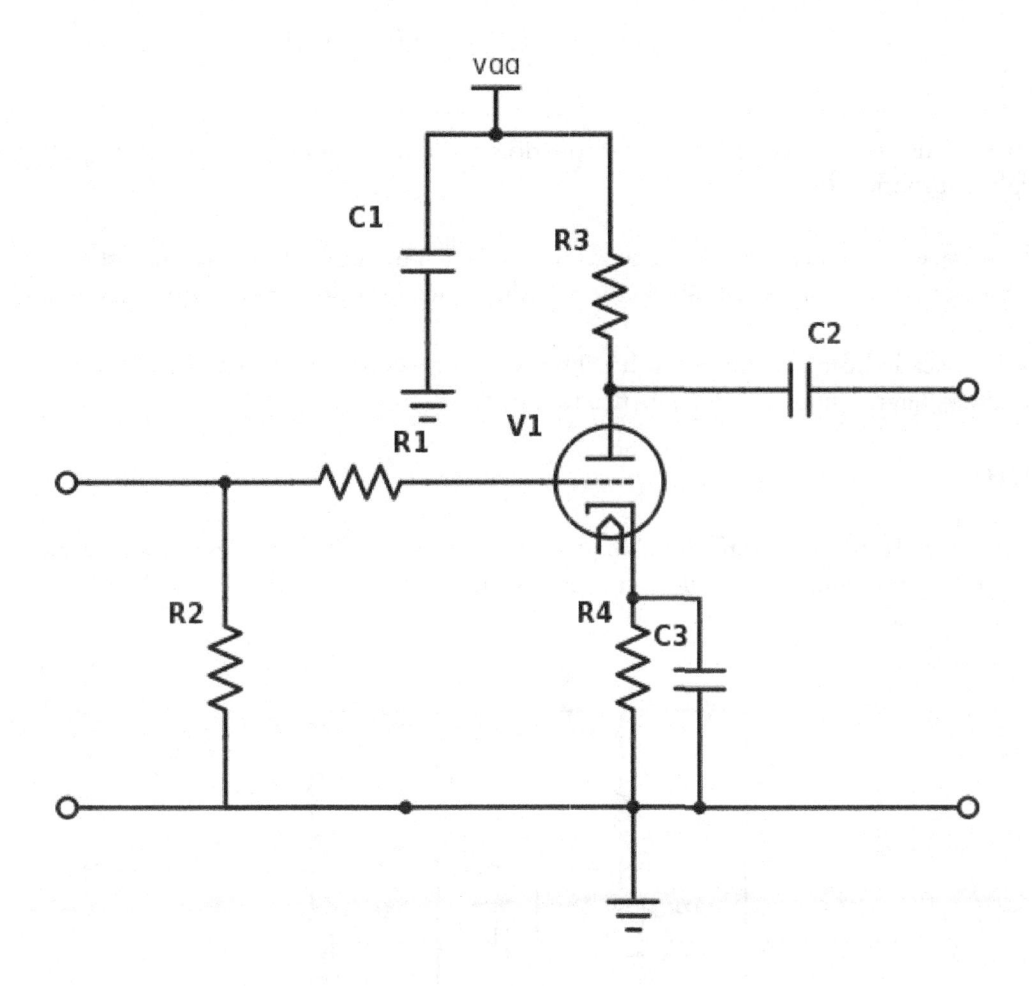

The output side is exactly the same, with the coupling capacitor C2, the load resistor R3, and the cathode resistor R4 and C3 bypass capacitor. But there are specific differences that becomes obvious. First, the input coupling capacitor is completely unnecessary because the vacuum tube grid terminal is DC-isolated. Second, the power supply capacitor C1 isolates this gain stage from other gain stages. If they are not isolated properly, we may hear "motorboating".

Just like the JFET, the cathode and plate resistors are what sets the output bias. The grid resistors sets the bias of the input and the control grid, plus the bandwidth of the amplifier. Just like the JFET, the full input voltage divider biasing network is not totally necessary.

The common source JFET and common cathode amplifiers have very similar characteristics (other than of course the issue of noise in vacuum tubes). They both have similar high input impedance, and operates similarly, plus having similar distortion characteristics. In fact, it is their distortion characteristics that gave the vacuum tube their specific sound that some audiophiles desire. Thus the solid state JFET amplifier is often a good substitute for those desiring the same vacuum tube sound... (Remember since vacuum tubes are not solid-state, it is sort of like car tires - has to be replaced every few years.)

Again, the common anode and common grid configurations are similar to the corresponding ones of the JFET and not further elaborated. Cascode tube mic preamps are very popular, for example. See next section for cascoding.

6.6 Cascade and Cascode

Here comes terms commonly confused.... How do we connect up more than one amplifier configuration to obtain higher amplification?

(A public service announcement from grammar police: be careful how you use them to qualify a noun; typically we say *cascaded* amplifier when they are in cascade, but *cascoding* amplifier when it is in cascode configuration.)
(The word "cascade" came from English. The word "cascode" was coined in 1939 as a portmanteau of "*cas*caded tri*ode*s having similar characteristics to a pent*ode*".)

Cascade

When we require more amplification than one common emitter configuration, we simply cascade more common emitter configurations called **stages** together. As an example, we have two common emitter amplifier stages below:

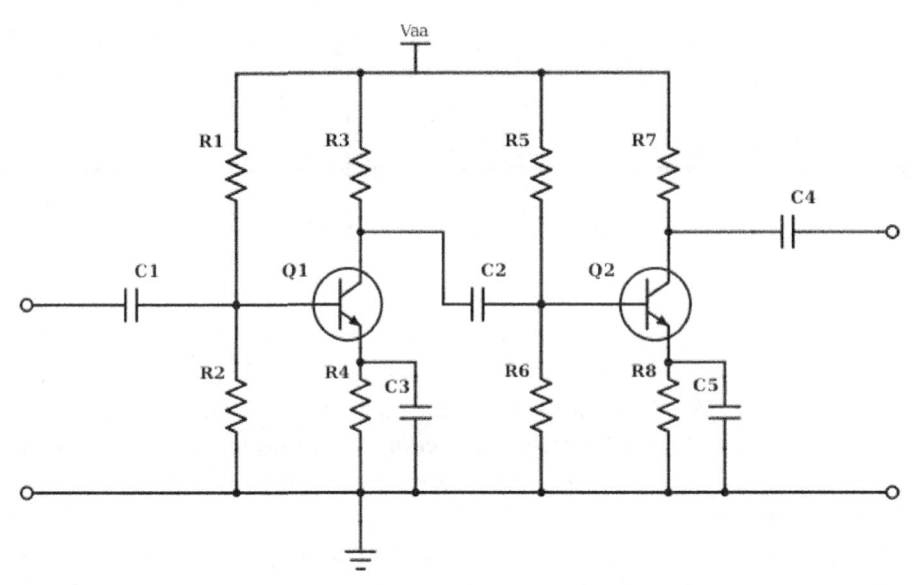

If the voltage gain of the first stage is -5, and that of the second stage is also -5, the resulting voltage gain is -5 x -5 = +25. Notice that having even numbers of common emitter stages results in a non-inverting output, and the amplification can go up fairly quickly, for example, four stages of -5 amplification will get an amplification of -5 x -5 x -5 x -5 = 625 !!! Notice also that the output coupling capacitor of the first stage is combined with the input coupling capacitor of the second stage.

Cascode

The common emitter configuration is very useful; one major issue is that it does not have the best frequency response, like the common base configuration. The cascode combines one common emitter configuration with a common base configuration in series and not in cascade. The result is a very good frequency response coupled with good amplification. This is often used in audio applications, especially for microphone pre-amplifiers.

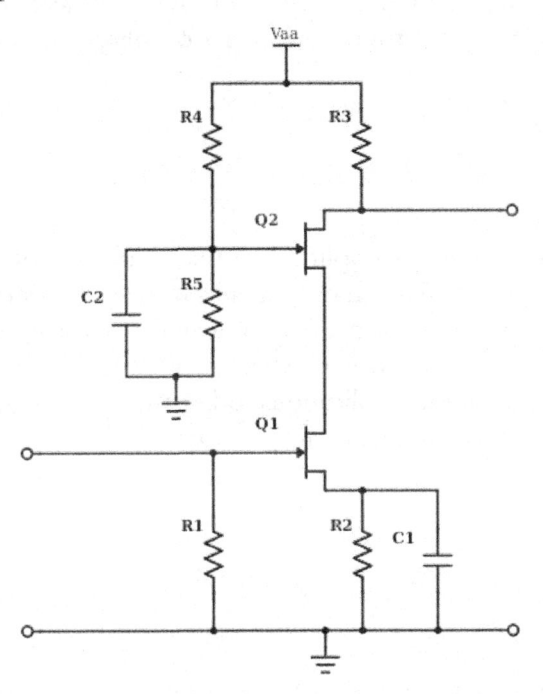

Note that in cascoding, the active device is typically drawn stacked one on top of one another, while in cascades, the active devices are drawn one on the side of another.

In this case, the input goes to the bottom active device Q1, which is configured as a common source amplifier, the most common configuration, with the load being the second active device Q2, and it is configured as a common gate amplifier, the least common configuration. It is common gate because R4 and R5 is a DC bias voltage divider, and it is essentially grounded through C2, and the input is the source which is connected to the drain of Q1, with the output being the actual output of the overall circuit. R3 is the drain resistor which limits current flowing. R2 and C1 are the standard source resistor and capacitor. R1 is a simplified bias resistor to ensure zero voltage output for zero voltage input.

Cascoding is very commonly used in pre-amplifiers, both in the form of integrated circuits, and in discrete tube pre-amplifiers, including inside microphones.

This configuration is useful because the output is well isolated from the input, both physically and electrically. In practice there is little feedback from output to input. It is also very stable. It retains the benefit of wide bandwidth and high voltage gain of the common base/gate configuration and high current gain and moderately high input impedance of the common emitter/source configuration. As for power gain, it is higher than all 3 configurations. Notice the part count is very low, because when the two configurations are combined many components are saved (while in the case of cascading only one coupling capacitor is saved). The output is inverted just as in common emitter/source because common base/gate is non-inverting. It has high slew rate.

The major disadvantage is that it requires a higher supply voltage than normal configurations. It also retains the high output impedance of the common base/gate configuration.

6.7 Amplifier Coupling

The most common coupling mechanism between amplifier stages or to loudspeakers is capacitive coupling, which was shown so far. Other alternatives include output transformer coupling, optoisolator coupling, and direct coupling.

Transformer Coupling

Historically this was the first coupling used for audio amplifiers; at that time transformers were expensive, but amplifier components were expensive also. Now transformers are just as expensive, but amplifier components have dropped tremendously, making transformers prohibitively expensive, relatively speaking.

In addition, transformers are not very linear, can be magnetized, and have many limitations reducing fidelity. Thus it is no longer commonly used in audio applications, other than for some classic microphone input coupling transformers. It is still used in radio frequency amplifier coupling.

Direct Coupling

It should be noted that **direct coupling** (or **DC- coupling**) was not needed for signal input in the days of vacuum tubes, because the grid of a tube essentially isolates and blocks DC. As mentioned earlier, in those days, the output was typically transformer-coupled. If it was not transformer-coupled, it was usually given the name of **OTL** (Output Transformer-Less). It could be implemented with either capacitive coupling (discussed in next section), or that it is also **OCL** (Output Capacitor-Less) and thus direct coupled.

Direct coupling has probably the best performance because there are no transformers or capacitors interfering in the way. However, the amplifier is difficult to design and manufacture to be stable, and thus seldom done that way in discrete amplifiers. In contrast, direct coupling is common in integrated circuit operational amplifiers (op-amps) that we'll discuss in the next chapter. In general, op-amps need to support DC and so must have direct coupling.

The opposite of DC-coupling is **AC-coupling**. Transformer and capacitive coupling are AC-coupling.

Capacitive Coupling

Thus this third option is used most often today, and also given in the earlier part of this chapter. The consequence is of course there would be no DC response. But precisely because of that it makes it much easier to set DC bias for individual active devices within the amplifier.

Optoisolator Coupling

This option is used when there is absolutely *no* coupling or connection allowable between the two stages. Transformer coupling has no coupling of DC bias, but the two systems are connected, and the impedance of the load is reflected on the amplifier. Likewise the impedance of the load is seen by the amplifier for capacitive coupling. Direct coupling often uses a servo-controlled circuit, which senses the DC of the output to change the DC bias on the amplifier. But for optoisolator coupling, the two sides are coupled through light, and so the impedances and other aspects of the stages do *not* matter. In fact, the two can be a long distance away, as long as optical connection can be maintained. It is thought that this may be the most secured means of communication between satellites and ground station. In contrast, a transformer-coupled circuit can be easily picked up by another coil coupled to an amplifier. Because of the lack of connection between the two stages, there is also not a possibility of electrical shock coming from one to the other.

All these benefits are negated by the lack of sufficient audio fidelity of the optoisolator, as it typically consists of a light-emitting diode (LED) and an opto sensor - neither perform with fidelity as well as a direct connection, or even capacitive coupling. Plus the optoisolator is relatively expensive, so it is seldom used in audio electronics. It is more often used in other scientific research and where complete isolation is necessary.

6.8 Amplifier Classes and Biasing

Amplifiers don't go to college. That is not the class we're talking about… (Nor is it about gender or racial bias…)

Are you telling me that there are 'class struggles' in the society of amplifiers? Could the cause be 'bias'? And with the class struggles, would a new class emerge, and which class becomes the winner?.

Bias

Remember the common emitter amplifier that we used as example? We talked about R1 and R2 as setting up the DC bias. But we had not yet explain how it may affect the amplifier. So let us look at the circuit again…

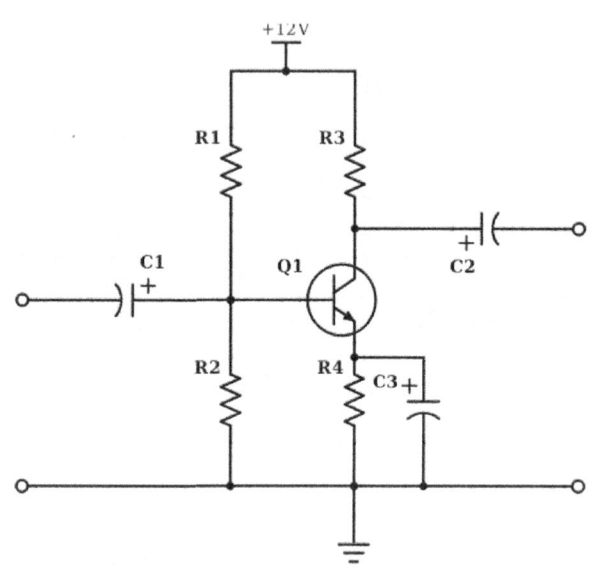

By changing R1 and R2, we can change the bias and thus establish new classes of amplifier. Let us illustrate when the amplifier is set to voltage gain of -2. They can be seen to have different input and output waveforms, as follows:

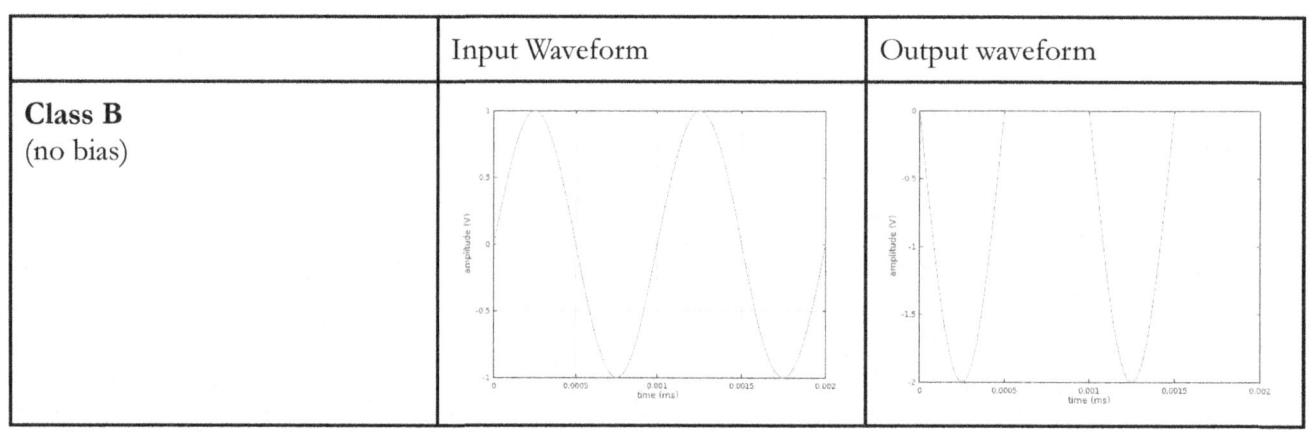

	Input Waveform	Output waveform
Class B (no bias)		

The input has no bias; it may look most natural, but look at the output. It is amplified by a factor of -2, except that the amplifier is conducting only when the output is negative, essentially clipping the output waveform and producing huge amounts of distortion. No one in a right mind will want this basic class B amplifier (but see push-pull later - often when people say class B they mean class B push-pull).

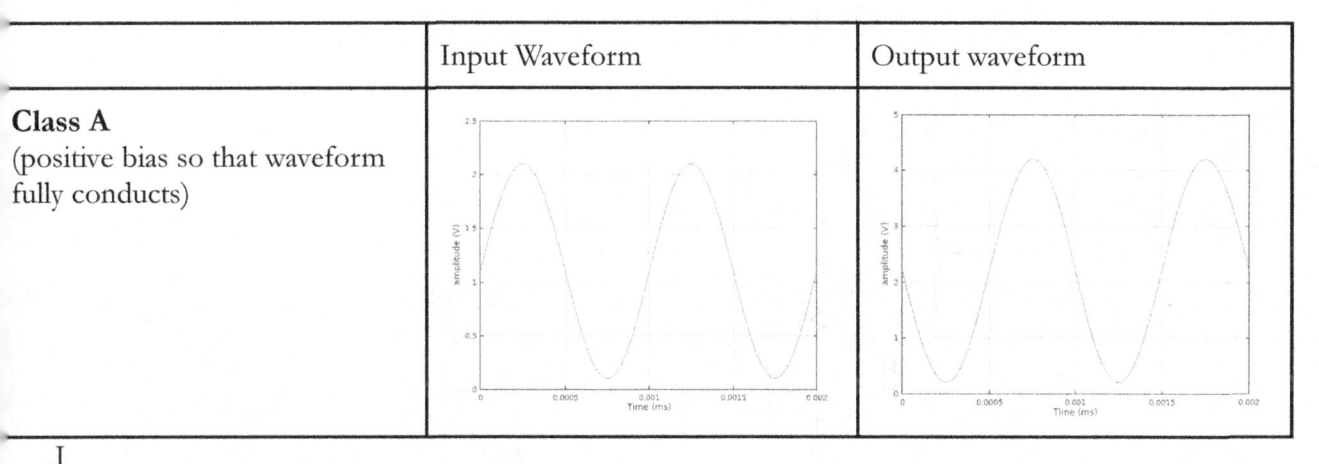

	Input Waveform	Output waveform
Class A (positive bias so that waveform fully conducts)		

In comparison, for class A amplifier, the input waveform is biased so that the largest waveform is still above zero. The output is likewise biased so that the transistor is always within its most linear conducting region, with the output being amplified by -2.

	Input Waveform	Output waveform
Class C (negative bias so that waveform is "barely above the water")		

For class C, the input is biased the "wrong way" deliberately, so that only peaks of the waveform are "above the water" or above zero. That means the amplifier is conducting only a fraction of the time, thus reducing the power dissipation in the amplifier. This is of no use in audio amplifiers, but in radio frequency amplifiers, we can add resonant circuit at the output which will regenerate the missing waveform at a known carrier frequency. This enables a more power-efficient RF amplifier.

Push-Pull

Recall the class B amplifier which can amplify on half of the cycle. What if we combine two such amplifiers, with one amplifying on the positive half, and another amplifying on the negative half? We can say one pushes, another pulls, and together they amplify the full waveform, whether positive or negative. The resulting amplifier is known as a **push-pull** amplifier. A symmetrical push-pull amplifier is when we pair a PNP with an NPN transistor (or P-channel with an N-channel FET) with both transistors operating in the same amplifier configuration. If we use the same type of transistors, or use different types of amplifier configurations, then it is asymmetrical push-pull amplification.

The previous configurations that are *not* push-pull are known as **single-ended**.

Here is an example circuit for push-pull:

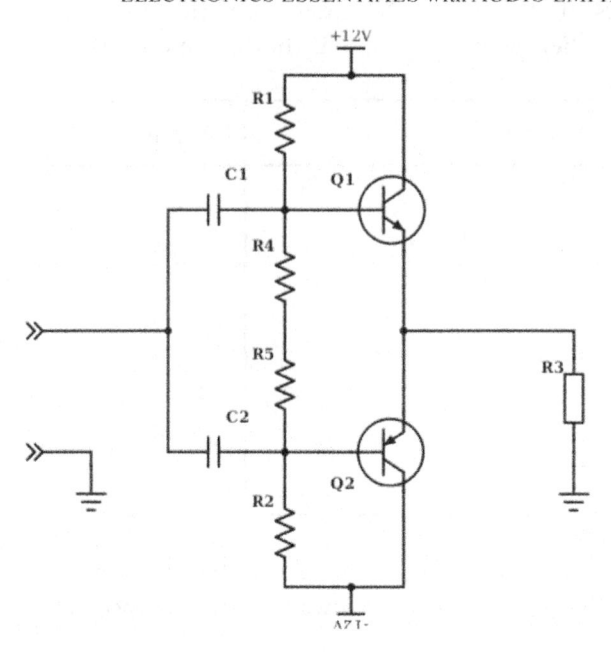

Notice that there are two rails of power supply, one labeled +12 V, the other -12 V, in this example. Actual voltages used may vary, but they would be providing symmetrical swing on the two transistors, and so expected to be the same in magnitude.

The upper transistor is NPN, while the lower transistor is PNP. Input signal is fed to the base of both. Output signal is derived from emitters for both, going to the load R3, making them operating in common collector mode, or emitter follower configuration.

C1 and C2 are the input coupling capacitors for each transistor. The biasing network is set up so that R1 = R2 and R4 = R5, thus ensuring that both transistors operate in the same manner and complementary to one another, consistent with a push-pull configuration. When R4 and R5 are zero, both transistors will have zero bias suitable for class B operation, with each transistor turning on during positive and negative cycles respectively, and when added together at the load they amplify both positive and negative cycles.

R4 and R5 are there to allow minor adjustments to the bias as needed.

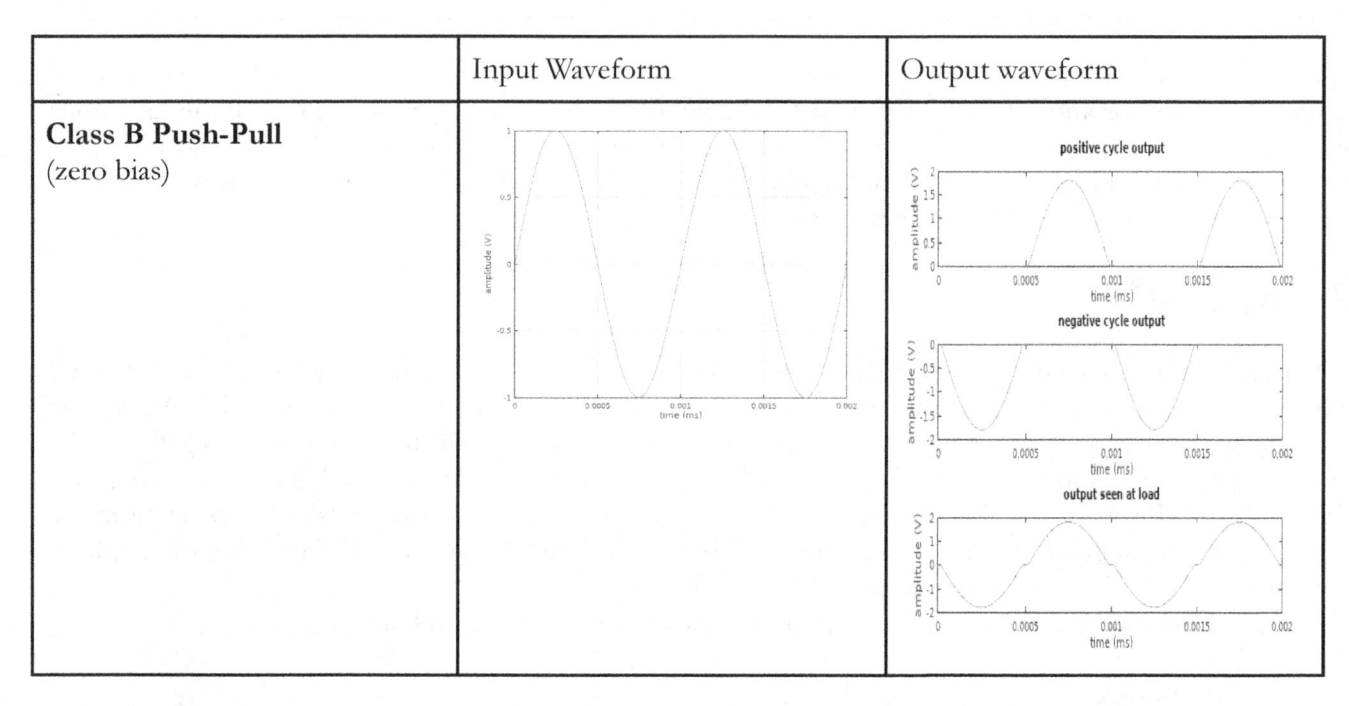

	Input Waveform	Output waveform
Class B Push-Pull (zero bias)		

Chapter 6 – Amplifiers

There is an obvious problem with this circuit, and that is the transistors do not turn on right at zero volts, but rather at about 0.7 V, introducing what is known as a **crossover distortion**, as we can see in the output waveform when the wave crosses zero.

R4 and R5 are there to adjust the bias to eliminate crossover distortion, as we move the bias slightly above the threshold of about 0.7 V, we can mitigate against crossover distortion significantly. Alternately, we can replace the two resistors with diodes, as shown below, and with the diodes chosen to match the respective transistors, will lead to an even better match to eliminate crossover distortion. When operating in this manner, it is usually called class AB push-pull. In many implementations, the designers play it safe and put in double the amounts of diodes in series, thus offering a definite but small bias on both transistors.

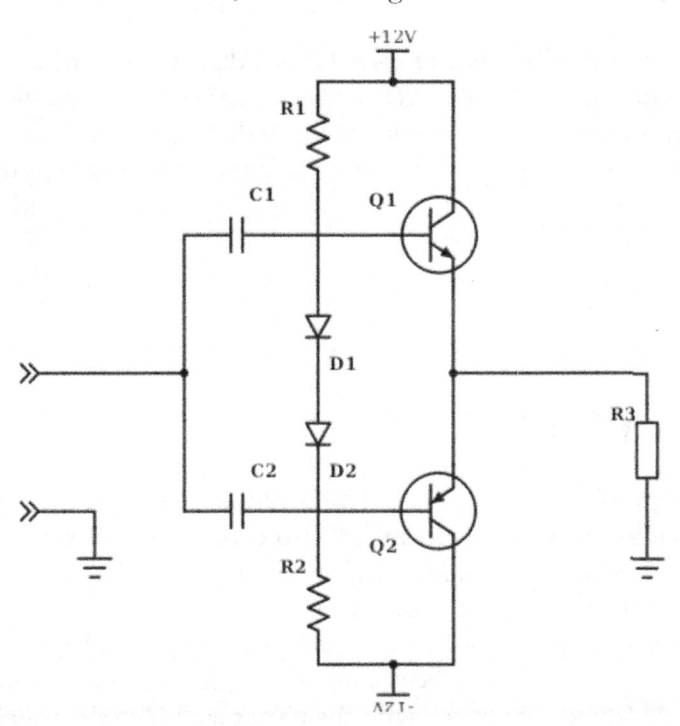

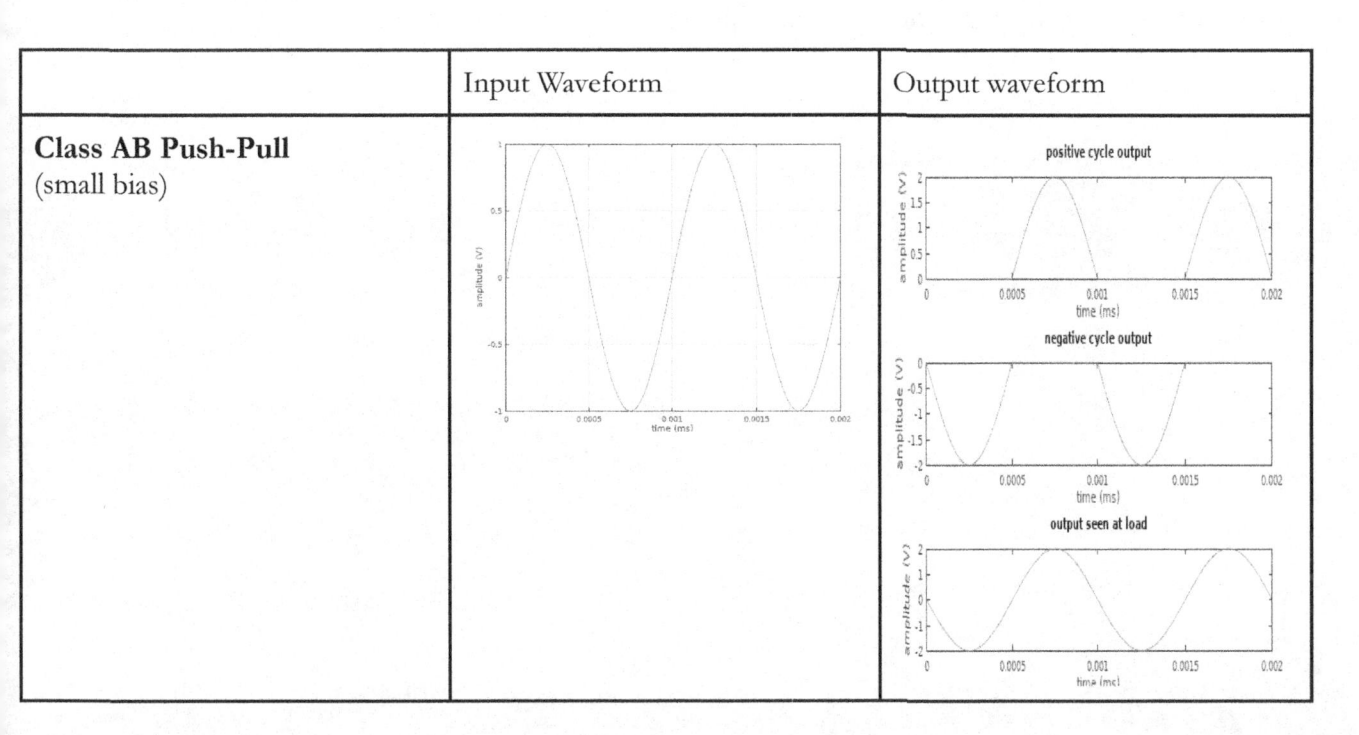

	Input Waveform	Output waveform
Class AB Push-Pull (small bias)		

6-19

Non-linear Switching Amplifier

This is also called PWM amplifier, where Pulse Width Modulation digital switching is used to achieve maximum power efficiency. Colloquially it is also known as digital amplifiers. The classical version is known as **Class D**. Class S and T as well as other so-called classes are variations on the same theme, although there are questions whether they should be considered distinct classes as they were simply marketing ploys to show that they were different.

The input signal is compared (in analog manner in class D, and digital manner in class S and T) with a ramp waveform and the result is fed to a switching power output of either plus or minus power supply. The result is a PWM waveform, which is then filtered to produce the audio signal. As long as the sawtooth and PWM waveforms have high enough frequency, say 10 MHz, the filtering is very effective to recover audio frequencies. Since the amplifier is either on or off, the power wasted is near zero. In other words, the power efficiency of class D reaches near 100%.

Supply Voltage Control

The bias is only *one* aspect of control of the amplifier. Eventually, even in class A, the output voltage will rise to the supply voltage, and then the output waveform will also clip. In contrast, if we use higher supply voltage all the time, then a lot of power is wasted when the input signal is low. To maximize power efficiency, we typically use class G or H coupled with class AB biasing.

Class G switches the supply voltage between two different supply voltages, depending on the instantaneous signal levels, so that larger signals will receive higher supply voltage to avoid clipping.

Class H is similar; instead of switching supply voltages, the power supply itself can change voltage levels instantaneously based on demand, thus avoiding clipping.

6.9 Amplifiers Bridged and Paralleled

The amplifier classes allow us trade-off with power consumption. What if we are already given the power amplifiers, or that our voltage supply is limited (for example in a car limited to a supply of 13.8 V DC)? How do we further increase the power to the loudspeakers?

Bridged Amplifiers

If we have two identical amplifiers with differential inputs, we can arrange them in this configuration:

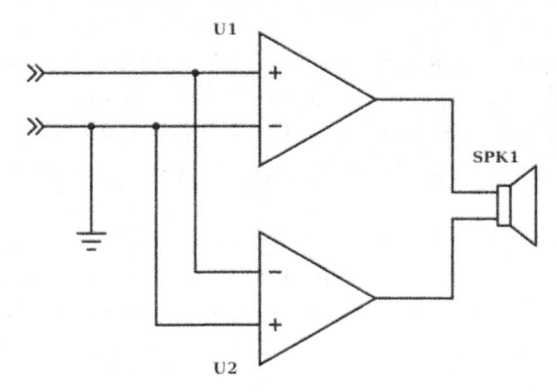

When one amplifier is driving positive, the other amplifier is driving negative, up to the maximum allowed by the power supply. The result is that the loudspeaker can produce **twice the voltage** possible with a single amplifier, and thus **twice the power**. Each amplifier would see only **half of the impedance** of the loudspeaker.

However, notice that the amplifiers must have the networks available to receive inverting or non-inverting inputs, and that the loudspeakers must be floating and not grounded at all.

Paralleled Amplifiers

If we have two regular identical amplifiers, we can arrange them in this configuration:

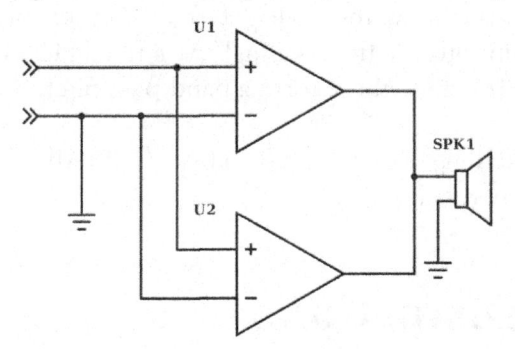

Assuming that they are perfectly matched, then the two amplifiers together can produce **twice the current** possible with a single amplifier with the same voltage, and thus **twice the power**. The output impedance of the system would be only **half of the impedance** of the regular amplifier.

However, the two amplifiers need to be perfectly matched, specifically with matched gains, or else one amplifier would end up driving the other. Also, both needs to have as little DC offset as possible, or else one would drive another with DC current which is not good. Since these are difficult to achieve, paralleled amplifiers are not commonly possible with commercially bought amplifiers, unless they were designed to serve that way.

6.10 Amplifier Distortions

Amplifiers are not perfect. Here are the common distortions that we need to control…

Amplitude Distortions

1) The amplifier may be nonlinear because it is operating in a non-linear region, meaning that a certain input may not produce a corresponding output.
2) The amplifier may clip when the output reaches the level of the supply voltage and cannot go any higher. Alternatively, improper bias also means that the active device may clip.
3) The active device may not conduct when it is reverse-biased and thus cannot go any lower.
4) The amplifier may also clip when it cannot supply the current necessary to drive the load, typically when the load impedance is too low.

Amplitude distortions result in measurable **harmonic distortion** and **intermodulation distortion (IMD)**. **Total Harmonic Distortion (THD)** is measured as a percentage of the RMS value of the generated harmonics to the RMS of the input signal. Likewise IMD is measured, both usually through international standards.

The crossover distortion discussed earlier was caused by 3) from both push-pull active devices. It is more audible than other amplitude distortions, say from soft clipping, for a comparable THD.

THD+N measurements are similar to THD but with Noise (N) included in the measurement.

Frequency Response Distortions

When we decompose the input signals into fundamentals and harmonics, and if the amplifier does not amplify the fundamentals and harmonics the same way, then we get frequency distortions. Capacitive-coupled amplifiers cascaded together have the capacitors acting as a high-pass filter. The intermediate stages act as integrators, which yields a low-pass filter. Together, these form a band-pass filter.

Frequency response is usually given as something like 5 Hz-90 kHz +/- 0.1 dB. Do not accept an "advertising" frequency range where the decibel part is not given.

Phase and Group Delay Distortions

When we decompose the input signals into fundamentals and harmonics, and if the amplifier delays the fundamentals and harmonics differently, then we get phase distortions. Reactances present in the circuit means that the components of the input signal are not amplified with the same phase shift. In "dispersive media," the phase velocity may vary with frequency. In a filter, the group delay tends to peak at the cut-off frequency (although different for different types of filter implementations).

Slew-Induced Distortions

This is different from the traditional amplitude and phase distortions. It is amplitude-like distortion, but it has to do with *not* the amplitude directly, but rather the rate of change of the amplitude, which is known as slew rate, or how fast the amplitude can change. Visualize it as the difference between distance and speed. Then visualize it as a sluggish train not able to change speed quickly due to its momentum. It occurs in the time domain, just like phase distortions, but with somewhat different results.

Slew rate is usually measured in V / μs or V / ms. It is is the maximum supportable rate of change in the output voltage caused by a step change on the input.

Slew-Induced Distortion (SID) can cause IMD when slewing or "clipping" at the amplifier's power-bandwidth product limit, effectively lowering the gain and thus amplitude modulating the other signal. When SID occurs for only mainly a portion of the signal, the result is called **Transient Intermodulation Distortion (TID)**.

Most high-end amplifiers are good enough these days to have minimum amplitude, frequency, and phase distortions. In a way, modern high-end amplifiers often distinguish themselves based on their slew-rate performance, plus their current drive capabilities.

Distortion Used as Effect

Rock music, heavy metal, and punk are among one of the first to take advantage of distortion in the electric guitar and used as an effect. Guitar pedals and other sound processors are now available to add distortions as needed to obtain such effects.

The vacuum tube sound is often subjectively described as "warm" and "rich." A famous (or infamous) research attributed that to the higher levels of second-order harmonic distortion common in single-ended tube amps, when compared with solid-state push-pull circuits where even-order harmonic products would cancel out, leaving only odd-order harmonics through. A single-ended amplifier, in contrast, produces both even and odd harmonic distortions. Regardless, with modern Digital Signal Processing (DSP), the distortions unique to various tube amps had been modeled and are available in some guitar pedals and processors, for those who wants the particular tube sound and yet desires the convenience of modern solid-state electronics.

6.11 Review Exercises

Chapter 6 Review Guide
Theory
- **concepts** of BJT, FET, vacuum tubes
- **concepts** of common emitter, collector, and base as well as their FET and tube counterparts
- **voltage gain** calculations for common emitter circuits
- **concepts** of bias and amplifier classes
- **concepts** of coupling
- **concepts** of bridged and paralleled amplifiers
- **concepts** of cascade and cascode
- **concepts** of Darlington pair and push-pull
- **concepts** of distortions and measurement
- **avoidance of wrong concepts**

Practice
- ability to **read** schematics of amplifiers and understand what each component does
- ability to **analyze** the pros and cons of each amplifier configuration

Exercise
For the circuit below, explain by filling in the blanks in the tables:

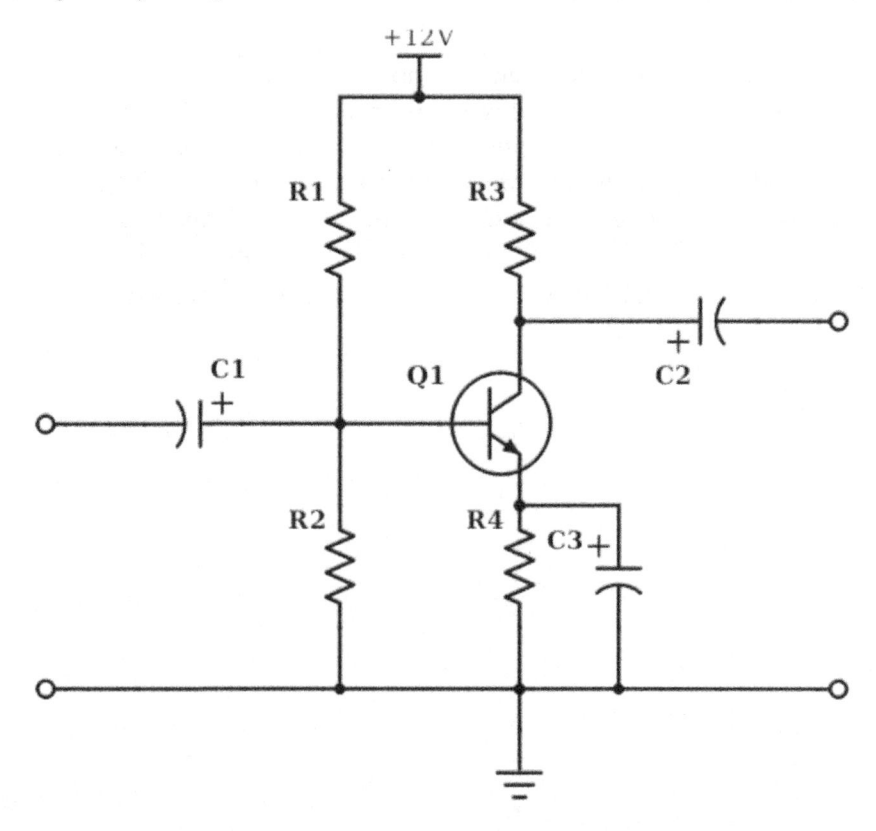

Chapter 6 – Amplifiers

1. Fill in the blanks:

Active device component type and configuration (e.g., PNP, common base)	
Inverting or non-inverting?	
DC or AC amplifier or neither?	

2. Identify and explain the functions of each component:

	Identify the Component Type (e.g., capacitor, resistor, PNP)	Function of the Component(s) (e.g., bias, output coupling, negative feedback)
C1		
C2		
Q1		
R1, R2		
R3		
R4		
C3		

3. Calculate the voltage gain of the circuit given values of the resistors, showing your steps:

R3	R4	Calculation of Voltage Gain
1k	100	
1.1k	220	
1k	500	

4. Calculate the DC bias at the base of the transistor given values of the resistors below, showing your steps, and identify the respective *probable* amplifier class:

R1	R2	Calculation of DC Bias Voltage	Amplifier Class
1k	1k		
9k	1k		
open	open		

For the circuit below, explain by filling in the blanks in the tables (note the power supply rail at the bottom is -12 V):

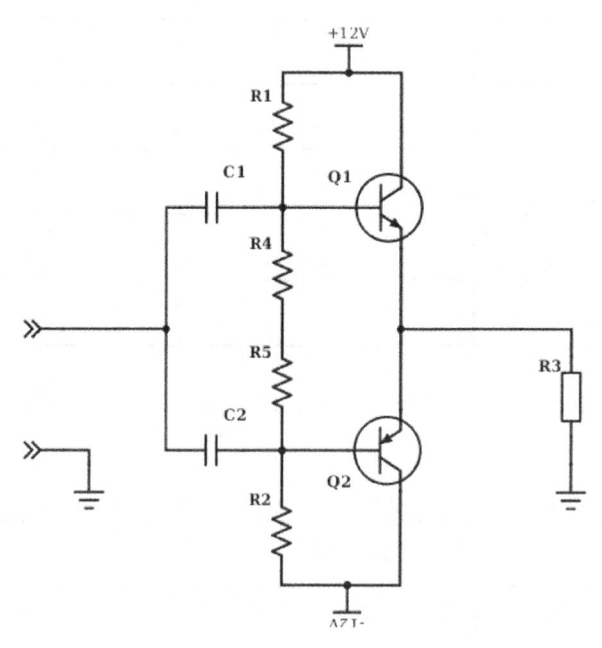

5. Fill in the blanks:

Q1 component type and configuration (e.g. PNP, common base)	
Q2 component type and configuration	
Analysis of Circuit type (*Explain its usage and configuration type.*)	

6-26

Chapter 6 – Amplifiers

6. Identify and explain the functions of each component:

	Identify the Component Type (e.g., capacitor, resistor, PNP)	**Function of the Component(s)** (e.g., bias, output coupling, negative feedback)
C1, C2		
Q1		
Q2		
R1,2,4,5		

7. Calculate the DC bias at the base of the transistor given values of the resistors, showing your steps, and identify the respective *probable* amplifier class:

R1	R4	R5	R2	**Calculation of DC Bias Voltage**	**Amplifier Class**
9k	1k	1k	9k	at base of Q1: at base of Q2:	
9k	short	short	9k	at base of Q1: at base of Q2:	

For the circuit below, explain by filling in the blanks in the tables:

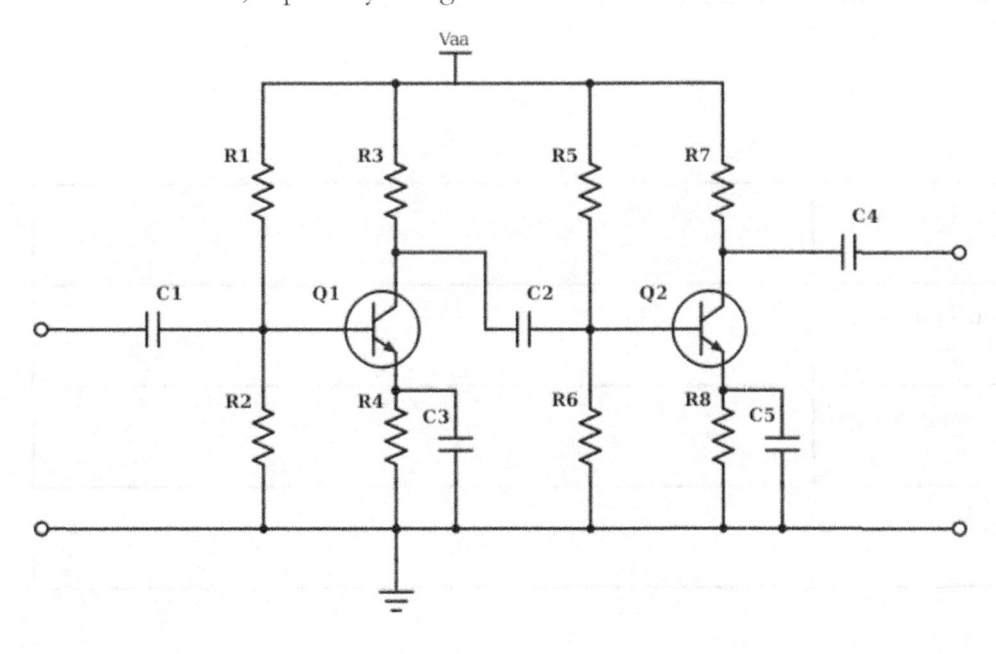

8. Identify the individual amplifier stages. (*draw circles or rectangles to carefully cover each stage*)

9. Calculate the voltage gain given values of the resistors, showing your steps:

R3	R4	R7	R8	**Calculation of Voltage Gain**
1k	100	1k	500	
1.1k	220	1.1k	220	

For the circuit below, explain by filling in the blanks in the tables:

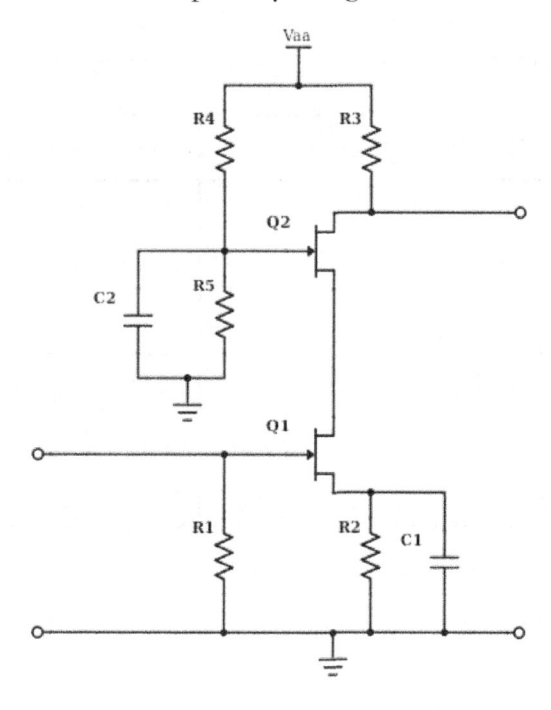

10. Fill in the blanks:

Q1 component type and configuration (e.g. PNP, common base)	
Q2 component type and configuration	
Analysis of Circuit type (*Explain its usage and configuration type.*)	
Function of R4, R5	

Chapter 6 – Amplifiers

11. Explain the various amplifier classes.

	Input waveform (include bias)	**Output waveform (*before* filter if used)**	**Key Characteristic**
Class A			
Class B Push-Pull			
Class AB			
Class D			
Class G or H			

12. Explain cascade and cascode.

13. Explain how to bridge and parallel amplifiers. Explain their differences.

Supplement Y - Project Tips: Order of Soldering, Height, etc.

So you have to complete a project on your own. How do you proceed? Where do you even start?

Most people immediately "just do it" - that's probably the worst approach; rather, we should "plan carefully".

Height of Components Above Circuit Board

The most-ignored consideration is: how high should we leave the resistors and other components at?

Some components are designed at a particular height: examples are potentiometers - even tiny preset potentiometers have a kink at theirs leads to indicate how far we should push it through. Likewise a few disk capacitors also have the kink at their leads. And IC's and transistors have a shaped lead so we should just push it as far as it can go...

Others, like resistors, have straight leads, which means we can have them all the way down the circuit board, intermediate, or as far up as possible. What should we do? Far up is not a good idea, as it increases the risk of potential shorts. Nor does it look good. As far down as possible looks good (in fact, many photos would be shot that way because it looks neater), but is not very practical for two reasons: 1) it does not leave enough lead for a re-soldering effort once the lead is trimmed, 2) the low position may block the silkscreen for component reference designators, making it difficult to see what that component is during troubleshooting in the future. Thus the pragmatic height is a little higher than the lowest position, enough for a re-soldering effort, and enough for one to see what is written underneath the component (typically the reference designator).

Occasionally, the components must be at the right height to interact with the enclosure. In that case we need to plan ahead. Example would be an LED or a mic - they need to fit a particular enclosure position.

Order of Soldering

One may think it is just our choice which to start with, and there is no difference. That is so far from the truth. There is a consequence in every action we take. In the case of soldering projects, the consequences may be severe...

Take for example what we learned from soldering the XLR cable: one must put the heat-shrink, or plastic insulator strip, or connector end-piece in early enough *before* the whole cable is soldered, because once it is soldered, we can't put the other pieces in.

Same is true for order of soldering. Some components may block the soldering of other components. So in general it may be a good idea to start soldering from the center and move outward, reducing the chance of items blocking the soldering iron. Consider especially the ultra-large and ultra-small components for special attention.

Transistors and IC's are usually soldered last, because they are usually more sensitive, and we don't want them damaged by heat or by static electricity as much as we can avoid.

Surface-mount ICs are often soldered first, because then they are not blocked by anything, as they require the utmost care and precise positioning for soldering.

Order Optimized for Testing

Often it is important to solder in such a way to help with checking along the way. We should check after every soldering step if possible. In that vein, the usual recommendation is to solder resistors first, followed by diodes and LED's.

Why? After soldering resistors, we can use the multimeter to check and see if the resistors are soldered correctly, and whether they are placed in the right place, by using the multimeter alone.

By the way, there is no polarity for resistors. But it is prudent to place all resistors in the same orientation, making it easier to read and re-check values.

Similarly, after diodes are placed and soldered, we can likewise check its value with the multimeter. In this case, we'd check with normal and reversed polarity, and we should get different readings - one high and one low.

Capacitors can't be easily checked, and once they are put in, they may interfered with the testing for resistors and diodes, and thus they are typically soldered a little later.

Component Orientation

It is important to know which components have polarity and which do not.

Polarized capacitors would have a strip on the cover to indicate polarity. Plus often the leads are different, or have different length.

IC's usually have a spot or notch and beveled edge to indicate location for pin 1. Learn to identify such position. The beveled edge is especially difficult to spot - you may use your finger to feel.

How to Know which Potentiometer to Use?

Given two preset potentiometers, one 10K and another 50K as shown on the schematic, but the values are not shown on the components, how do we figure out which is which?

In such a case, the multimeter is again your friend! There are three leads in a potentiometer. The two outermost leads are fixed, while the middle lead is associated with the wiper and will give variable values when the potentiometer is changed. Measure the outermost leads, and that'll give the values of 10K and 50K…

Also, it should be noted that potentiometers designed for volume control may be tapered differently than those designed for bass and treble. So one should not think that the three 50K potentiometers are exchangeable. In buying a replacement, always double-check the taper. If it is not indicated in the schematic, it is fair to assume that the volume control would be "**audio-taper**" or "**log-taper**" (the two means the same thing). On the other hand, tone controls are usually "**linear-taper**". They are sometimes labeled as LIN or LOG, or with code like A and B. In the latter case, we need to check the exact manufacturer to know what that code means - A means audio-taper in Asia and America, but linear in Europe; B means linear in Asia and America; while C means audio-taper in Europe.

Which position should we use on the multimeter?

Depending on the specific multimeter, there may be up to five positions for resistance measurement. Which one should we use?

The general idea is to guess what value we are working with, and set the position accordingly (larger than the expected value). If the reading is too high or too low (or too little significant figures), we change to the next setting, until we get the best significant figures and values out of it.

Read the Instructions

Sometimes the instructions are poor. But given the above hints, it may make more sense how to execute your project. If they recommend something unusual, figure out why first before proceeding further…

Matched Components

For high-end audio electronics, occasionally components are matched. In that case make sure you identify them at the beginning, and keep pairing them up. Otherwise you may lose all the benefits of the kit. With those labeled, you should not consider that any component with same values can replace them. This is especially true for output transistors - they are usually packaged separately in different bags - keep them isolated and not try to organize things on your own, as you may lose the matching.

Heat Transfer Considerations

High power transistors may need good heat dissipation. Heat sinks plus a good heat conduction path through heat pastes and other methods would be needed. Otherwise the temperature may be kept too high during operating, thus reducing operating life. The heat sinks are usually anodized black for good heat dissipation.

(this page deliberately left blank)

Chapter 7   OP-AMPS & ACTIVE FILTERS

> The superpower of negative feedback and op-amps introduce us to the various configurations of op-amps, and how they can be used to implement amplifiers and audio mixers. Then op-amps are used to implement integrators and state variables as well as voltage followers for various kinds of active filters, leading to implementation of crossovers, tone controls, equalizers and other audio electronics.
>
> *Expected Learning Outcome*: The student will be able to connect op-amp characteristics and configuration with typical audio applications. The student will also be able to explain how to mitigate against noise via differential signaling and balanced cables.

CONTENTS

Differential Amplification, Op-Amps, Analog Computing	7-3
Positive about Negative Feedback	7-6
No fair? Op-Amp as Comparator	7-8
Can You Hear Me Now? Op-Amp as Amplifier	7-9
♪ Why not Y? 'Cause Bus is Better: Op-Amp as Summing Amplifier	7-11
I'll Follow You: Op-Amp as Voltage Follower	7-13
♪ Balanced and Differential Signaling in Pro Audio, Noise of the 3rd Kind	7-14
Op-Amp as Integrator, Intro to Active Filters	7-19
Virtual Lab: State Variable Active Filter Synthesis	7-25
Do You Want to Dance? Op-Amp as Gyrators	7-29
♪ Practical Circuits: Crossovers, Parametric Equalizers, Tone Controls	7-33
Classifying Active Filters	7-39
♪ More Building Blocks: VCA, OTA, Analog Multipliers	7-42
Review Exercises	7-44

This chapter is divided into background plus two main parts: op-amps and active filters.

Background:
1) analog computing
2) superpower of negative feedback

Op-amp configurations:
- no negative feedback: comparators
- some negative feedback: inverting and non-inverting amplifiers
- some negative feedback: summing amplifiers
- full negative feedback: voltage followers
- differential-inputs
- fully-differential amplifiers
- integrators
- gyrators

Active filters:
- state variable synthesis
- feedback topologies
- gyrator-based simulation of inductors
- implementation examples in audio circuits

Chapter 7 – Op-Amps/Active filters

7.1 DIFFERENTIAL AMPLIFICATION, OP-AMPS, ANALOG COMPUTING

Brief History of Analog Computing

Some credits the south-pointing chariot in China during the first millennium BC as the first analog (analogue) computer, using mechanical gears to keep the south-pointing direction even when visibility is low.

In the West, the mechanism discovered off a wreck near the Greek island of Antikythera, dated to around 100 BC, was considered the first analog computer, again using mechanical means to calculate astronomical positions.

Starting from World War II (and all the way to the Vietnam and Korean Wars), gun fire controls and bombsights were mechanical analog computers.

Slide rules can also be considered analog calculators; it is still a backup option for aircraft personnel.

The first general-purpose fully-electronic analog computer was built in 1941. Heathkit produced the first educational fully-electronic analog computer in the 1960s. (Photo credit: Computer History Museum, Michael Holley.) As can be seen on the front panel, nine (9) operational amplifiers are available, and the analog computer can be programmed using patch cords. My first "programming" was done on this **analog computer**.

Meaning of Analog

In these cases, **analogy** was used to enable the calculations, sort of like the use of "equivalent circuit" in electronics. It is this concept of analogy that gave rise to the name **analogue** (or analog) **computer**. The analog computer also works in the **analog domain**, where time and value are continuous. Notice that "*analog*" was used in different ways and contexts here: *analog* computers use *analog* circuitry and works in the *analog* domain... :)

Likewise, analog computing in most cases are linear (with some exceptions operating non-linearly). Analog computers typically use **linear circuits**. But not all analog circuits are necessarily linear. Students often mistakenly equate analog with linear, and it is important to understand the definitions clearly.

The Operational Amplifier

Operational amplifiers (from now on simply called **op-amps**) enabled the "rise" of analog computing. For mechanical analog computers, the "programming" was done during construction, which took a while. That means each mechanical analog computer have a specific purpose. There were no general-purpose analog computers until the advent of operational amplifiers. With a set of patch cords, programmers can reprogram the problem in just hours or even minutes (for simple problems).

Initially, op-amps were constructed using vacuum tubes. Then transistors followed. And then integrated circuits. The first widely successful integrated op-amp was the 741 introduced in 1968. Now it is difficult to find the 741 parts, although we can still implement 741 circuits using the 747 which is simply a jumbo twin-741 chip. (Memory aid: recall that the first widely successful jumbo jet is the 747.)

Op-amps are sometimes also called "continuous current amplifiers" because they have no low frequency limitation. In other words, **op-amps are DC amplifiers**, capable of amplifying DC.

For its successful operation, one other characteristic is that it must have a relatively high open-loop gain and relatively wide frequency response. When the two factors are multiplied together, it is known as the **gain-bandwidth product** (abbreviated **GBW**). **The 741 has a gain-bandwidth product of one million!**

> **How to use GBW**
> The 741 op-amp has GBW=1000000.
> If we implement it for 0-20 kHz, then we can have a maximum gain of 50!
> (50 x 20000 = 1000000)
> The more common usage is for it to have a gain of 10 (or less) - that means a frequency response of 0-100 kHz! (10 x 100000 = 1000000)

The symbol for a basic op-amp is shown below, consisting of a triangle shape with three terminals, one output and two inputs called the **differential input**, designated as non-inverting and inverting with the symbols + and - (plus a reference designator here of U1):

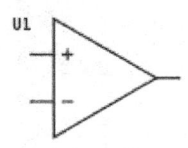

The op-amp requires two or more power supply terminals for operation, but is omitted above by convention. When it makes sense to include them, the symbol is shown below:

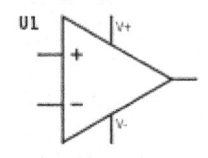

Op-amps using only one power supply is available, although relatively rare.

Differential Amplification

For comparison, the symbol for an amplifier is shown below:

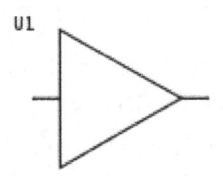

Hopefully one difference is now obvious. Op-amps provides **differential amplification**, whether we use it or not. In other words, **op-amps have a non-inverting and an inverting input terminal**.

> **Differential signaling** is a method for electrically transmitting information using two *complementary* signals.
>
> An **operational amplifier** (**op-amp**) is a 1) DC-coupled 2) high-gain 3) electronic voltage amplifier 4) with a differential input.
>
> A **differential input** consists of an inverting and a non-inverting terminal, the two signals being complementary.

Pro's and Con's of Analog Computing

Analog computers operate in **real-time**, meaning there are no discernible time lags at all. In contrast, digital computers always encounter a time lag, no matter how small, and so cannot be used in time-sensitive or "real-time" applications. In the recording studio, microphones going through an A/D converter and a digital signal processing unit will encounter sufficient delays that some recording artists will complain about "latency" sourcing a headphone monitor from such output; thus, usually a special analog headphone output is provided that have only negligible analog delay.

It is for the same reason that we use op-amps in creating active filters in audio electronics - the resulting circuits operates in real-time!

Non-ideal effects like limited voltages allowed, limited gain and frequency response limits restrict the use of analog computing. Noise and resolution are key challenges as well. That explains why the audio electronics industry have gone primarily digital - except for real-time needs like in monitoring.

The Ideal Op-Amp

1. Infinite bandwidth, starting from 0 (DC) and going as high as possible.
2. Infinite open-loop gain, or as large as possible.
3. Infinite input impedance, meaning essentially zero input current.
4. Infinite output voltage range (of course limited by power supply).
5. Infinite slew rate.
6. Infinite common-mode rejection ratio and power supply rejection ratio.
7. Zero input offset voltage or as low as possible.
8. Zero output impedance or as low as possible.
9. Zero noise or as low as possible.
10. Zero phase shift or as low as possible.

7.2 Positive about Negative Feedback

What enables the op-amp's ideal characteristics are its high open-loop gain and the use of negative feedback in most of its applications. So, it is important to *first* look at what is **negative feedback**.

The human body, and most of nature, uses negative feedback; for example, the regulation of sugar in blood using insulin. In the 1600s, we saw the rise of thermostatically-controlled ovens and centrifugal governors in windmills - all examples of negative feedback. James Watt patented a form of governor for the steam engine in 1788. The term "feedback" was well established by the 1920s in reference to boosting the gain of amplifiers (which was severely limited then). The negative feedback amplifier was invented by Harold Stephen Black at Bell Laboratories in 1927, and a patent was applied the following year and granted in 1937. (In other words, it is fairly recent - less than a century old!) What is significant is that Nyquist and Bode then built upon Black's work to develop a theory of amplifier stability, leading to the rise of modern audio electronics...

A Model of Negative Feedback

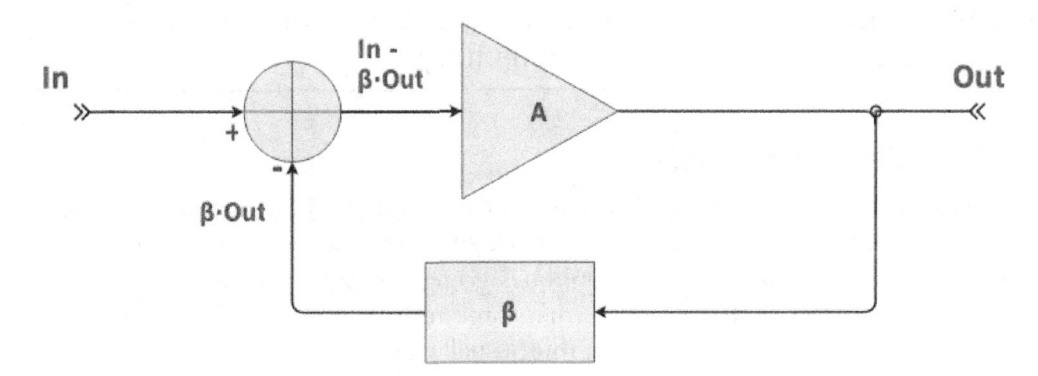

We have an input signal, **In**. The output signal is **Out**. That output is fed back through the feedback network with gain of β, resulting in signal $\beta \cdot$**Out**. That signal is mixed together (notice the + and - sign at the mixer input) with the result being (**In** - $\beta \cdot$**Out**). The open-loop gain of the amplifier is **A**. Thus, the output of the amplifier is:
A(In - β·Out). This is the same as **Out**. Thus:
    Out = A (In - β·Out)
Rearranging:
    Out + A·β·Out = A·In
    Out/In = A/(1+A·β)

Out/In is the **closed-loop gain**, or the **transfer function** of the system.
When A·β is large and much greater than 1, we can approximate and say:
    Out/In ≈ 1/β

In other words, the *system level* gain is independent of the *amplifier* gain! This is significant because variations in manufacturing result in significant variations in open-loop amplifier gain, sometimes in the range of 10-30%. In contrast, the feedback gain is usually due to passive components, which have much closer tolerances down to 1%. **Thus, use of negative feedback allows us to have much less variations in the overall system.**

Chapter 7 – Op-Amps/Active filters

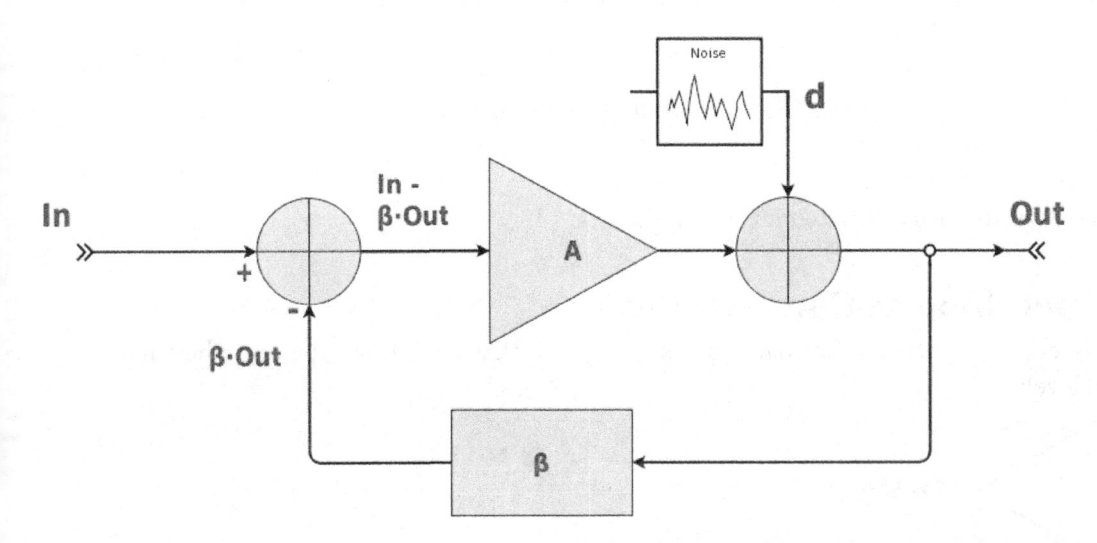

Now let us add a disturbance d to the system. That disturbance could be the result of noise from the environment or noise from the amplifier itself (noise of the first or third kind), or distortion coming from the amplifier. If there were no feedback, the result would be:
   Out = A·In+d

With feedback, Out = A(In - β·Out) + d
Rearranging,
   Out = A·In - A·β·Out + d
   Out + A·β·Out = A·In + d
   Out = A·In/(1+ A·β) + d/(1+ A·β)
In other words, with feedback, the disturbance is reduced by a factor of 1/(1+ A·β)!
When A·β is large and much greater than 1, we can approximate and say the improvement factor ≈ 1/A·β
Thus noise and distortion performance are improved significantly by negative feedback!

Please go through the math again to make sure you understand how to derive the formulas above.

Example
1. Calculate the system gain if the amp has gain A=10000 and the feedback network has nominal gain of 0.5
Since A·β = 10000 · 0.5 is much bigger than 1, we can apply the approximate formula Out/In ≈ 1/β and get:
   gain = 1/0.5 = 2
We can apply the fundamental and more exact formula just as a check:
   Out/In = A/(1+A·β) = 10000/(1+10000 · 0.5) = 1.99960007998400319936012797440051
Thus we see the result of gain =2 is accurate to *four* decimal places, all because of the high gain of the amplifier, even though its gain does not show up in the final result of 2, which is determined by the passive element…

2. If there is noise disturbance of 1 mV equivalent at the amplifier, what is the effect on the final system output?
The effect is d/(1+ A·β) = 1m/(1+ 10000 · 0.5) = 2 nV
In other words, the effect of the noise disturbance was reduced by 5000 times!

7.3 No Fair? Op-Amp as Comparator

An op-amp used ***without negative feedback*** is a comparator.

Op-Amp in Open-loop as Comparator

The case with an open loop, as the potential loop is an open circuit, is shown below when it is used to compare to ground level:

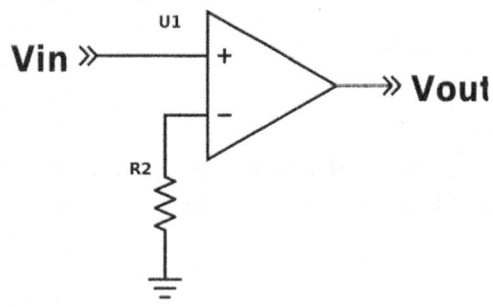

If Vin is higher than the compared signal (which is ground in this case), it is amplified by the high open-loop gain which is close to a million, thus Vout would have been very high, limited only by the output voltage range (which itself is limited by the power supply), to a value we call maximum positive as the amplifier goes into saturation (meaning no longer acting linearly).

If Vin is lower than the compared signal (which is ground in this case), it is amplified by the high open-loop gain which is close to a million, thus Vout would have been very negative, limited only by the output voltage range (which itself is limited by the power supply), to a value we call maximum negative as the amplifier goes into saturation.

Thus, the op-amp without negative feedback acts as a **comparator**: if Vin is higher, it produces maximum positive; if it is lower, it produces maximum negative. The *magnitude* of the voltages are irrelevant, only whether it is relatively higher or lower.

Op-Amp with Positive Feedback as Schmitt Trigger

The circuit with positive feedback is shown below, with similar result other than action is faster and has *hysteresis*, thus earning a special name of Schmitt Trigger:

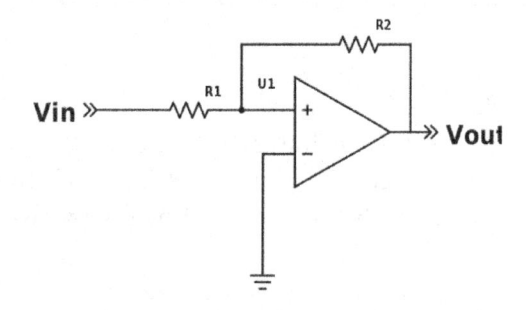

Comparators are not common in audio electronics in studio or live sound, but may be used in synthesis. It is used in our 555 circuit for synthesizing a waveform.

7.4 CAN YOU HEAR ME NOW? OP-AMP AS AMPLIFIER

We now know negative feedback is necessary to produce an amplifier. (With positive feedback or no feedback, the op-amp becomes a comparator.) There are again two cases, depending on whether the Vin goes to the inverting or non-inverting input.

Op-Amp in Negative Feedback as Inverting Amplifier

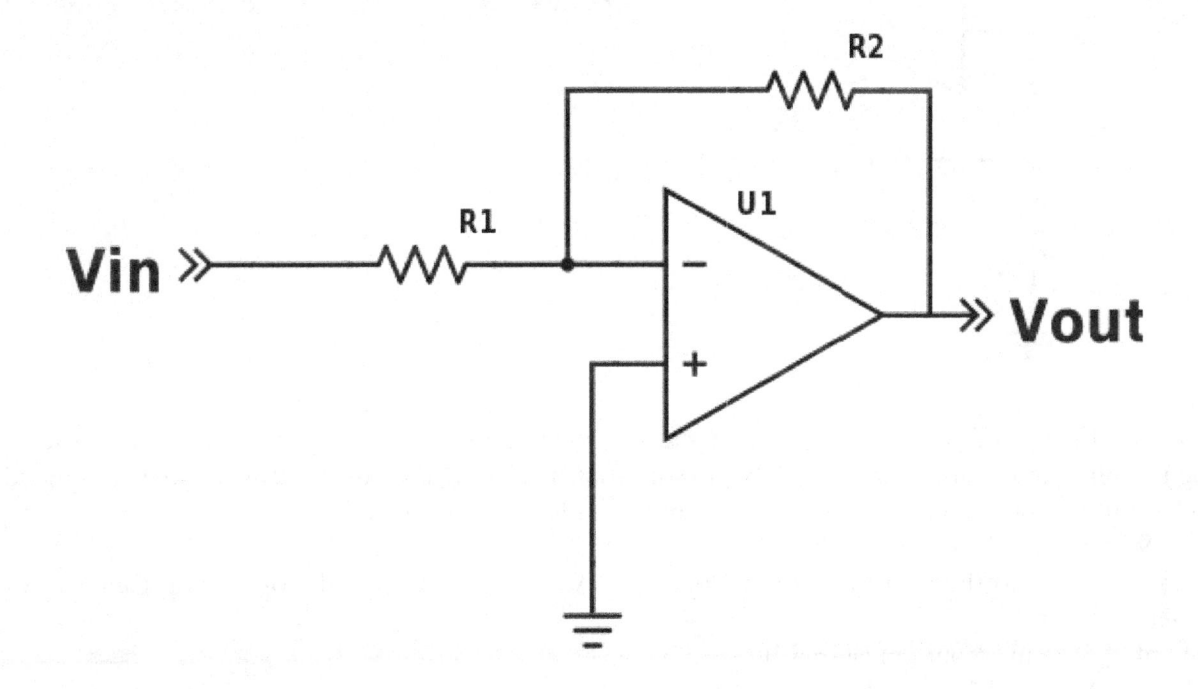

Using the notation of V_- as the voltage at the inverting terminal, and V_+ as the voltage at the non-inverting terminal, and seeing that V_- is a superposition of Vin divided by the voltage divider of R1 and R2, and Vout divided by the same voltage divider, and the ideal op-amp has infinite input impedance and zero output impedance, thus:

V_- = Vin (R2/R1+R2) + Vout (R1/R1+R2)

Substituting this into the gain formula of Vout = A (V_+ - V_-) where A is the open loop gain and V_+ = 0, we get:

Vout = - A [Vin (R2/R1+R2) + Vout (R1/R1+R2)]
Vout = - Vin (A R2 / (R1 + R2 + A R1))

Since A is very large, the **gain = Vout / Vin ≈ - R2 / R1**

Often a similar resistance to ground is added at the non-inverting input to reduce input offset voltage.

Example

1. If R1 and R2 are equal, what is the gain?
   Gain = - R2/R1 = - R/R = -1. The gain is unity, the negative sign indicates it is inverting.

2. If R1 is 10 kΩ, R2 is 20 kΩ, what is the gain? What is the bandwidth if GBW is 1 million?
   Gain = - R2/R1 = - 20k/10k = -2 .
   Since Gain · BW = 1000000 · |gain| = 1000000/2 = 500000 = 500 kHz

Op-Amp in Negative Feedback as Non-Inverting Amplifier

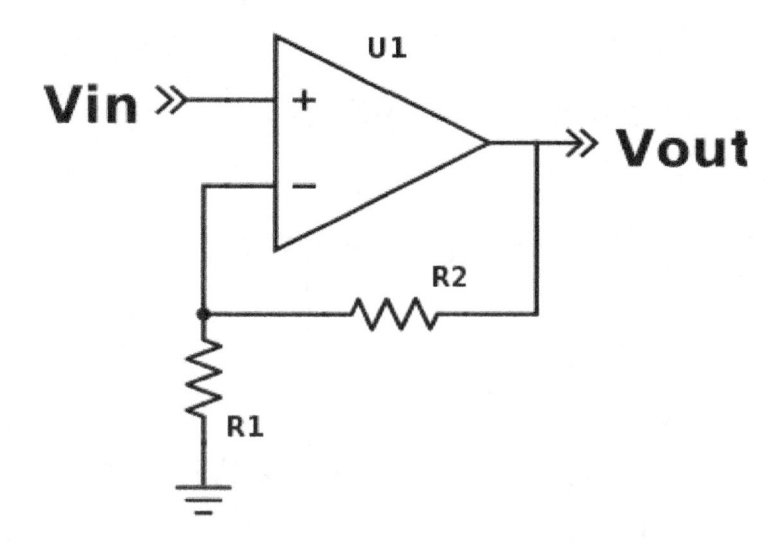

Using the notation of V₋ as the voltage at the inverting terminal, and V₊ as the voltage at the non-inverting terminal, and seeing that V₋ is a function of Vout divided by the same voltage divider, and the ideal op-amp has infinite input impedance and zero output impedance, then we get:
V₋ = Vout (R1/R1+R2)
Substituting this into the gain formula of Vout = A (V₊ -V₋) where A is the open loop gain and V₊ = Vin, we get:
Vout = A [Vin - Vout (R1/R1+R2)]
Vout = Vin [A / 1 + A · R1/(R1+R2)]
Since A is very large, the **gain = Vout / Vin ≈ 1 + R2 / R1**
From the formula, we can deduce that **for a noninverting amplifier, the voltage gain is always more than unity and never less than one, and it is always positive**. In contrast, **for the inverting amplifier, the absolute value of the gain could be bigger or smaller than one, but the gain is always negative**.

The student may be attracted to the non-inverting configuration because the derivation is much simpler than the inverting; however, in actual implementation the non-inverting configuration is trickier, because the non-inverting input requires a path for DC to ground. If that is not provided by the signal source, then a resistance to ground is needed. In that case, the resistance needs to be balanced to reduce offset voltage. Same consideration is necessary if a different input impedance is desired. Thus, you may see extra components in a non-inverting configuration. Also, the distortion is usually higher than the inverting configuration. That explains why the **inverting configuration is more common**.

Example

3. If R1 and R2 are equal, what is the gain?
Gain = 1 + R2/R1 = 1 + R/R = 2. The gain is two; the positive sign indicates it is non-inverting.

4. If R1 is 10 kΩ, R2 is 20 kΩ, what is the gain? What is the bandwidth if GBW is 1 million?
Gain = 1 + R2/R1 = 1 + 20k/10k = 3 .
Since Gain · BW = 1000000, gain = 1000000/3 = 333000 = 333 kHz

♪ 7.5 Why Not Y? 'Cause Bus is Better: Op-Amp as Summing Amplifier

Invariably at some occasions, audio engineers may find a need to combine two inputs into one output. The obvious instinct is to use a Y-cable to combine them together. But there are a few reasons why we should not Y:
1) The two output circuits may harm one another (as they are not isolated by virtual ground) ...
2) The frequency response may suffer...
3) There is a better way...

Examine the inverting amplifier circuit again:

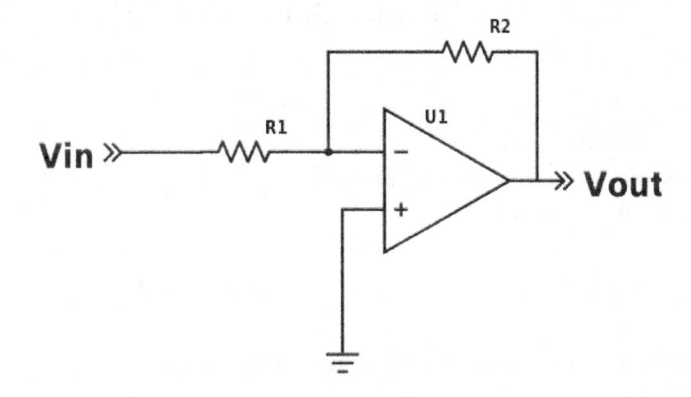

There are two important observations to be made about the inverting amplifier configuration: 1) We can assume **no current flows into the inverting input terminal** (because of the high input impedance); 2) We can assume the **voltage difference between the two input terminals to be negligible** (because A is large and Vout is limited, so $V_+ - V_-$ is small). In other words, $V_+ = V_-$), and since V_+ is ground, we call V_- a **virtual ground** (or "virtual earth" across the pond).

At the virtual ground, we also call it a summing point or **summing junction**. because no current flows into the inverting input terminal, thus all currents flowing into the summing junction added together will equal the current flowing out to the feedback resistor. If we implement a circuit like this:

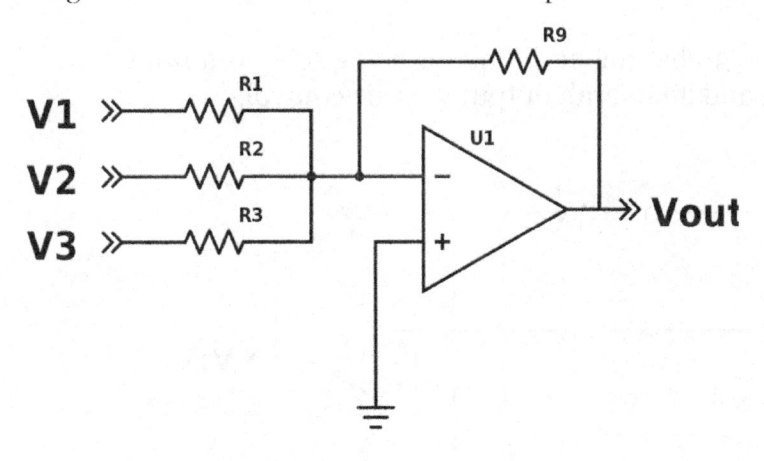

Because of 1) and 2), we can get: $i1 + i2 + i3 = i9$,
thus: $V1/R1 + V2/R2 + V3/R3 = -Vout/R9$ ………………………….. (1)
If $R1 = R2 = R3$, then we arrive at the **summing amplifier equation**:
   Vout = - R9/R1 (V1 + V2 + V3)

What is important to recognize is that because of the two properties of the summing junction, **the input signals are effectively isolated from one another**. Each input signal sees a high input impedance, which will not reduce or affect the input signals. Thus, they will not have the unfortunate effect of the Y junction mentioned earlier.

When implemented in an audio mixer, the summing junction is known as a **bus**. As many input signals as needed can be connected to the bus. Expansion modules can be added to the bus to further expand the number of signals accessible.

The op-amp in the inverting amplifier configuration, with more than one input signals and equal input resistors, is usually known as a **summing amplifier**.

When the input resistors are not equal, the configuration is known as a **scaling summing amplifier**, with this equation derived from (1):

$$V_{out} = - V_1 (R_9/R_1) - V_2 (R_9/R_2) - V_3 (R_9/R_3)$$

The scaling factors are R_9/R_1, R_9/R_2, and R_9/R_3 respectively for the three input signals.

Examples

1. $R_1=R_2=R_3=R_9$. What is the output?

    $V_{out} = - R_9/R_1 (V_1 + V_2 + V_3) = V_1 + V_2 + V_3$

    The output is simply the sum of the inputs.

2. $R_1 = 1\ k\Omega$, $R_2 = 2\ k\Omega$, $R_3 = 3\ k\Omega$, $R_9 = 6\ k\Omega$, and $V_1 = 1\ mV$, $V_2 = 2\ mV$, $V_3 = 3\ mV$. What is the output?

    Let us try to calculate using first principles, without using formula…

    $A_1 = R_9/R_1 = 6k/1k = 6$
    $A_2 = R_9/R_2 = 6k/2k = 3$
    $A_3 = R_9/R_3 = 6k/3k = 2$
    $V_{out} = A_1 V_1 + A_2 V_2 + A_3 V_3 = 6 \cdot 1 + 3 \cdot 2 + 2 \cdot 3 = 6 + 6 + 6 = 18\ mV$

    We need to check that the purported output is less than the supply voltage (in other words, the amplifier is not yet saturated). Since 18 mV is relatively small, we can assume that it is the case, and that is the output voltage.

    As a separate exercise, you can try substituting into the scaling summing amplifier equation.

3. Implement a basic 3-channel audio mixer using one op-amp, with an input gain control for each channel, and an overall output volume control.

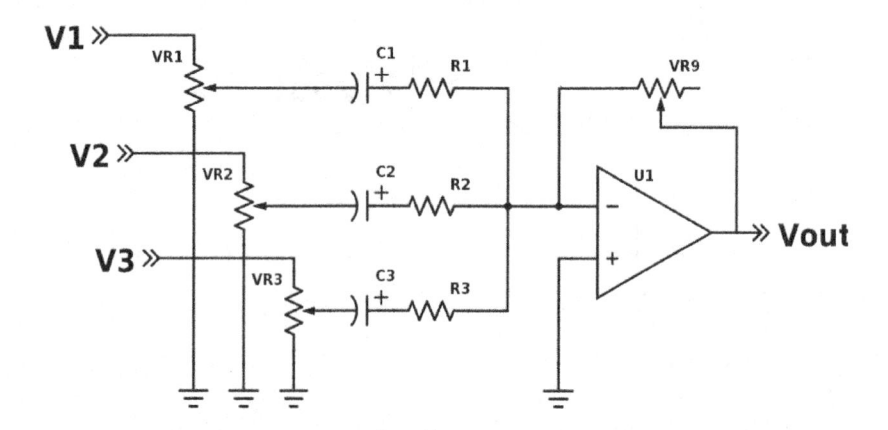

    Gain pots VR1, VR2, VR3 acting as voltage dividers, and volume control VR9, with equal input resistors R1, R2, R3 going to the summing junction. Blocking capacitors C1, C2, C3 blocks potential DC coming from the source.

7.6 I'll Follow You: Op-Amp as Voltage Follower

One extremely simple op-amp circuit frequently used in audio electronics is shown below....

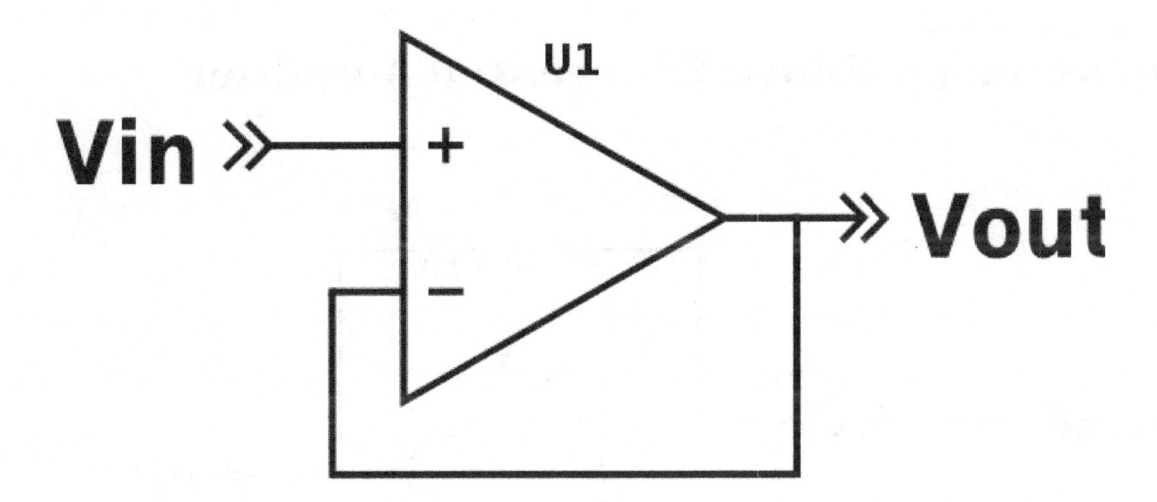

The voltage gain is one - we call that **unity gain** - because the full output is fed back to the inverting terminal to reduce it; the high open-loop gain of the op-amp makes sure the Vout follows Vin precisely. That is why this op-amp configuration is also known as **voltage follower**.

Why bother using this configuration if there is no voltage gain?

The reason lies in the other name for this circuit: **voltage buffer amplifier**. A buffer isolates the input circuit from the output circuit, so that the signal source will *not* be affected by the output load. This is especially important for buffering passive filters in audio electronics.

This behavior is possible because the input impedance of the op-amp is high while the output impedance is low, thus creating an effective **impedance bridging** condition.

Even though the *voltage* gain is essentially unity, the current gain is *not* - in fact usually it is quite considerable. Thus, even though there is no voltage gain, **there is current gain and with it power gain** as well.

It may sound superfluous, but voltage followers are essential in many applications, some being shown in active filters in section 7.9.

♪ 7.7 Balanced and Differential Signaling in Pro Audio; Noise of the 3rd Kind

Thus far we have used only *one* of the inputs of the op-amp. To complete our collection of audio electronics op-amp configurations, let us use *both* inputs.

Op-Amp as Single-Ended Differential Amplifier

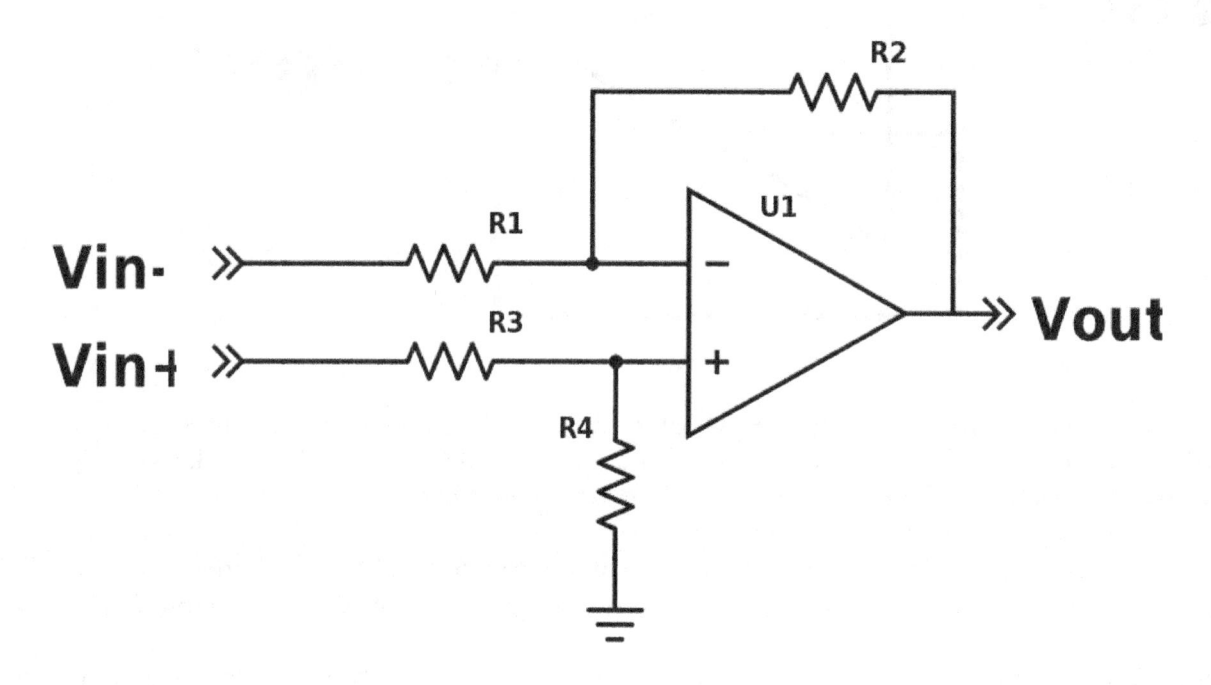

It looks *very* familiar: the inverting path is *exactly* the same as that of the inverting amplifier. It is also very much like that of the non-inverting amplifier except the non-inverting input is connected to ground, and the non-inverting input has an input resistor, and the path to ground (R4) that was already alluded to earlier. In other words, this circuit is a practical implementation of the non-inverting amplifier when Vin- is connected to ground!

The derivation of the Vout can be obtained by superposition of the inverting amplifier and the non-inverting amplifier (i.e. the two added together assuming Vin- and Vin+ equal to zero respectively). It'll be left to the reader for practice. Thus,

  Vout = - Vin- (R2/R1) + Vin+ (R4/R3+R4) (R1+R2/R1)

When we choose R1=R3 and R2=R4, the equation simplifies to:

  Vout = (Vin+ - Vin-) (R2/R1)

In other words, the differential amplifier is a *difference* amplifier, computing the difference of the differential input signals and multiplying it by the feedback ratio of R2/R1. Note that the differential amplifier configuration here has a differential *input* but a single-ended *output*.

Differential Output

So far, we have used single-ended op-amps. The symbol for a **fully differential** op-amp (with differential output which consists of two complementary outputs, one complements or negative of the other) is shown below, with the ball-shaped pin indicating a complement by convention (the minimum required pins are the 4 pins of differential input and output, others are optional):

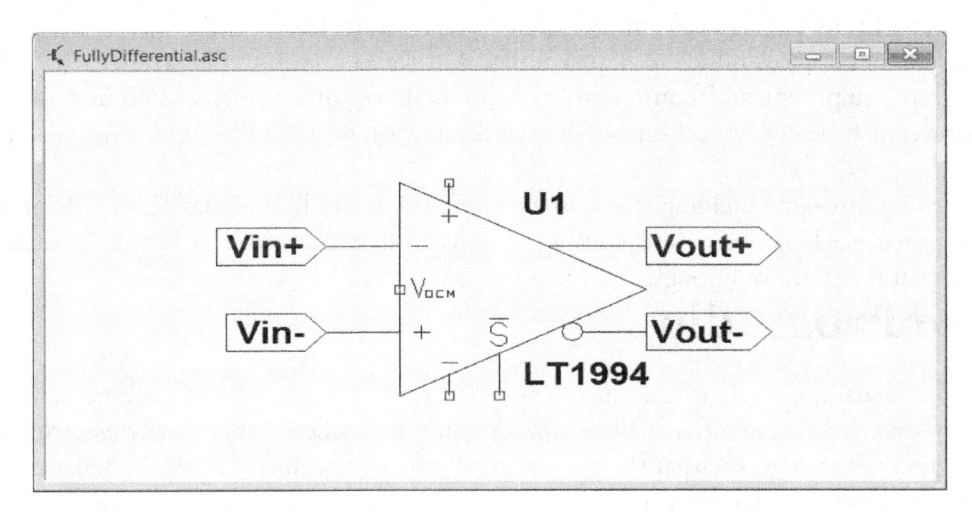

In addition to the voltage supply pins, we also typically have other pins, for example here Vocm is a pin designed to be connected to a precision ADC to supply a reference voltage for offset correction. But the key we should note is that in addition to the differential input we now have differential output as well. That makes a **fully differential amplifier**. A typical usage is shown below:

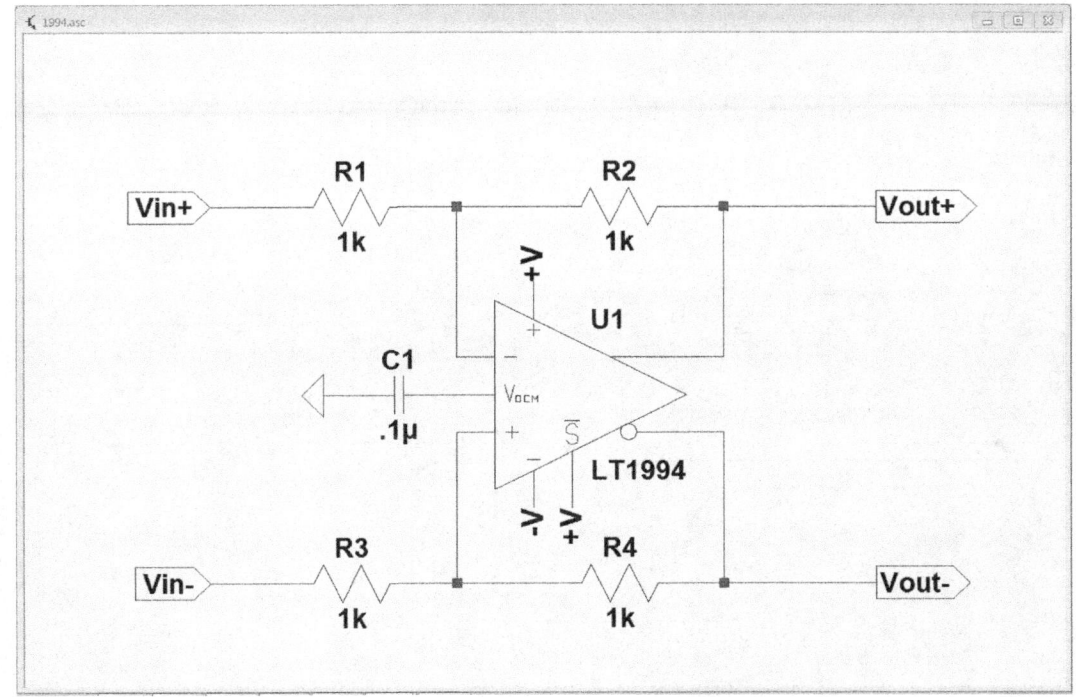

The schematic is drawn deliberately to emphasize its **balanced** nature. In use, it is important to make sure that R1 is precisely balanced with R3, and R2 with R4. One may guess that the amplification factor (when R3=R1 and R4=R2 as before) is the same as that of the differential amplifier mentioned in previous page:

$$Vout_+ = (Vin_+ - Vin_-)(R2/R1)$$
$$Vout_- = (Vin_- - Vin_+)(R2/R1) = -Vout_+$$

Noise of the Third Kind

Sometimes in the studio we may just be connecting cables to amplifiers, but suddenly we can hear a local radio station. That is because of electromagnetic interference (EMI). The electromagnetic waves from the radio station get caught in the wires, and then at certain junctions (typically poorly soldered joints), the signal got rectified and amplified, and thus the signal from the radio station can be heard.

In the early days of consumer audio, and even of pro audio, that was a common occurrence; thankfully it seldom occurs with modern equipment and configuration. But EMI of other forms, such as from utility power connections, fluorescent tubes (a typical hum that is at a pitch of roughly Bb), and other electronics, are still potential issues.

One approach is of course through shielding, the concept being that of the Faraday Cage. By shielding each audio electronics cabinet, we have less noise coming to affect other electronics. Likewise, each cable can also be shielded. But that is often not enough...

Application in Pro Audio

You may have heard that "balanced cable is used in pro audio to reduce noise." In this age of "alternative facts," we may have to say that the statement is at best *partially* true. **Balanced cable is *necessary* but not sufficient.** A stricter engineer may even say that the statement is wrong, because it is *not* the balanced cable that reduced the noise, it is the differential signaling that did it, but balanced cable made it *feasible* as balance is essential in differential amplifiers as we had just seen.

Had we use an "unbalanced cable" consisting of only two conductors (one signal, one ground), the result would be as follows:

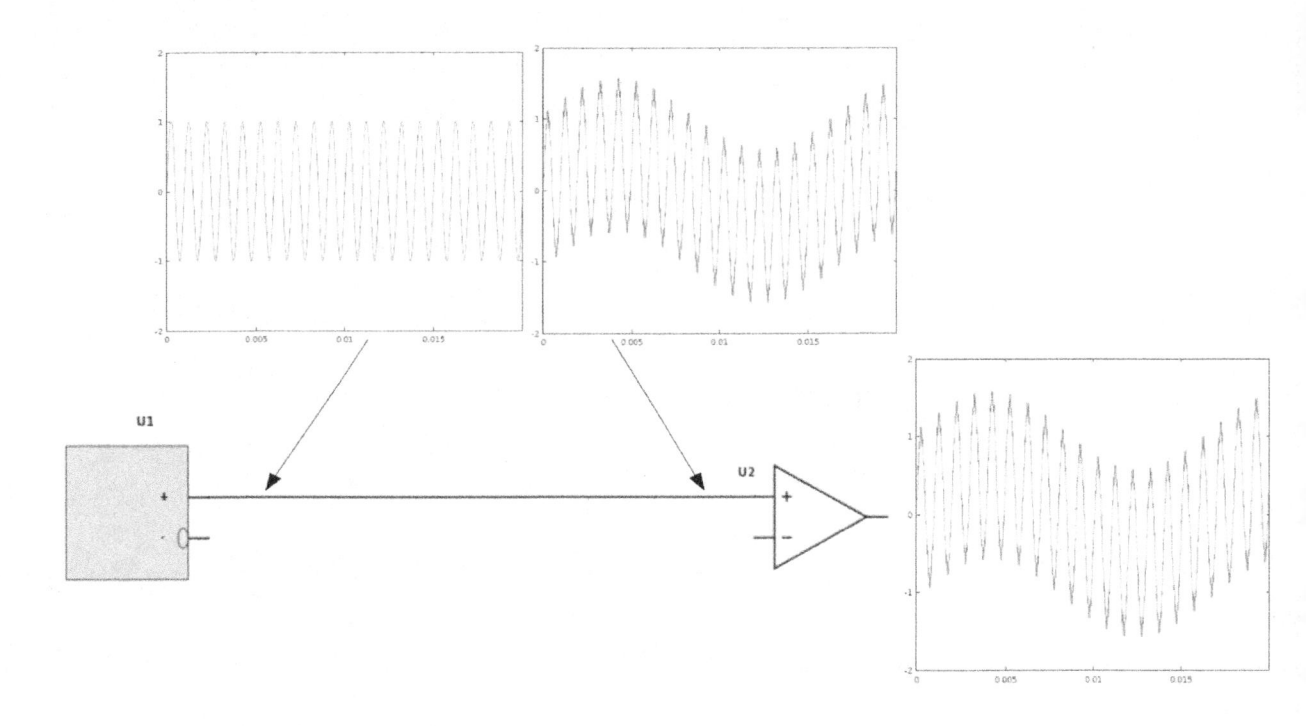

The input was travelling along the cable, and then noises consisting of 60 Hz and 15 kHz EMI (exaggerated) were encountered along the way. The 60-Hz contributed to the slow undulations, while the 15-kHz contributed to the porcupine-like character of the waveform (quills coming out more frequently near the peaks). Notice the output is *not* double that of input (as we shall see later for balanced cable with differential input), and none of the noises were rejected at the final output.

Fortunately, that is *not* what we do in pro audio. The "balanced cable" consists of at least three conductors, two carrying the signals in complementary form shown in the schematic below, and one being the ground conductor. This is what happens when a 1 kHz signal traveling on a balanced cable using differential signaling caught 60 Hz and 15 kHz EMI (exaggerated) along the way, and the resulting waveform:

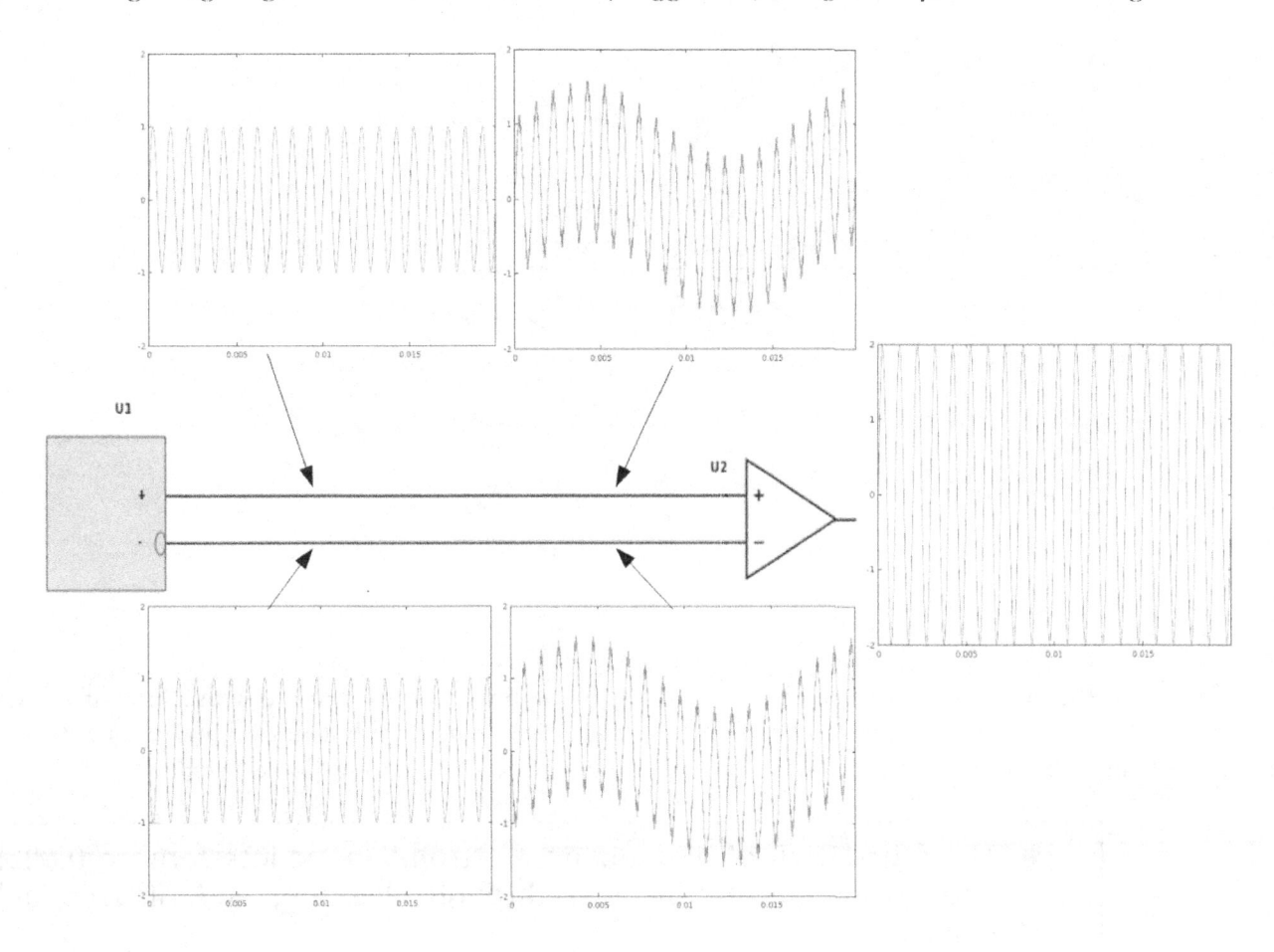

Notice that the inputs may look the same, but are complementary, i.e., one moves upward while the other downward from zero. The noises encountered along the way (notice they are the **same** shape on the two conductors, and *not* complementary of one another) cancelled out one another at the differential amplifier. Note that the output waveform is just like the input, except twice as large - that is a seldom-mentioned benefit of balanced differential signaling. We'd lose the extra headroom had we use an unbalanced cable.

Differential signaling is also used in your cell phone or tablets, when the processor communicates with the display: it is a very noisy (electromagnetically speaking) environment, and thus differential signaling helps greatly. (Balanced cables are not used in that case because of expenses.)

As alluded earlier, differential signaling is typically also used when signals go to a precision ADC (analog digital converter). Again, balanced cable may or may not be needed, depending on the configuration.

A "balanced" cable can become slightly unbalanced quickly when someone "coil" the cable improperly, thus exerting stress on the cable, resulting in strains in the crystalline structure of the copper. This is what happens when a 1 kHz signal traveling on an imperfect "balanced cable" caught 60 Hz and 15 kHz EMI (exaggerated) along the way, and the resulting distorted waveform:

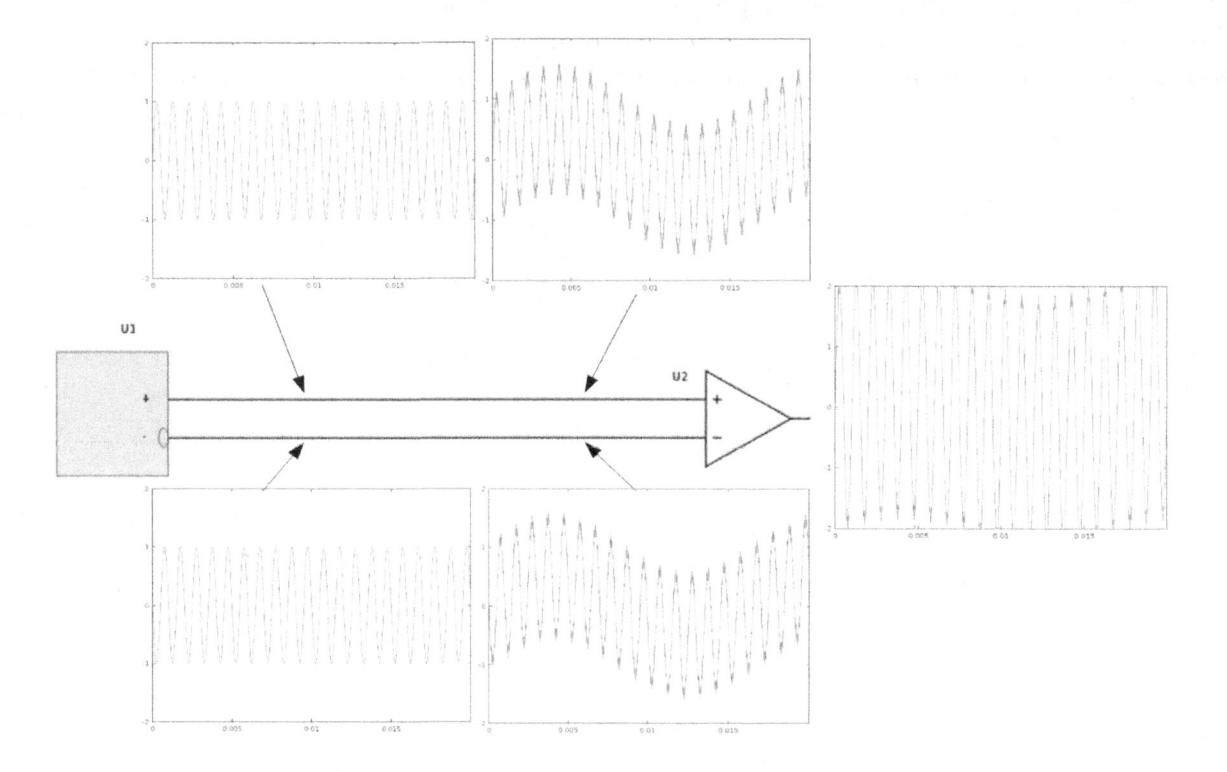

Notice that the output signal is also twice that of the input, but it is contaminated with a little of the EMI noise - *most* of the noise had been removed by the differential amplifier, but noise removal is *not complete* because of the imbalance in the cable - even though the imbalance in the noise "looks" slight.

- The foundation of minimizing electromagnetic interference (EMI) and noises caused by it is through shielding and differential signaling.
- Differential signaling is possible with **differential output** (sender) driving **differential input** (receiver) through a cable which is balanced.
- Fully differential amplifiers are one way to achieve the differential inputs and outputs on both ends, although the minimum requirement is a differential output circuit at the sender, and a (partially) differential amplifier at the receiver.
- For maximum noise immunity, the full differential signaling path needs to be carefully balanced. Properly **balanced cable** is essential to help achieve maximum immunity.
- When the cable is slightly imbalanced (for example with regular incorrect coiling of the cable), the EMI immunity is reduced accordingly.
- Implementing differential inputs and outputs are expensive, thus it is typically used only when 1) signals are low enough that interference becomes critical; 2) signals are run outside the electronics box and so is less under control (inside a box we can use box shielding). That explains why balanced cable is typically the place where we see differential signaling.

7.8 OP-AMP AS INTEGRATOR, INTRO TO ACTIVE FILTERS

There are tons more use for op-amps; we'll just mention one more by way of introduction to active filters. By changing the feedback resistor into a capacitor for the inverting input configuration, we have changed the op-amp into an integrator…

Basic Op-Amp Integrator

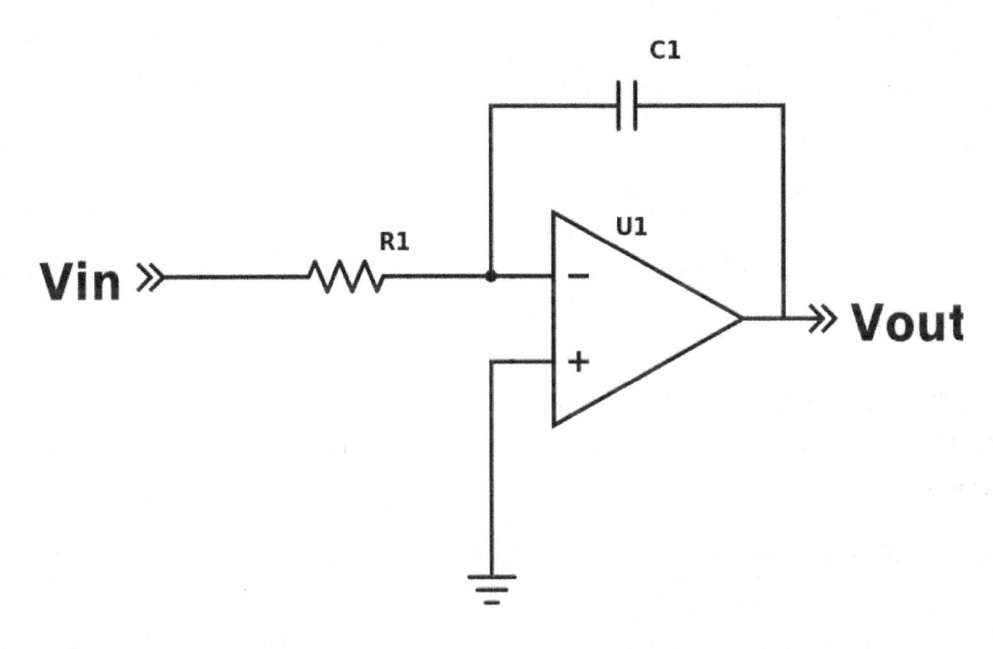

The term integrator sounds very mathematical and complicated - in fact it performs the function of integration. However, all we need to know is that it performs just like the RC lowpass filter we had discussed in chapter 5, which also performs (imperfect) integrator function. This circuit has the benefit of typical op-amp circuits, a low output impedance that isolates the output, and a virtual ground that isolates the input. Thus, this can be cascaded easily (and that's what we'll do in the next section), while in comparison passive filters can't be cascaded easily. That explains why passive filters are used in very simple applications not requiring cascading, while active filters are used in more complex situations where cascading is necessary.

If this is an integrator, one may suspect that by reversing the R and C, we would get a differentiator. That is correct: and the behavior is like that of a RC highpass filter also. However, we shall not dwell on such circuits.

Let us now make a quick trip to our virtual lab…

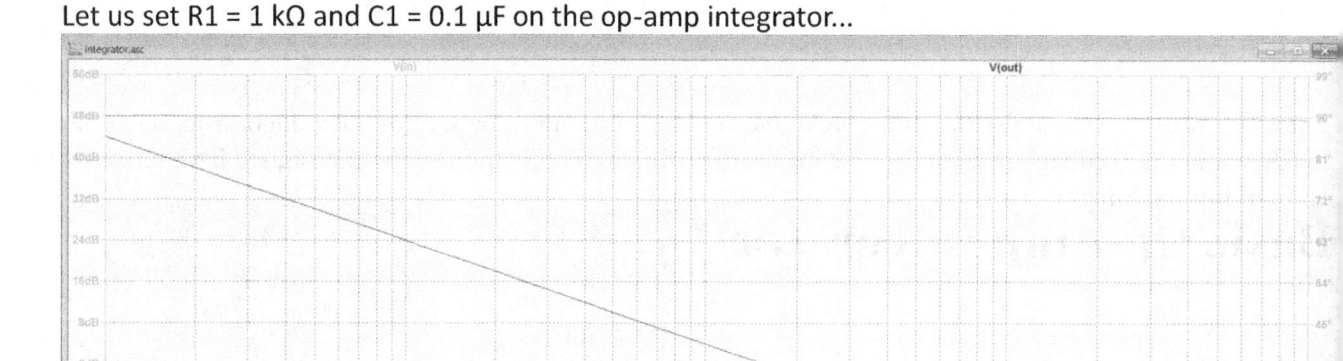

Let us set R1 = 1 kΩ and C1 = 0.1 µF on the op-amp integrator...

Note that:
1) it has a perfect rolloff of -20 dB/decade (measure it on the chart to be sure!)
2) the 0 dB crossover point (indicated by the crossing of the rolloff and the 0 dB line) is 1.6 kHz; this case is for R1 = 1 kΩ and C1 = 0.1 µF

Let us change the RC values and see what happens. Voila, C1 = 0.01 µF...

Note that:
1) it has a perfect rolloff of -20 dB/decade (measure it on the chart to be sure!)
2) the 0 dB crossover point is now 16 kHz; this case is for R1 = 1 kΩ and C1 = 0.01 µF

Can you figure out a potential relationship (or formula, which represents relationships) for the 0 dB crossover point?

Let us do one more experiment, this time changing the resistor value to 10 kΩ and capacitor back to 0.1 µF...

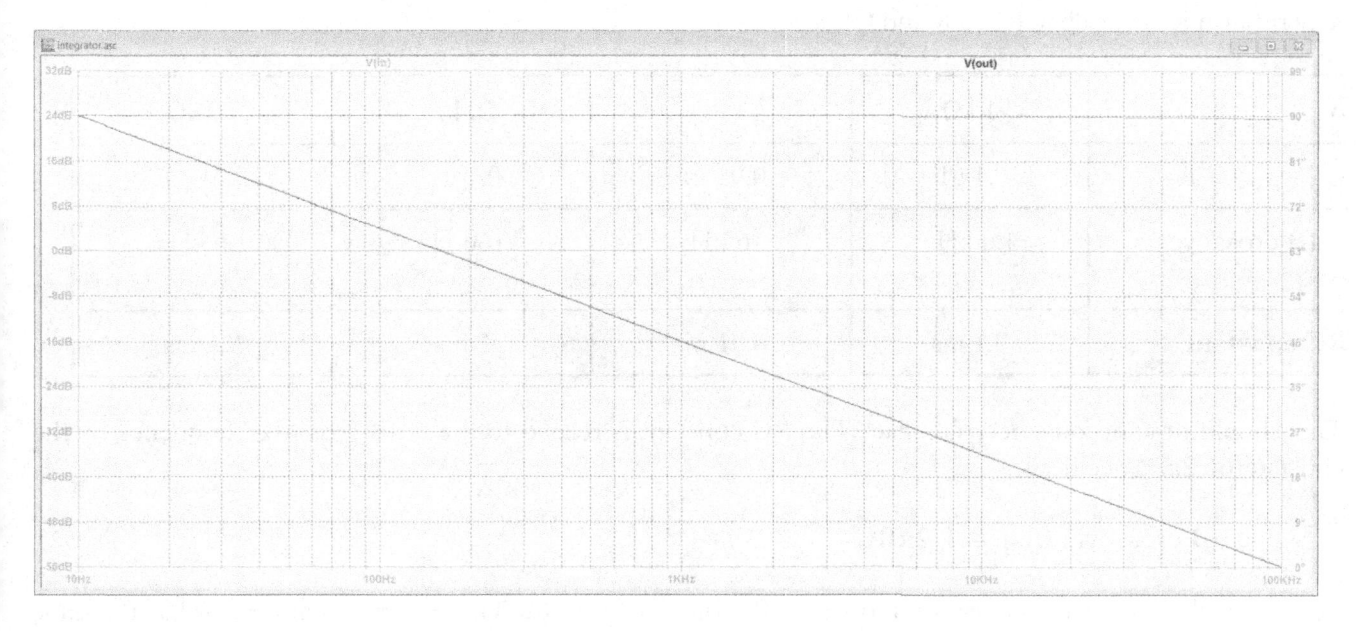

Note that:
1) it has a perfect rolloff of -20 dB/decade (measure it on the chart to be sure!)
2) the 0 dB crossover point is now 160 Hz; this case is for R1 = 10 kΩ and C1 = 0.1 µF

Just to round out the experiment, let us let have R1 = 10 kΩ and C1 = 0.01 µF...

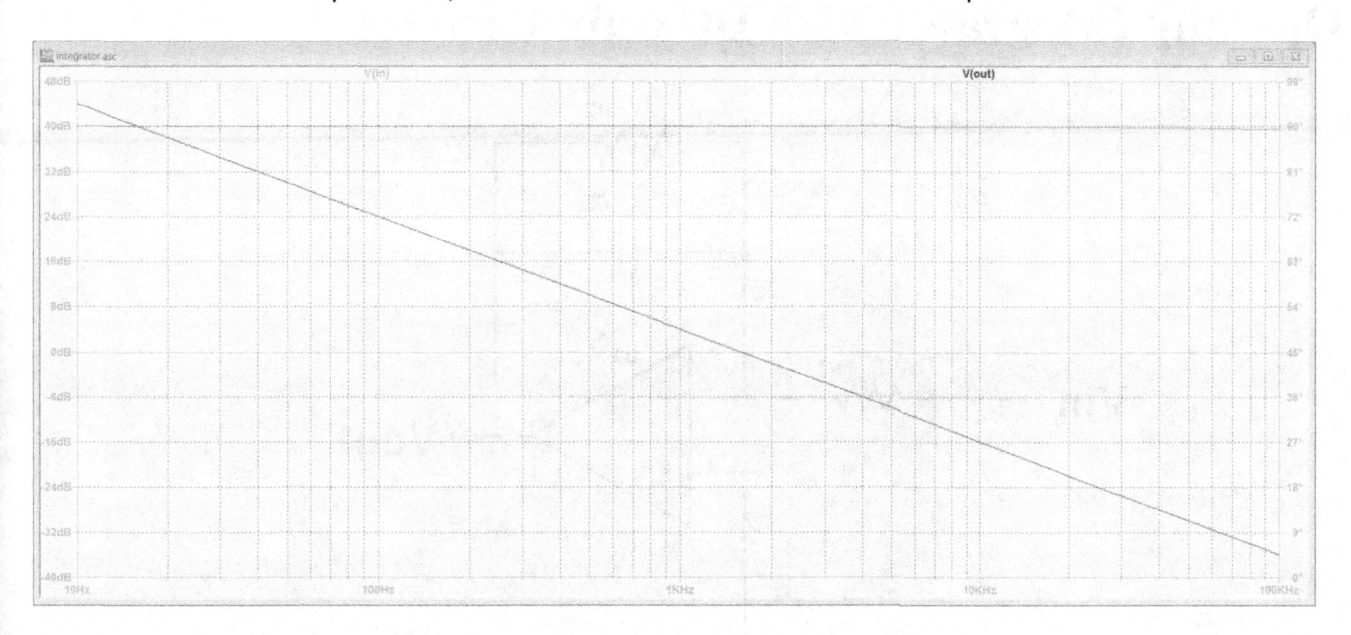

Note that:
1) it has a perfect rolloff of -20 dB/decade (measure it on the chart to be sure!)
2) the 0 dB crossover point is now back at 1.6 kHz; this case is for R1 = 10 kΩ and C1 = 0.01 µF

7-21

Let us summarize our findings in a table; I have added a row of RC constant (which is R1 times C1 as you may recall, and I'm calling R1 as R and C1 as C):

R	1 kΩ	1 kΩ	10 kΩ	10 kΩ
C	0.1 μF	0.01 μF	0.1 μF	0.01 μF
0 dB crossover freq.	1.6 kHz	16 kHz	160 Hz	1.6 kHz
RC constant	0.1 ms	0.01 ms	1 ms	0.1 ms

This should now become very familiar. The RC constant is related to the 0 dB crossover frequency by the constant of 2π. Specifically,

0 dB crossover freq. = $1/2\pi RC$

Recall that the corner frequency for a low pass RC filter is $1/2\pi RC$. The op-amp as integrator has the same rollover slope, and instead of a *corner* frequency, the $1/2\pi RC$ formula predicts the 0 dB *crossover* frequency instead.

Op-Amp Integrator with DC Gain Control

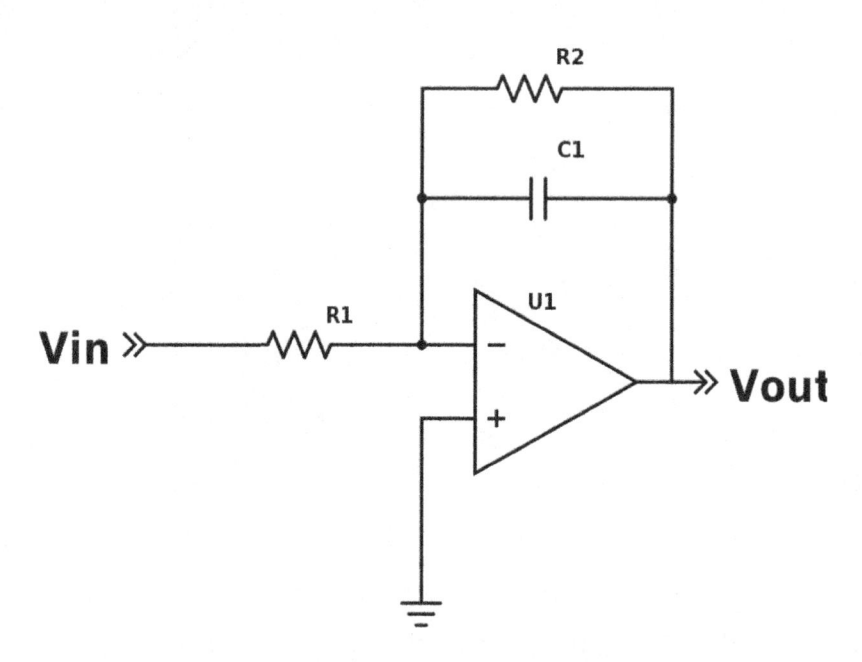

With the addition of one resistor R2 parallel to the capacitor, we gain DC control.

Study its response carefully: it is one of the simplest **active filter**:

[Bode plot showing DC gain = -R2/R1, -20 dB/decade rolloff, (-3 dB) corner frequency = 1/2πR2C1, (0 dB) cross-over frequency = 1/2πR1C1]

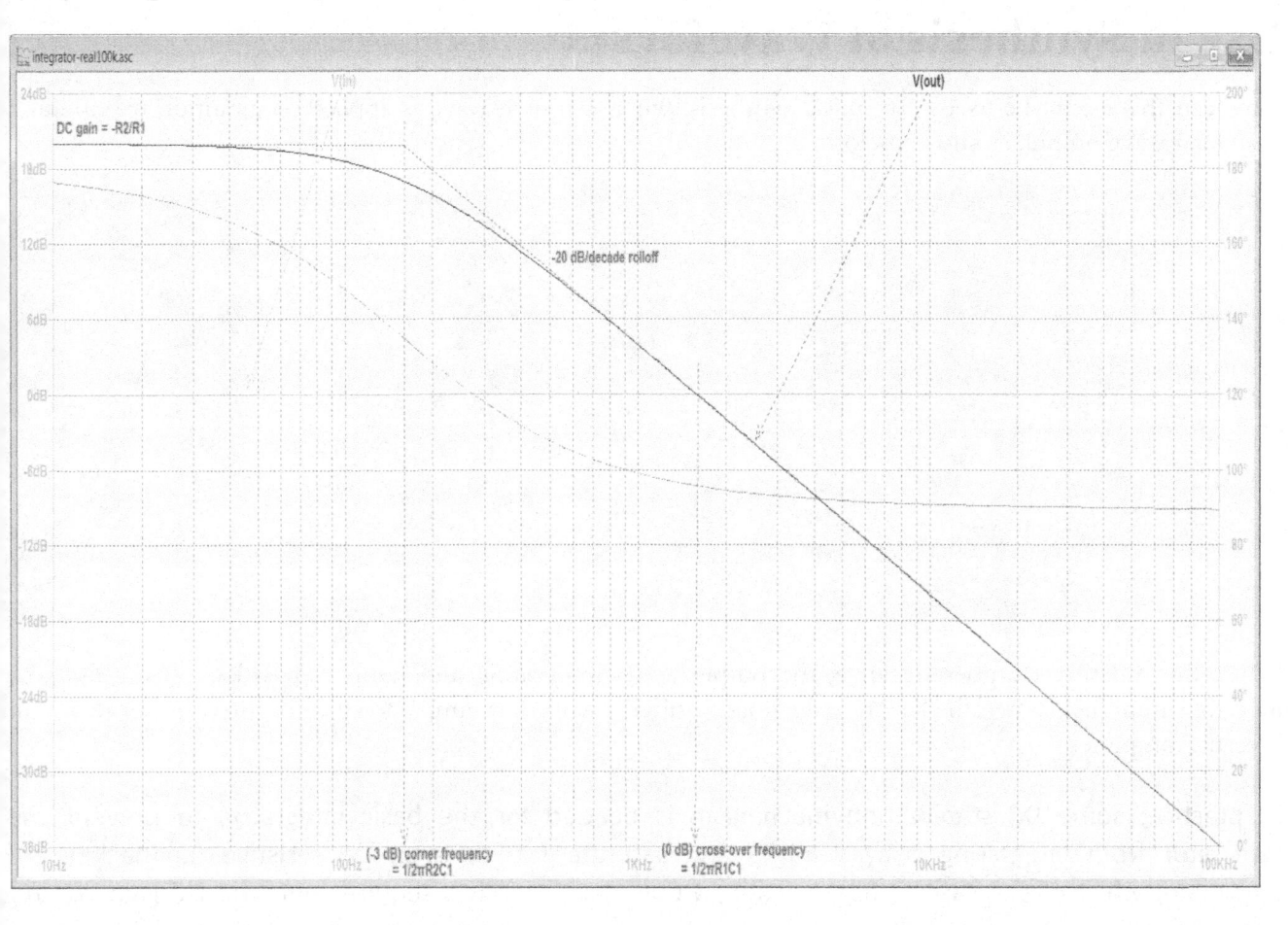

Its response is *exactly* like that of the basic RC lowpass filter (except of course it behaves more ideally).
The **DC gain = - R2/R1**.
The -3 dB **corner frequency = 1/2πR2C1**, controlled by the RC constant of the two parallel elements.
The **rolloff is -20 dB/decade**, same as before.
The 0 dB **crossover frequency = 1/2πR1C1**, controlled by the RC constant of the two series elements, is the only additional element.

Do not get *corner* and *crossover* frequencies confused! (Haha, they are both *critical* frequencies.)
Exercise: imagine the response when R2 becomes infinitely large (it degenerates into the basic op-amp integrator).

Example
Calculate the Bode plot parameters for the op-amp integrator with DC gain control when R1=10 kΩ, C1=0.01 µF, R2=100 kΩ.

DC Gain = -R2/R1 = - 100k/10k = -10, the negative indicating inverting output.
DC Gain in dB = 20 log (10/10) = 20 dB.
Corner frequency = 1/2πR2C1 = 1/2π·100k·0.01µ = 0.159 kHz = 159 Hz
Crossover frequency = 1/2πR1C1 = 1/2π·10k·0.01µ = 1.59 kHz
There is a -20 dB rollover slope.
The plot is exactly the one shown earlier.

Use in Synthesis of Waveforms

How can this circuit be used? In music synthesis, when a square wave is applied to its input, we obtain a triangular wave output, as shown below:

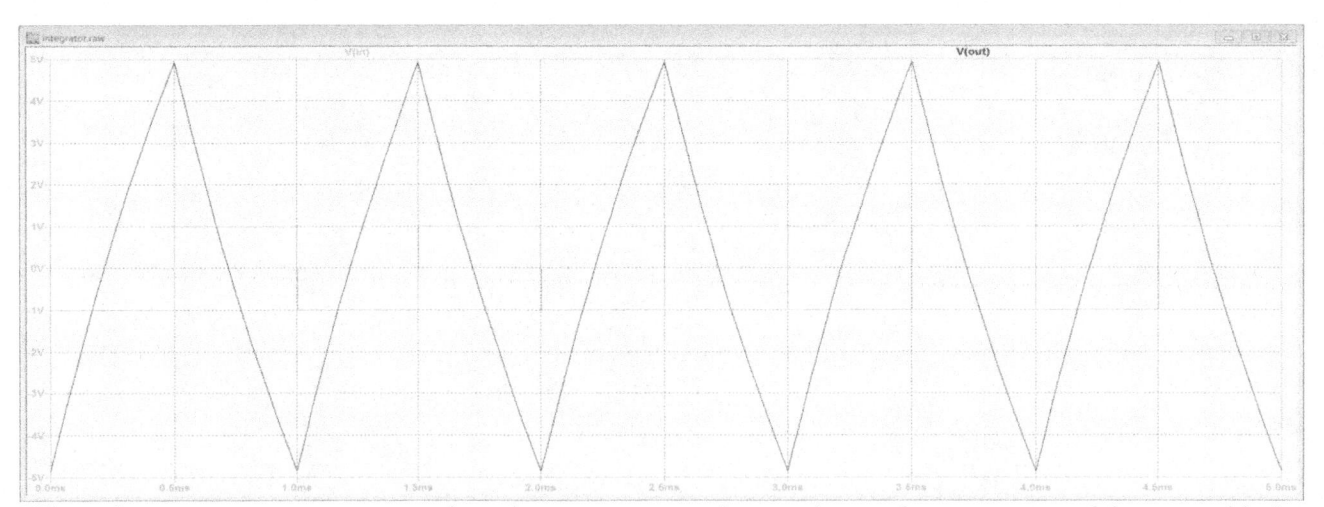

When the square wave goes negative, the output starts integrating and ramps upward because this is an inverting amplifier. When the square wave goes positive, the output ramps downward. (Remember this is an inverting amplifier.)

In practice, some DC stabilization mechanism is needed for the basic integrator, to prevent the integrator from integrating stray voltages and saturate the op-amp. A resistive feedback (from anywhere, which is the case in next section) would satisfy such a requirement; the DC gain control above happens to provide such also. The above waveform was generated from this modified circuit, with R1=5 kΩ, C1=0.01 µF, R2=100 kΩ. The bigger R2 is, the straighter would be the edge of the triangular wave, but also the bigger DC gain, which would thus rob bandwidth response from the gain bandwidth product. Plus, a bigger DC gain may overload and saturate the op-amp in practice. (In this case it peaked at 5V which would still be OK with a +/-15V power supply; but if the gain is just 4 times higher it would saturate and overload the op amp.)

Another application of the op-amp integrator circuit will be shown in the next section.

7.9 VIRTUAL LAB: STATE VARIABLE ACTIVE FILTER SYNTHESIS

Let us visit our virtual lab one more time; this time we want to find out what op-amps can do for us in active filters. There are many kinds and categories of active filters; a very powerful and general-purpose technique is presented here. This method can be used to synthesize low-pass, high-pass, bandpass, and band-reject filters - with the same circuit! In fact, some mixer boards use this topology for their parametric equalizers...

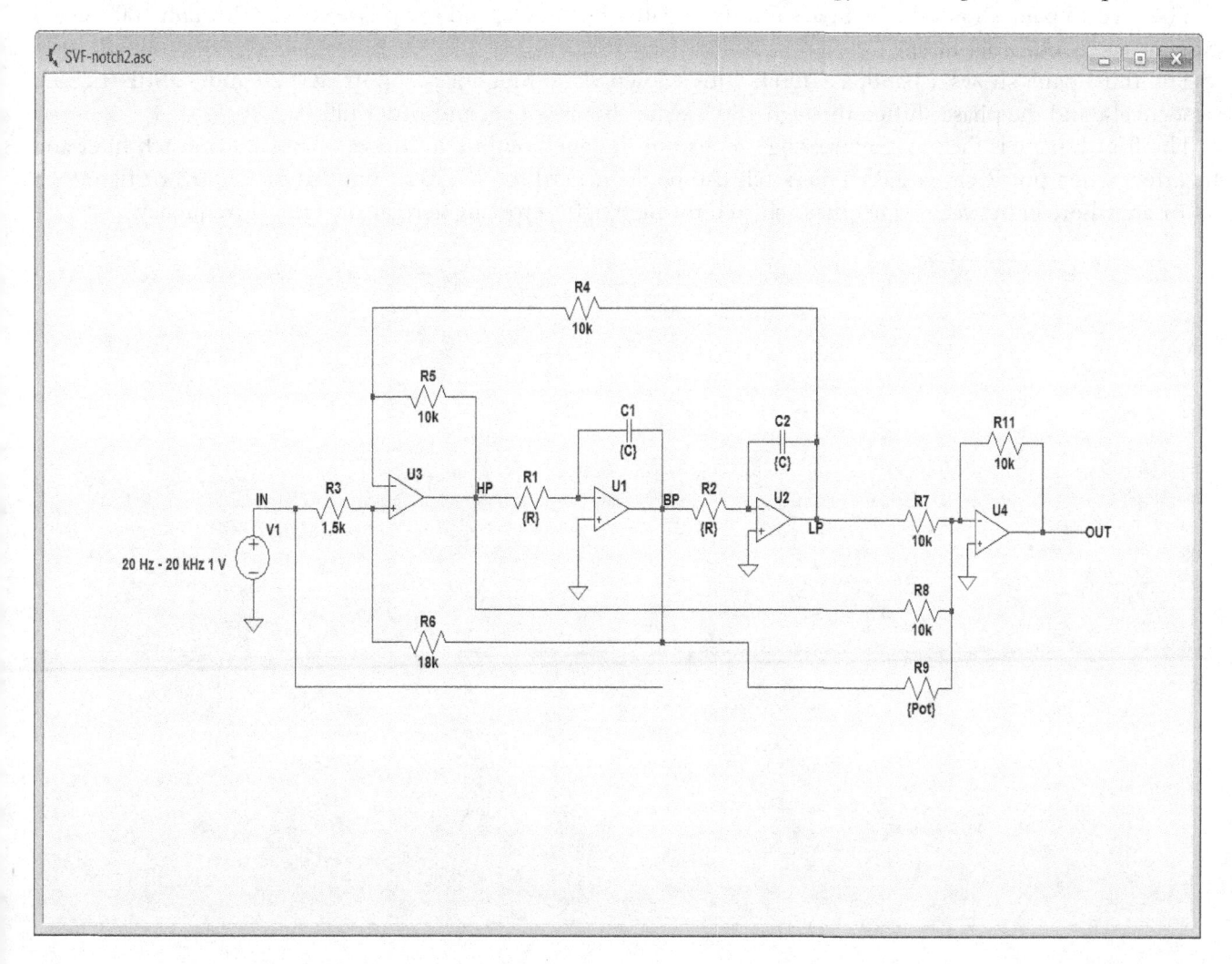

This circuit uses four op-amps which can be conveniently available in *one* package. U1 and U2 are the integrators, producing state variables BP (for BandPass) and LP (for LowPass). U3 is a mixer which combines the input with feedback from LP and BP, producing HP (for HighPass). U4 is an audio mixer combining LP with BP and either HP or the input signal through a pot (variable potentiometer acting as variable gain).

The line from input going nowhere is *not* a bug, but rather the engineer leaving the schematic so that it is easily switched by editing to show that R9 can be connected to either BP or input.

By varying R1 and R2 (on U1 and U2 respectively), the critical frequency can be adjusted. (Alternatively, C1 and C2 can be adjusted for different critical frequencies.) R1 and R2 should be ganged together so that changes will track.

Critical Frequency at 1 kHz

Right off the bat, one may guess that this circuit can produce LowPass, HighPass, or BandPass outputs from the input signal. This can be seen in the next chart, where the first three panes indeed show the three outputs, with the critical frequency set to 1 kHz.

The first pane shows the lowpass rolloff at -40 dB/decade, and the phase shifted through 180°, showing that it is a second-order filter, as the two state variables in U1 and U2 would indicate.

The second pane shows the highpass rolloff at +40 dB/decade, and the phase shifted through 180°, again showing a second-order filter.

The third pane shows a bandpass filter, with a lowpass and highpass rolloff at +20 and -20 dB/decade respectively, and the phase shifted through 180°, again showing a second-order filter.

The fourth pane is the most interesting - a custom-designed output, in this case showing a notch filter and the effect when pot R9 is varied. The notch can be downward to -10 dB or upward to +20 dB or flat at +5 dB, or anywhere in between! The phase shifted through 360°, crossing zero at the critical frequency.

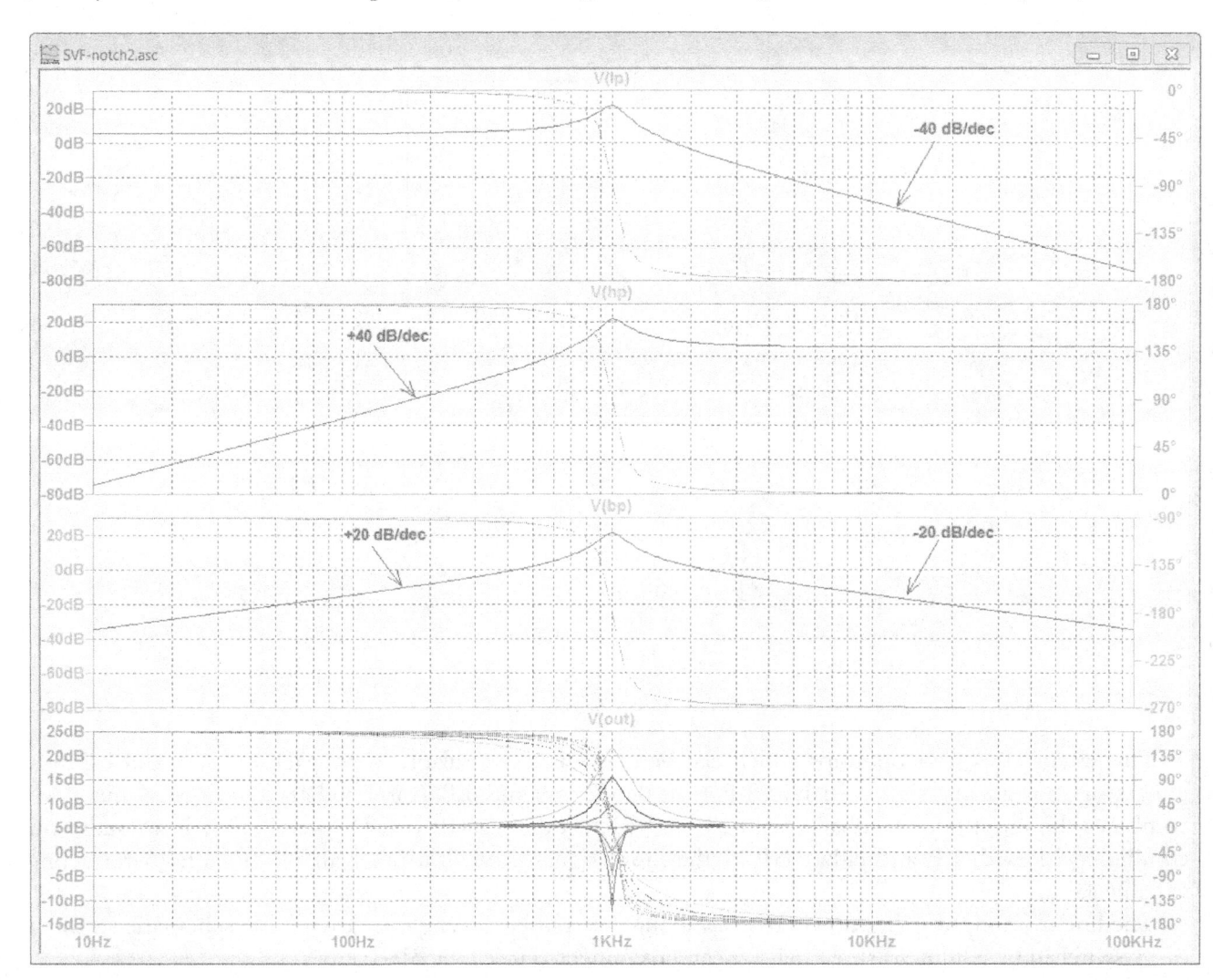

Critical Frequency at 80 Hz

Let us now change R1 and R2 so that the critical frequency is at 80 Hz.

The first pane shows the lowpass rolloff at -40 dB/decade, and the phase shifted through 180°, showing that it is a second-order filter, as the two state variables in U1 and U2 would indicate.

The second pane shows the highpass rolloff at +40 dB/decade, and the phase shifted through 180°, again showing a second-order filter.

The third pane shows a bandpass filter, with a lowpass and highpass rolloff at +20 and -20 dB/decade respectively, and the phase shifted through 180°, again showing a second-order filter.

The fourth pane is the most interesting - a custom-designed output, in this case showing a notch filter and the effect when pot R9 is varied. The notch can be downward to -25 dB or upward to +25 dB or flat at +5 dB, or anywhere in between! The phase shifted through 360°, crossing zero at the critical frequency.

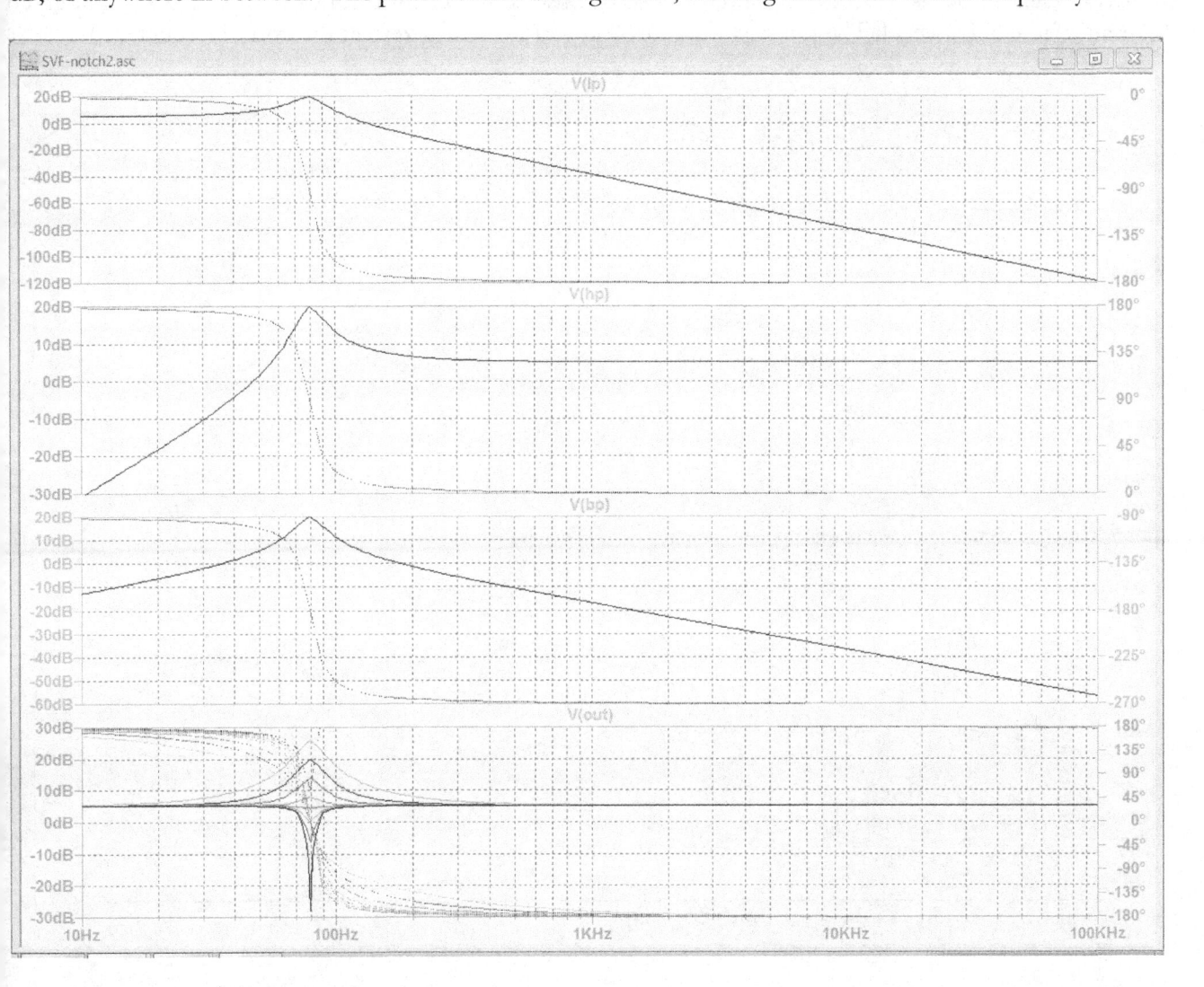

This is useful for mitigating against room modes in audio recording.

Critical Frequency at 6.2 kHz

Let us now change R1 and R2 so that the critical frequency is at 6.2 Hz.

The first pane shows the lowpass rolloff at -40 dB/decade, and the phase shifted through 180°, showing that it is a second-order filter, as the two state variables in U1 and U2 would indicate.

The second pane shows the highpass rolloff at +40 dB/decade, and the phase shifted through 180°, again showing a second-order filter.

The third pane shows a bandpass filter, with a lowpass and highpass rolloff at +20 and -20 dB/decade respectively, and the phase shifted through 180°, again showing a second-order filter.

The fourth pane is the most interesting - a custom-designed output, in this case showing a notch filter and the effect when pot R9 is varied. The notch can be downward to -15 dB or upward to +25 dB or flat at +6 dB, or anywhere in between! The phase shifted through 360°, crossing zero at the critical frequency.

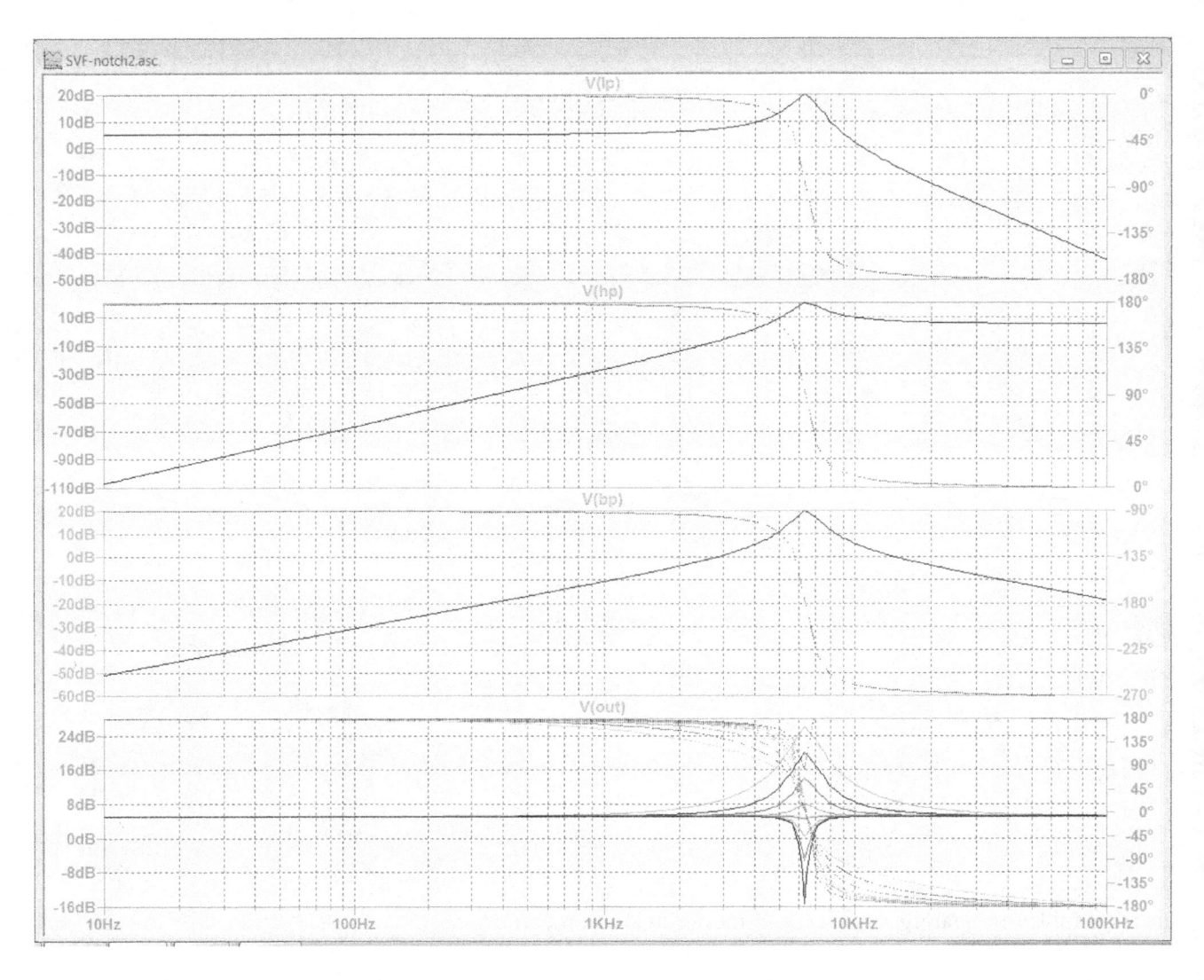

This is useful for mitigating against excessive sibilance in audio recording.

The take-away is that by suitably mixing lowpass, highpass, and bandpass signals (together with the input) and changing the mix ratios, one can custom-build general-purpose second-order filters at will - in this example we built notch filters which are very useful in audio recording to correct problems like sibilance and room modes.

7.10 Do You Want to Dance? Op-Amps as Gyrators

Gyrators are **impedance converters**: capacitors are "gyrated" into inductors, and vice versa. **Gyrators provide inductance without magnetism**. Op-amps can implement gyrators. But unlike real inductors, such implementations can absorb only small amounts of current.

There are at least two implementations of single op-amp gyrators and one common one using dual op-amps. The single op-amp gyrators suffer from thermal resistor noise. But if we lower the resistor values, then it will make the "parasitics" large for the emulated inductor. So, we can't win with single op-amp gyrators. (Gyrators can also be implemented with OTAs, which we'll discuss later in the chapter.) All of these implementations require the inductance to be grounded. To make a floating inductance, more op-amps are needed... :(

The Simplest Gyrator

For simplicity, we shall use a single op-amp implementation for illustrative purposes. R, C, and an op-amp - that's all it took was three components as shown within the dotted lines. Remember this will be a grounded inductance being emulated, with the equivalent circuit shown on the right side.

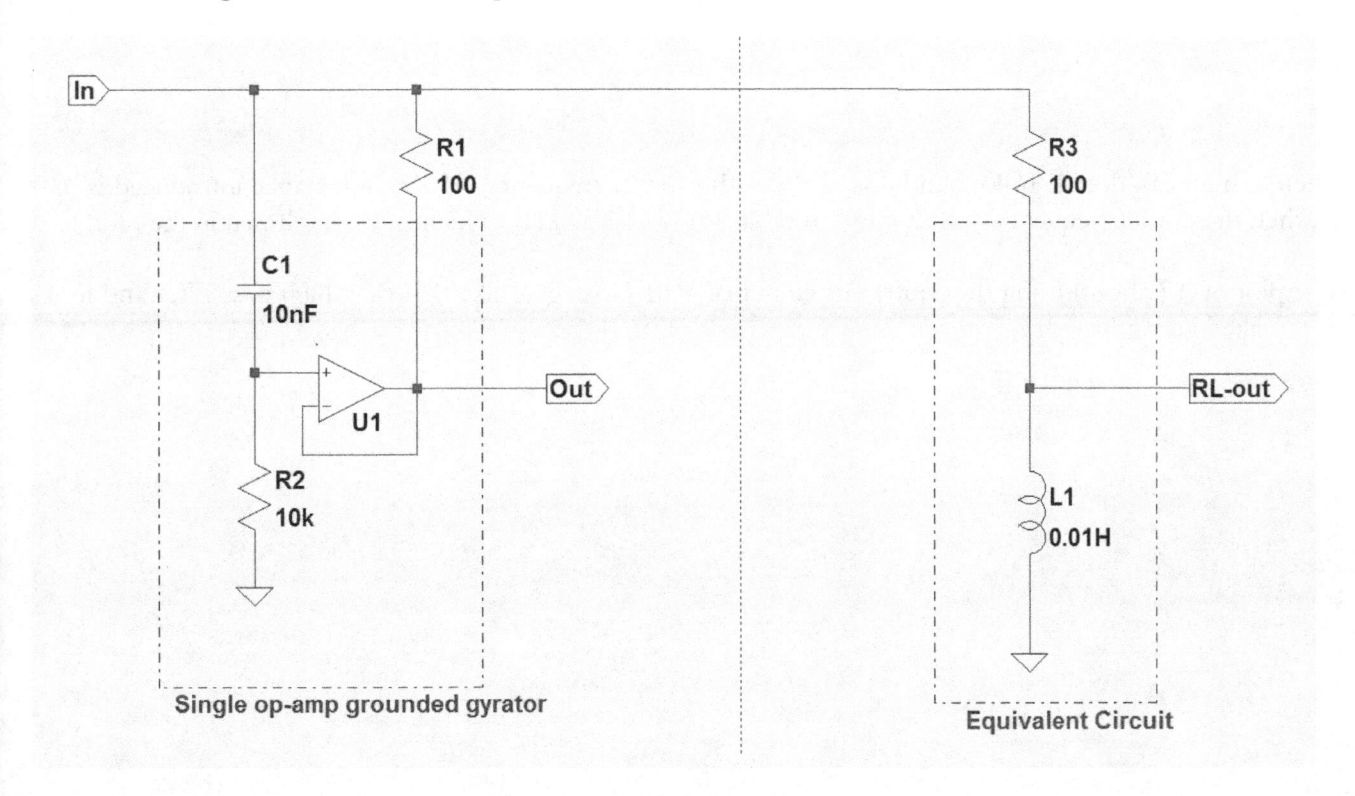

Single op-amp grounded gyrator | *Equivalent Circuit*

The equivalent inductance is given simply by **L1 = R1 · R2 · C1**
Thus L1 = 100 · 10k · 10 n = 0.01 H

Note that these component values are probably the minimum recommended (the op-amp tend to misbehave on smaller values). Yet the gyrator still produces 0.01 H which is a comparatively large inductance, which would have been expensive and large and heavy were it a real inductor. By increasing the component values, we can easily obtain inductances of a couple of Henries (which would be "impossible" to obtain from real inductors, and if we can make it, its serial resistance would have been about 100 ohms which would render it useless in a filter).

The plot of V(out) using the gyrator implementation is just like that of a high pass filter:

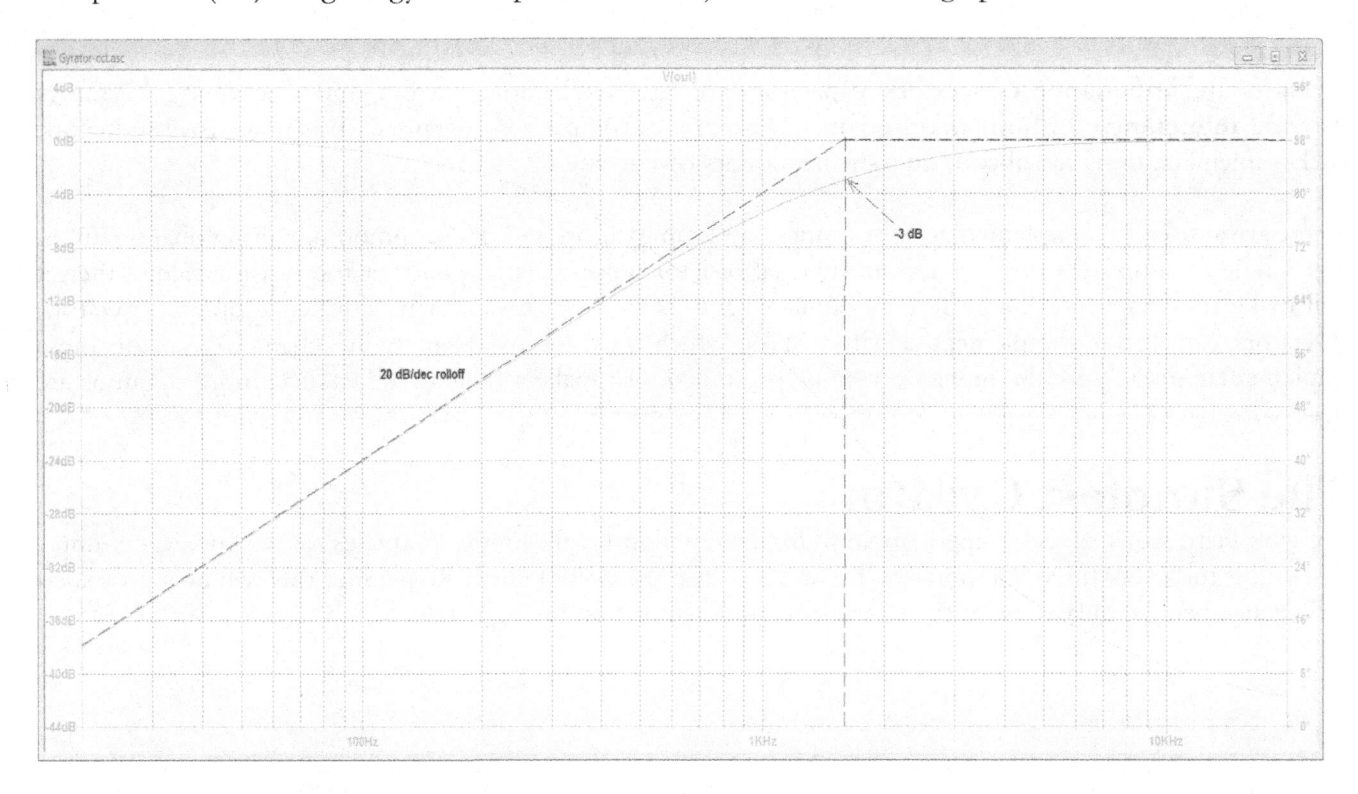

There is a 20 dB/decade rolloff, and it is -3 dB at the corner frequency. If the inductance introduced is 0.01 H, then the corner frequency = R/2 π L = 100/2π * 0.01 = 1.59 kHz, which is just as shown in the plot.

The plot of V(RLout) using the equivalent circuit of 0.01 H is again like that of a high pass filter and looks *exactly* like V(out):

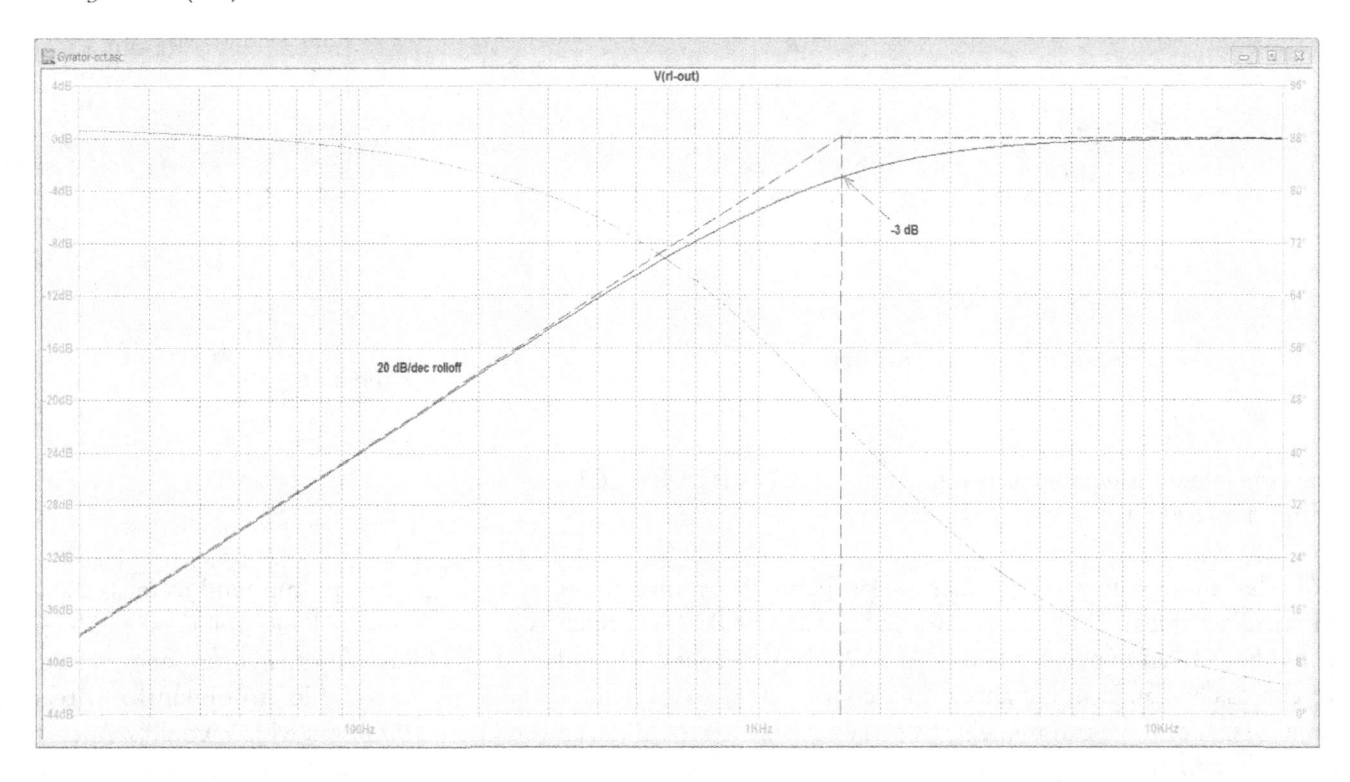

The outputs from the gyrator and the equivalent RL circuit looks identical. How can we quantify how different they are? We can simply ask for the plot of V(out)-V(RLout): if the difference is small, then they are essentially identical…

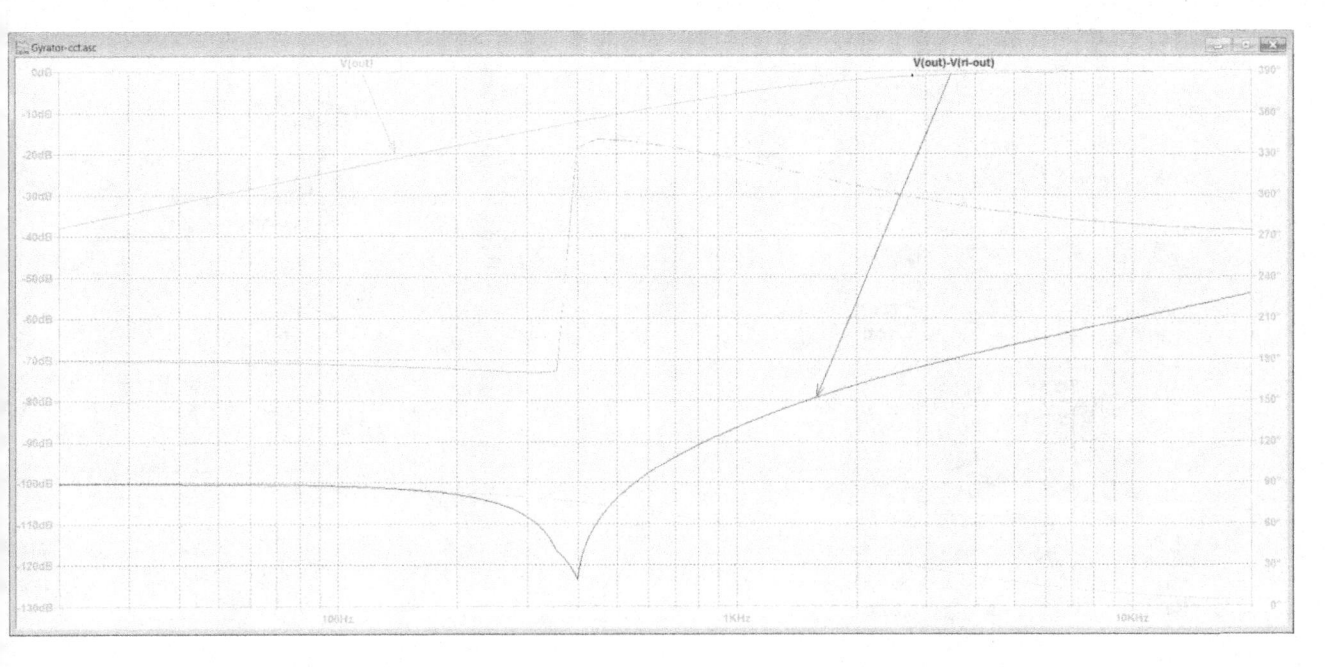

The shape of V(out) looks different, but only because the scale has changed; its value at 20 Hz is still around -38 dB. But notice the difference signal: it is -100 dB at low frequencies - which means the difference is negligible. The difference signal increases as frequencies increase past 400 Hz, but is still small (less than -50 dB) even at 20 kHz.

It turns out that gyrators are not suitable at high frequencies like radio frequencies anyway, but it is OK for audio frequencies. Use polypropylene or polystyrene capacitors where available; if not (as they are often difficult to get), then the more readily available (and cheapest) polyester/mylar film capacitors are OK for audio use.

Again, notice the minimal component count achieved when the single op-amp gyrator is usable (and remember again the gyrator is not usable in all situations because of its limitations). That makes it favorable for DIY community. Unfortunately, that configuration is typically only used in entry-level pro or consumer grade audio electronics, specifically because of that attribute. Many graphic equalizers use this configuration for simplicity, and even a few parametric equalizers do the same.

Which leads us to our next example, an RLC series resonant circuit… wait, although conventionally we always say RLC in that order, let us call it RCL circuit to emphasize that the L must be grounded. In other words, we can't produce an RLC-equivalent gyrator circuit using the simple gyrator.

Gyrator RCL Circuit

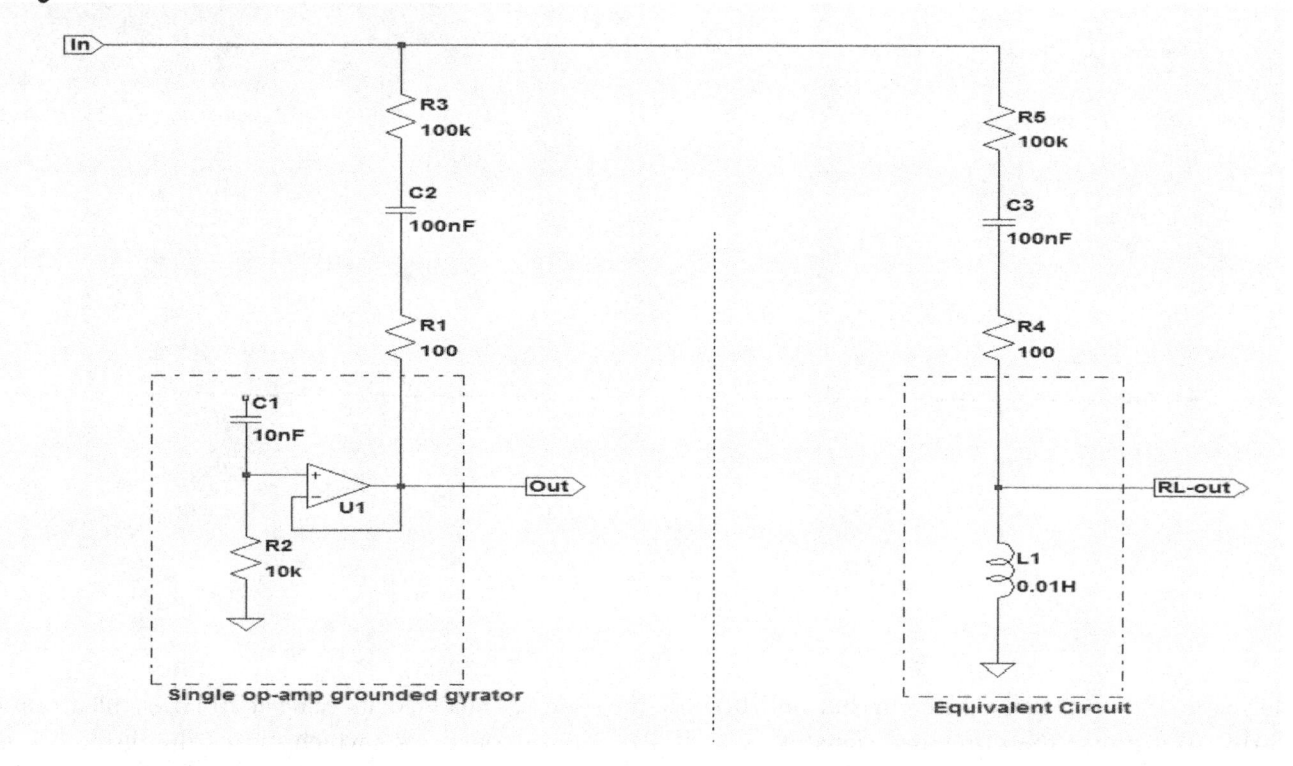

Single op-amp grounded gyrator

Equivalent Circuit

An even more accurate summary of this circuit is that it emulates an R-CRL network; as shown on the right side and reading down, it is R, with the output tab, then C, R, and finally L grounded. Note that we *cannot* lump the second R with the first, because it is on a different (and wrong) side of the output. This second R emulates the serial resistance of a large 1 H inductor, meaning it is an *imperfect* RCL network.

When R2 is varied from 10k to 100k, the following plot results; it is a "sweepable" notch filter:

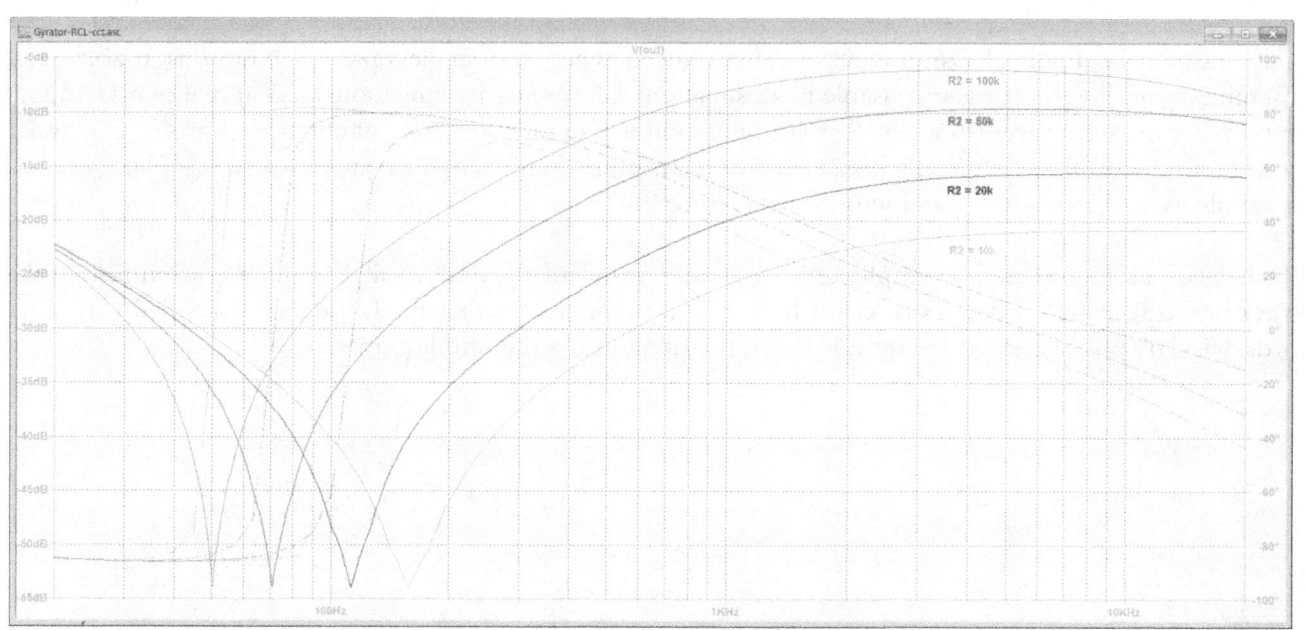

Unfortunately, when the critical frequency changes, so does the Q and the gain - that is the typical issue with passive network-based filters. Quality factor Q will be discussed 5 pages later.

Chapter 7 – Op-Amps/Active filters

♪ 7.11 Practical Circuits: Crossovers, Parametric Equalizers, Tone Controls

We have kept this chapter as simple as possible; likewise, the circuits were presented with fewest components possible. In reality, one may find practical circuits with a few more components. There may be components used for compensation, for bypass, or for DC path or balance for biasing purposes.

We have waited until this moment, near the end of the book, to present some practical audio shaping circuits used in the industry, specifically in pro audio mixer boards. Why? Because: 1) the basic filter types aren't the typical controls we desire for use in studio or live sound; 2) the controls we desire typically require active circuits like the op-amps shown in this chapter; 3) some of the circuits first appeared *only* in the 1980's (fairly recently in terms of the long history of audio electronics!), and that is because the prevalence and reduced cost and improved performance of op-amps finally made it feasible in a cost-effective manner; 4) as you may guess, the circuits are significantly more complex than the simple circuits we had presented so far...

Specifically, the fully-adjustable parametric equalizer *first* appeared in the 1980's; the now-common bass/treble tone control *first* appeared 30 years earlier.

"Traditional" 3rd Order Two-Way Active Crossover

The traditional **Sallen-Key** unity gain filter has a characteristic T-shaped network (often used in pairs, in specific configurations would be twin-T); it performs well, but have odd component values. It does require few and minimal components for a 3rd order crossover (which provides 3x 20 dB/dec or 6 dB/oct, meaning 18 dB/oct or 60 dB/dec crossover), and thus is shown first:

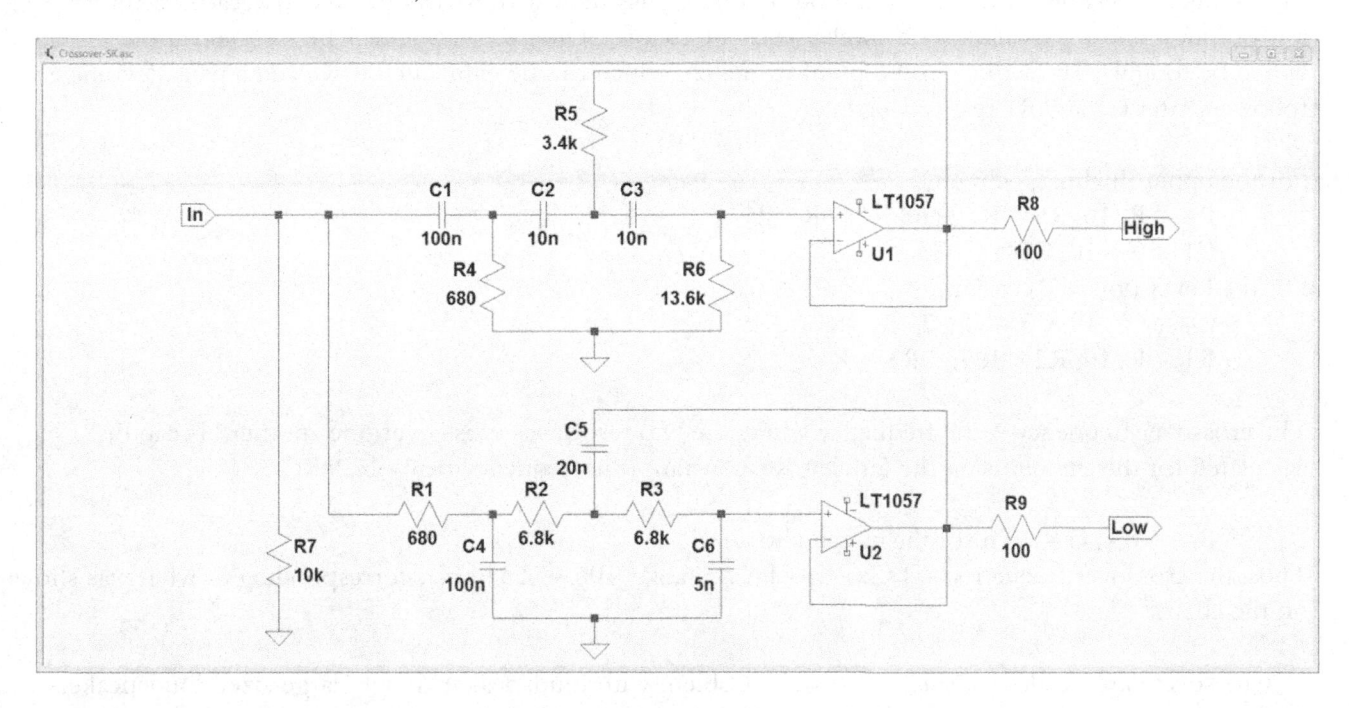

A two-way crossover circuit splits an input signal (In) into two signals (High and Low), to be sent typically to two circuits eventually to a tweeter and a woofer respectively. The op-amps act simply as voltage followers. The upper and lower circuits show some symmetry; in fact, you will see that resistors and capacitors are swapped in the key (tee) part of the circuit. There is feedback from the op-amp output directly into the tees. Because there are three capacitors in each circuit, we can expect 3rd order filter with 60 dB/decade rolloff performance. R7 is there to ensure a DC return path to ground (and can be omitted if there is a return somewhere else).

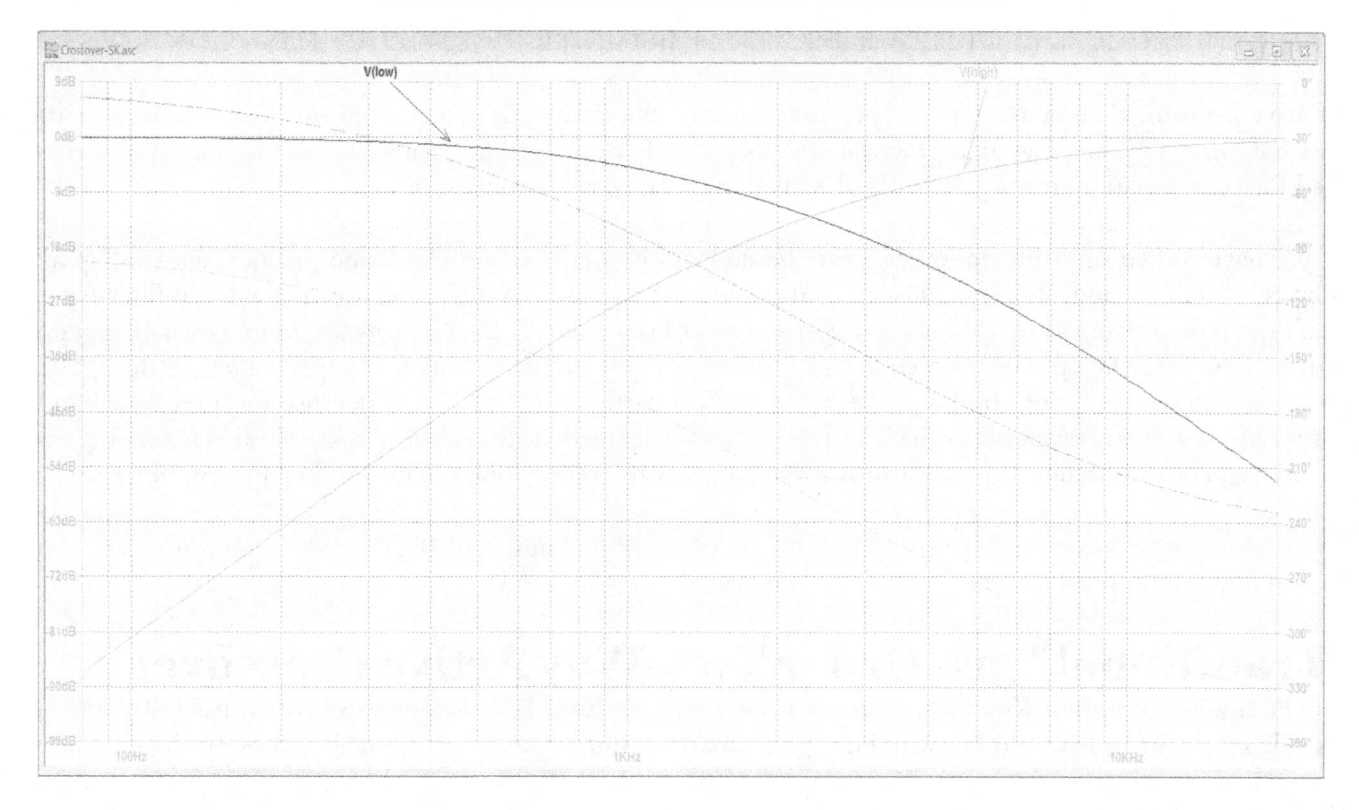

The frequency response for the high output is a high-pass filter with rolloff of 60 dB/decade. Similarly, the low output is a low-pass filter with similar levels of rolloff. The two outputs will be summed at our ears, and would be roughly flat (with a small dip unfortunately, which can be eliminated if we add a pair of voltage followers after C1 and R1 respectively).

For the upper (highpass) circuit,
    R4 = R/10, R5 = R/2 and R6 = R * 2
    C1 = C * 10, C2 = C, C3 = C
For the lower (lowpass) circuit,
    C4 = C * 10, C5 = C * 2, C6 = C / 2
    R1 = R/10, R2 = R and R3 = R

The crossover frequency is the frequency where the two responses cross over one another. It can be calculated for this circuit using the familiar RC constant (the frequency being $1/2\pi RC$).

    R = 6.8 k, C = 10 n for the circuit shown.
Thus, the crossover frequency = $1/2\pi RC = 1/2\pi \cdot 6.8k \cdot 10n$ = 2.34 kHz, corresponding to what was shown on the chart.

Crossovers are needed because of a basic problem with loudspeaker design: large-sized loudspeakers can generate good bass, but not treble; small-sized loudspeakers can generate good treble, but not good bass. By splitting the signal into high and low through the crossover circuits, and then feed the low signals to the large-size loudspeaker, and the high signals to the small-sized loudspeakers, we get the best of both worlds. When the signal is split two ways, the crossover is known as a two-way crossover.

State Variable 3rd Order Two-Way Active Crossover

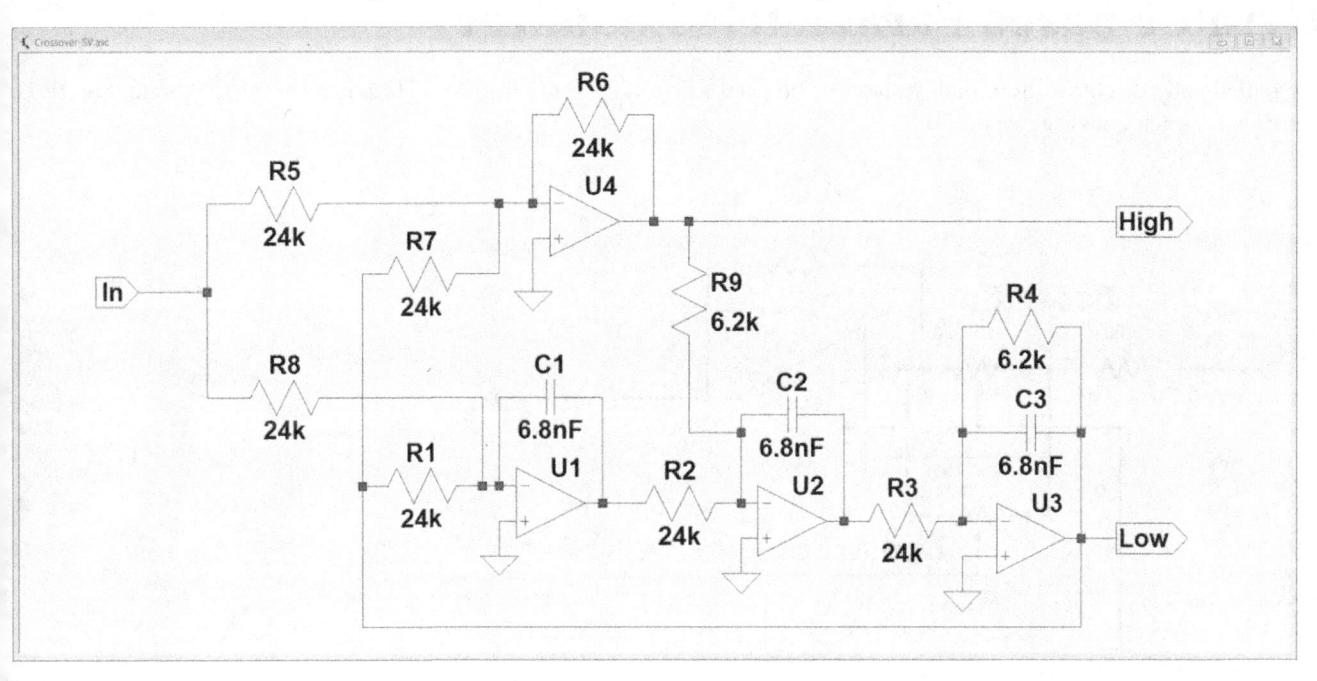

Doesn't this look familiar? Yes, it is like the state variable filter we had seen earlier, except in this case we have three stages of integrators instead of two (op-amps U1, U2, U3), and two outputs instead of one. This shows the flexibility of the state variable synthesis approach - it can be used to create crossovers as well!

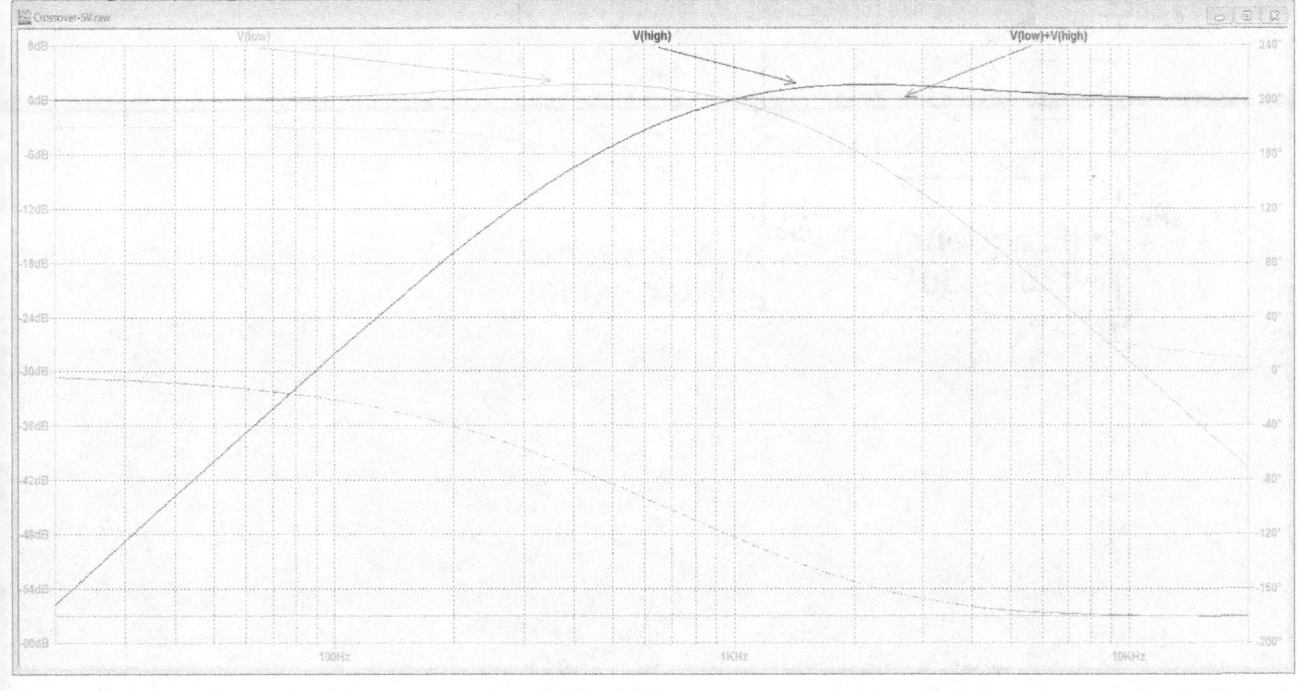

The frequency response can be seen below:

Doesn't this look familiar? Yes, it is very similar to the chart for the traditional crossover. Except in this case the curves are *near perfect*. Notice that the highpass and lowpass outputs are higher than first order, so their response peaks slightly (by 2 dB) before the crossover point. The result is that the sum of V(low) and V(high) is perfectly flat, and it also will be when summed at our ears. No dips like in the traditional crossover.

Gyrator-based Parametric Equalizer

We had already discussed how the gyrator can be used to produce a notch filter. Here is a circuit to provide the three controls for each equalizer section…

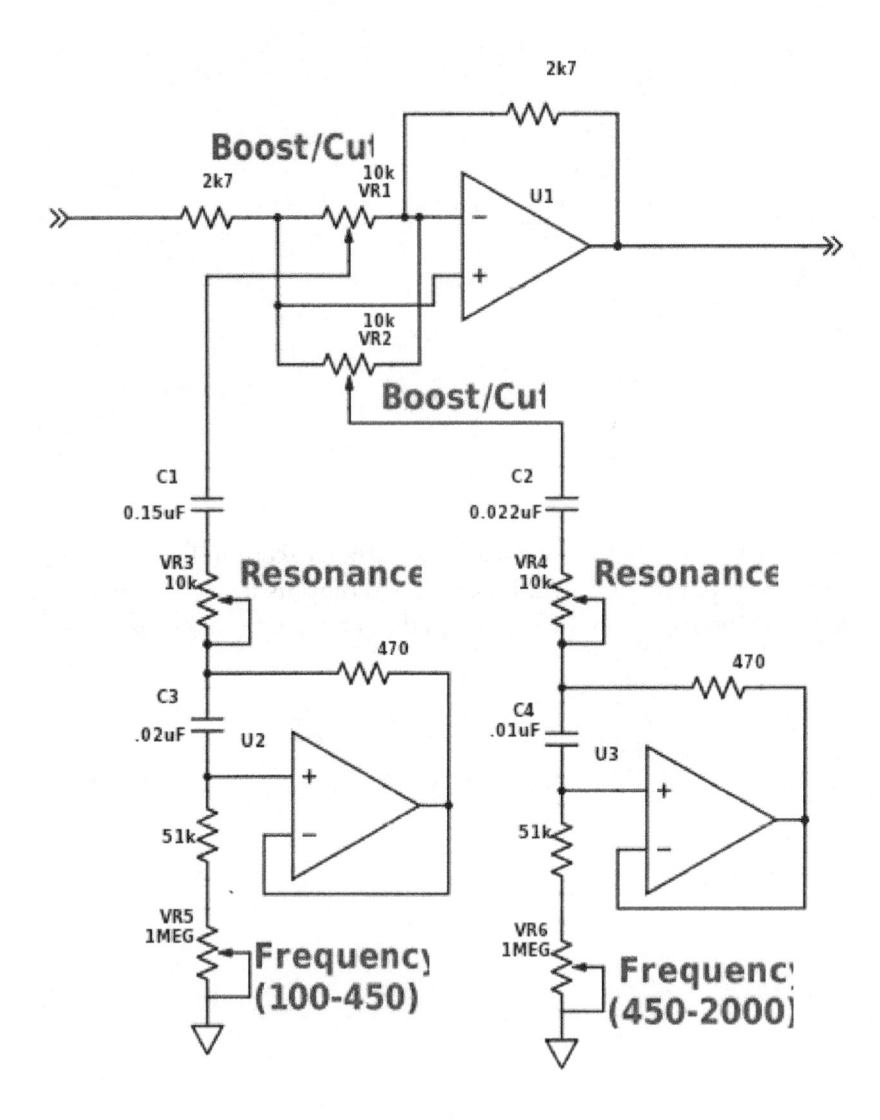

State Variable Parametric Equalizer

As you may expect by now, the parametric equalizer can be implemented using something like the Sallen-Key approach, using twin-T circuits. Or, we can use the state variable synthesis method…

Guess what the *first* implementation of the parametric equalizer was, historically speaking, in the early 1980's?

Well, the first *two* implementations widely known (and one patented) were in fact based on twin-T and state variables. *Both* became reference designs that audio mixer boards since then continued to emulate. (There is a third method, based on the use of op-amps as gyrators, which was also mentioned in one of the original papers.) Which one came "first" was subjected to debate, so we'll simply leave it like that…

The theory had already been presented before when we show you the notch filters, so here it is a matter of simply getting more pots into the circuit so that the frequency, the bandwidth, and the gain can all be independently varied. It will be left for the reader to implement such, based on the above two ways. The major difference is that the notch filter has a higher Q, while the parametric equalizer has a Q that starts at very low and reaches near the lower end of the notch filter. So, in a hurry the parametric equalizer when turned to the highest Q, can perform like that of a notch filter with its low-range Q. It is sufficient to handle most room modes and sibilance problems. More severe room modes need to be handled with a separate notch filter.

> **Quality Factor (Q)**
> Q is a dimensionless parameter that describes the resonance behavior of an underdamped oscillator or resonator. Higher Q means lower rate of energy loss relative to stored energy. Higher Q also means resonating at higher amplitude at the resonant frequency, but have a smaller range of frequencies around which they resonate.
> We shall use the latter approach to calculate Q. First, we find the peak amplitude at the **resonant frequency f_c**. Then, we find the two surrounding frequencies at which the amplitude is -3 dB. The difference between the two surrounding frequencies is the **bandwidth BW**.
> Then, we define $Q = f_c / BW$

Bass/Treble Control

This circuit was first published by Peter Baxandall in 1952. He used a vacuum tube triode as the active device, and referred to the virtual ground provided, just like we now understand it for the op-amp. The passive tone control circuit inside the feedback loop is known as the James Network, and is still often used as tone control guitar pedal, its main advantage as passive circuit being that no power supply is required.

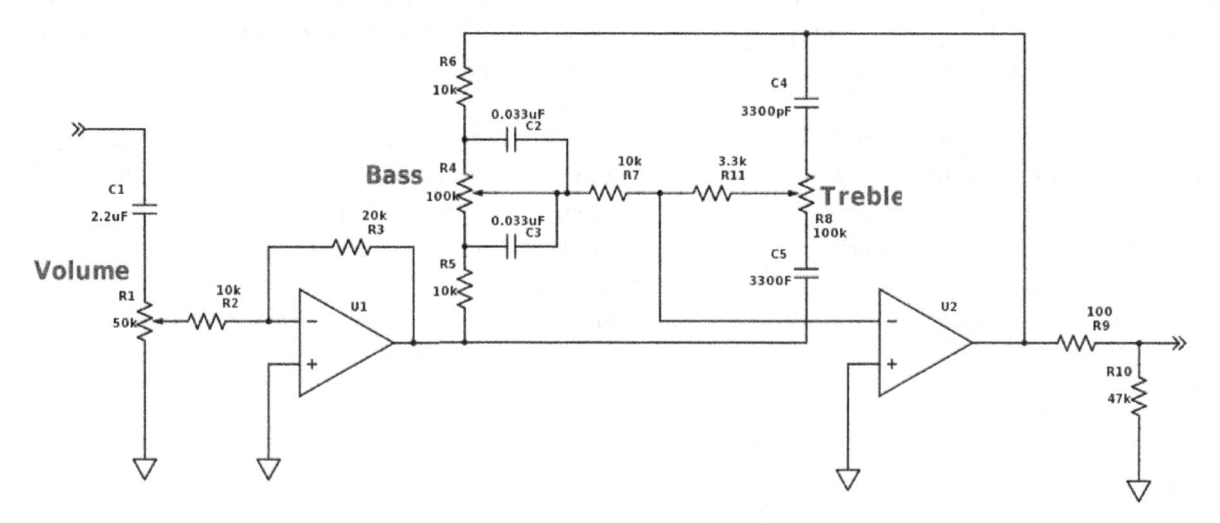

This circuit starts with the volume control (which uses audio taper potentiometer) going into a buffer amplifier providing a gain of 20k/10k = 2 or 6 dB. The tone control circuit is located in the feedback loop of the second op-amp. Mid-range frequencies are not affected by the tone control. When the tone controls are in middle positions, the overall response is flat.

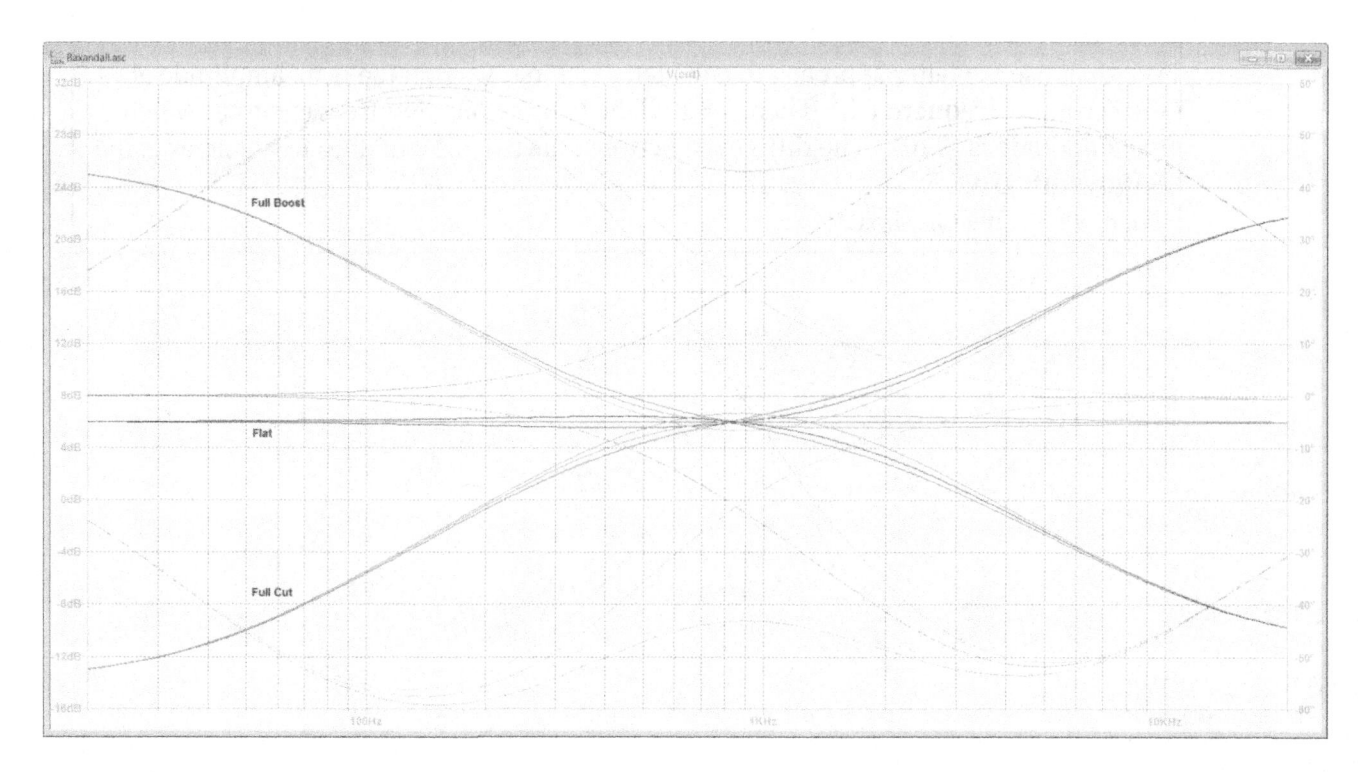

7.12 CLASSIFYING ACTIVE FILTERS

Here's a recap...

Passive filters or networks use R, L, C. In contrast, **an active filter or network uses only R, C, and active elements** like op-amps. Why the discrimination against inductors? Because inductors are physically large, heavy, often expensive, and often dissipates more energy compared with capacitors of similar size. Dissipation (or resistance) in inductors cause problems in filter implementation; in other words, passive filters containing inductors are *not ideal*. But active RC networks can do everything that passive networks can do, plus more. So, let's be active and stamp out inductors where possible... (just kidding)

There are so many kinds and ways of implementing active filters that it is useful to categorize them. Here is a brief approach:
1. **High Gain Single Feedback**: High Gain refers to the ideal op-amp with infinite gain, or probably 60 dB gain or more. Common examples use **bridged-T** or **twin-T** RC networks (which use *more* RC elements than other active filter techniques, making it harder to adjust also) in the op-amp input or feedback network. But it is stable, having a low output impedance, has a summing junction, relatively high gain available, and high Q is possible.
2. **High Gain Multiple-Feedback**: Its multiple feedback means the summing junction is *not* available, nor high gain or high Q. But it is also stable and has low output impedance. The network design procedure is a little complicated, and so is seldom used.
3. **Low Gain Controlled Source**: Low Gain refers to the ideal op-amp in a low gain configuration, 20 dB or less. Common example would be the **Sallen-Key** networks using unity-gain op-amp. It has characteristic that the pass band gain is independent of the element values, and the gain appears in the transfer function equation. So it is typically used when we want to use the gain to help control parameters of the filter, like bandwidth or Q or frequency, say for parametric equalizers or tone controls. It has low output impedance, relatively high gain and high Q is possible.
4. **Negative-Immitance Converter**: Another name is negative-impedance converter (remember immitance is the inverse of impedance), which converts a positive impedance to a negative one, or vice versa. It does *not* have low output impedance (which means cascaded voltage followers are needed for isolation) nor summing junction, and stability is a concern. But high Q is possible, and minimum circuit elements are used, plus Q and frequency are easily adjusted (because of the few circuit elements involved).
5. **Gyrator or related Converter**: A gyrator converts an inductor to a capacitor, or vice versa; it also converts a voltage source to a current source, or vice versa. Frequency-Dependent Negative Impedance would be another conversion mapping technique for implementing active filters. The most common usage is to convert capacitors into inductors when using simple RLC passive network as an initial blueprint. In such usage, the gain would be the same as the equivalent RLC passive network (which will be less than 1). Because it is lossless it should be very stable. And because the emulated inductor is more ideal than real inductors, the result should better than the corresponding RLC passive network, yet using relatively small and low-valued circuit elements which are less expensive.
6. **State Variable Synthesis**: Common example would use typically two (or at most three) stages of integrators with multiple feedback, limiting it to second or third order filter design. We have used many examples of it in this book to illustrate its benefits Just like the high gain multiple-feedback situation, the network design procedure is a little unfamiliar to most engineers, and so unfortunately is seldom used.

Recall that the approach the audio engineer often takes is to visually identify the circuit topology, and then from it use reference books or memory to find out the formula for that topology, and then the behavior can be produced through the language of audio electronics that we had already studied. Here are some further topologies to recognize…

Sallen-Key Topology

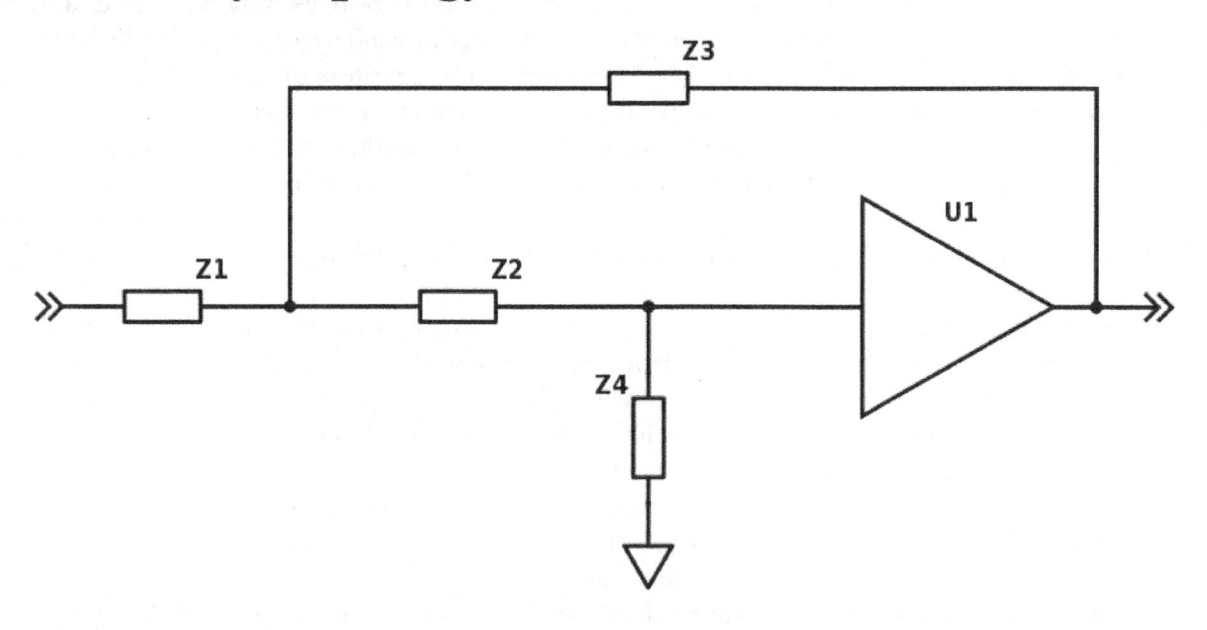

The critical frequency would be:
$1/(2\pi\sqrt{R_1 R_2 C_3 C_4})$ **for low-pass filter** implemented with the corresponding R1, R2, C3, C4;
$1/(2\pi\sqrt{C_1 C_2 R_3 R_4})$ **for high-pass filter** implemented with the corresponding C1, C2, R3, R4.
Easy enough? This was introduced in 1955 by R. P. Sallen and E. L. Key of MIT Lincoln Laboratory, long time ago in terms of audio electronics years.
Note again that the amplifier U1 has to be either unity-gain or have a low gain, and a differential input is not even required.

Bridged-T Topology

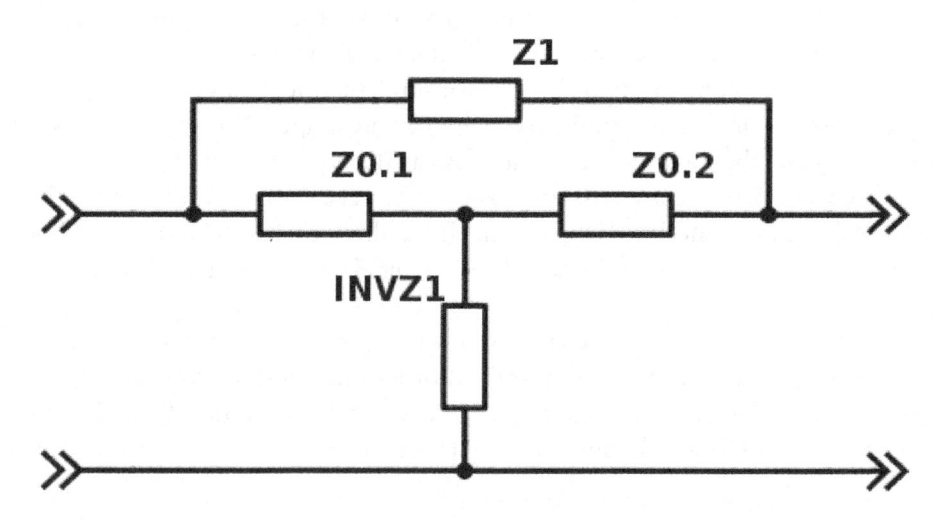

Note that Z0.1 and Z0.2 have equal values, say Z0. On the other hand, Z1 and INVZ1 are inverses, meaning:
INVZ1 = 1/Z1.
Then simply, **transfer function = Z0/(Z1+Z0)**

Although the transfer function is simple, it is no simpler than the voltage divider. In fact, it is *exactly* the same! (Hint: that should make it easy to remember!) So why another topology? This one is unique in that the input and output impedances are the same! (*Not* the case for the voltage divider.) In fact, the impedance at both input and output are simply Z0.

Because of the analogy with voltage divider, we already know how to create the basic low-pass and high-pass filters with this configuration, although of course more variations are possible.

Twin-T Topology

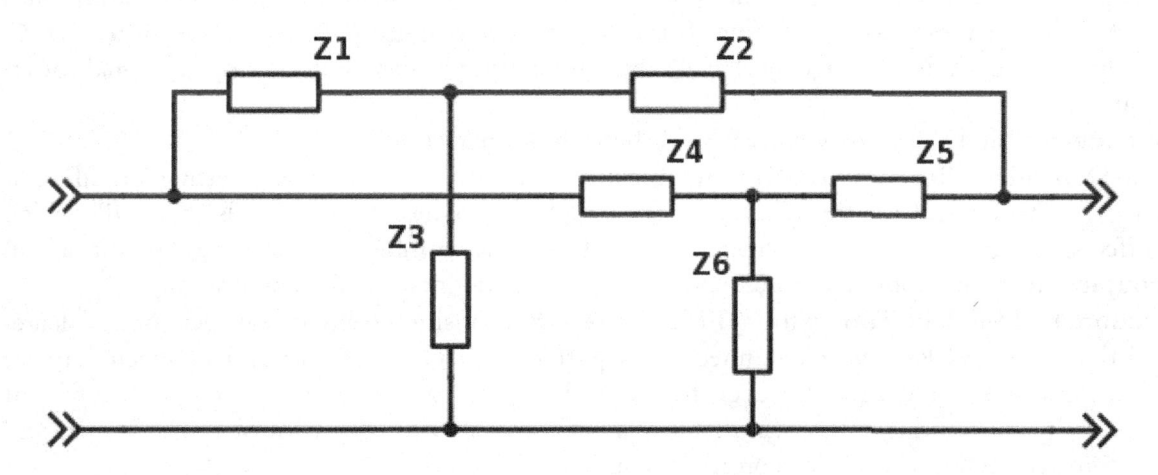

In use, typically the two T sections consists of an R-C-R T-network and a C-R-C T-network.
One example:
R1 = R2 = R; C3 = 2C
C4 = C5 = C; R6 = R/2
The **critical frequency = 1/2πRC**

See how simple it is? We already knew the answer if we had remembered the importance of the RC constant...

In general, the engineers designed these topologies so that the result are easy to use.

Q = ¼ in the above circuit. But we can add op-amps to provide feedback to control the Q.

♪ 7.13 More Building Blocks: VCA, OTA, Analog Multipliers

There are other building blocks important to music and audio that are similar to op-amps that deserves a mention before we end the chapter, such as compressors, automatic gain control (AGC), ADSR (Attack-Decay-Sustain-Release) envelope control for music synthesizers, mixer console automation, and even AM radio. VCA, OTA and analog multipliers can all be bought as modules or ICs, and are used like op-amps.

Variable Gain Amplifiers

So far, we have varied the gain of an op amp by changing its resistor values; the resistors can be implemented as potentiometers. In contrast, an amplifier with a variable gain that varies depending on a control voltage (CV) is a **variable gain amplifier**. It is often called a **Voltage-Controlled Amplifier** (**VCA**) on the assumption that the CV is a voltage level, even though historically and often the control signal for an VCA is a current.

Multiple technologies can be used to achieve a VCA; here are some examples:

- A **Light-Dependent Resistor** (**LDR**) in the feedback loop of an op-amp as inverting amplifier, lit by an LED. The control for the LED affects its brightness, which then controls the LDR, which then affects the gain of the amplifier. The LED-LDR combination can be bought as an **optocoupler** unit. This approach is quite popular in DIY and vintage audio compressors.
- A **Junction Field-Effect-Transistor** (**JFET**) biased with a simple circuit can act as a voltage-controlled resistor (VCR). When connected as part of a voltage divider and connected to an amplifier, the result is a VCA. However, the control voltage band is narrow; there are significant variations in manufacturing; even within the narrow control band the control is not linear, and requires some correction if accurate control is required.
- An **Operational Transconductance Amplifier** (**OTA**) is an amplifier whose differential input voltage produces an output *current*. (An op-amp is an amplifier whose differential input voltage produces an output *voltage*.) Its characteristics and usage are like op-amps, except that it can often be used in open-loop when a suitable limiting resistor is connected at the output, as the limiting resistor prevents the OTA from going to saturation. See inset box on next page.
- An **Analog Multiplier** multiplies two analog signals together to form the output; thus, if one input is the control voltage, then we obtain a VCA! It may be an overkill, as analog multipliers are usually more expensive than typical VCA chips, and analog multipliers are symmetrical (the two inputs are identical) while VCAs are usually not (the VC has more limited bandwidth than the signal) - in other words, analog multipliers can be used as VCAs, but VCAs usually cannot be used as analog multipliers! Analog multiplier circuits are similar to op-amps; it is usually based on the Hall effect; but it is often more susceptible to noise and offset voltage issues when compared to op-amps, because in the case of the analog multiplier those effects are multiplied (pardon the pun)!
- A **digital potentiometer** can be used as part of the op-amp gain control circuit (the R1/R2 circuit) as **digitally-controlled resistors**, with the digital potentiometer controlled digitally (implementing logarithmic control and calibration in the digital domain) but originated with a control voltage which was then digitized through an ADC. A **multiplying DAC** likewise can be used like a digital potentiometer coupled with an amplifier, but eliminate problems like poor tolerances; it functions like the analog multiplier, except using a digital control signal.

David Blackmer founded dbx in 1971, first creating a VCA, then an RMS detector, and with those two in place, an audio compressor in 1976. The company soon became famous for noise reduction systems based upon those circuit blocks, by essentially compressing the signal on recording, then decompressing on playback, thus reducing the effect on noise.

Blackmer's historic VCA has exponential control, and is categorized as **logarithmic VCA**. The gain control thus mimics how the ear would hear loudness. That gain control is achieved via current. Thus, the circuit uses an OTA (with current control) followed by a current-to-voltage converter which is a simple op-amp with a resistor in front converting that current from the OTA to voltage and the buffered and amplified.

Recall our discussion on waveform RMS and peak? Note this general rule:
- to make an audio compressor we should best use an **RMS detector**
- to make an automatic level control we should best use a **peak detector**

7.14 Review Exercises

Chapter 7 Review Guide
Theory
- **concepts** of analog computing, analog circuits, analog domain, linear circuits, negative feedback
- **concepts** of op-amp, gain bandwidth product, differential input/output, fully differential amplifier
- **configurations** of typical op-amps: open-loop, positive feedback, negative feedback (non-inverting input, inverting input, differential input), voltage follower
- **gain** calculations for various op-amp circuits
- **concepts** of virtual ground, summing junction, and bus, and how it is used in audio mixer
- **concepts** of balanced and differential signaling, and how it may mitigate against noise
- **concepts** of basic op-amp integrator and addition of DC gain control, and how it relates to basic RC lowpass filter
- **concept** of synthesizing general-purpose filters through state variable synthesis
- **concept** of op-amp, VCA, OTA, analog multiplier as building blocks for audio electronics
- **avoidance of wrong concepts**

Practice
- ability to **estimate** gain and bandwidth and function by looking at schematics
- ability to produce a standard op-amp schematic based on a particular need in audio

Exercise

1. Explain the terms analog computing, analog circuits, analog domain, and how they may not be the same. Likewise explain linear and why they may not be the same.

2. Explain negative feedback. Assuming you have an amplifier with open-loop gain of 5000 and a feedback loop gain of 1/5, calculate the system gain. Explain the amount of contribution to the system gain that the amplifier and the feedback network has.

3. List some major characteristics of op-amps and ideal op-amps:

Characteristic	Practical Op-amp	Ideal Op-amp

4. Fill in the table, assuming the op-amp is a typical 741 (gain bandwidth product of 1 million), using your best engineering judgment (explain how you interpret the resistors if needed); *show all calculation steps*. Fill in "N.A." if it is not applicable, or *short* or *open* for zero or infinite impedance. The two impedances correspond to R1 and R2.

	feedback impedance	non-feedback impedance	gain	bandwidth
Inverting amplifier	499 Ω	499 Ω		
Inverting amplifier	2 kΩ	1 kΩ		
Inverting amplifier	10 kΩ	2 kΩ		
Inverting amplifier	1 kΩ	10 kΩ		
Non-inverting amplifier	499 Ω	499 Ω		

Non-inverting amplifier	2 kΩ	1 kΩ		
Non-inverting amplifier	10 kΩ	2 kΩ		
Non-inverting amplifier	1 kΩ	10 kΩ		
Voltage Follower				
Open Loop				

5. Explain differential input, output, differential amplifiers, and fully differential amplifiers. Show schematic diagrams illustrating each (you can use one schematic to illustrate multiple points).

6. Show schematic diagrams for each of the following implementations, and provide gain equations:
Open-loop:

Positive feedback:

Voltage follower:

Negative feedback: Non-inverting input:

Negative feedback: Inverting input:

Differential input:

Fully differential amplifier:

7. Show the circuit for op-amp integrator with DC gain control. Show its Bode Plot with formulas for defining its response. Explain how its response degenerate into that of the basic op-amp integrator.

8. Show the Bode plot for a basic RC lowpass filter with 2.2 kΩ resistor and 0.1 µF capacitor. Label the axis and provide rough values. (*This is a review of chapter 5.*) Draw the schematic of an op-amp integrator with the **same** response. Explain your reasoning. (*This tests your understanding of this chapter 8.*)

9. Explain corner and crossover frequencies for an Op-Amp Integrator lowpass filter. For a two-way crossover circuit, show the corner and crossover frequencies.

10. Show how you'd implement a simple four-channel mixer with input gain pots and overall output volume control. Explain the purposes of each part used. Point out where is virtual ground and summing junction, and explain what they mean.

11. Explain how balanced cables can be used in pro audio to mitigate against noise. Is the balanced cable the reason for the success? Explain with a schematic of the circuits involved.

12. Give several ways a voltage-controlled amplifier (VCA) can be implemented.

13. Analyze the following circuit:

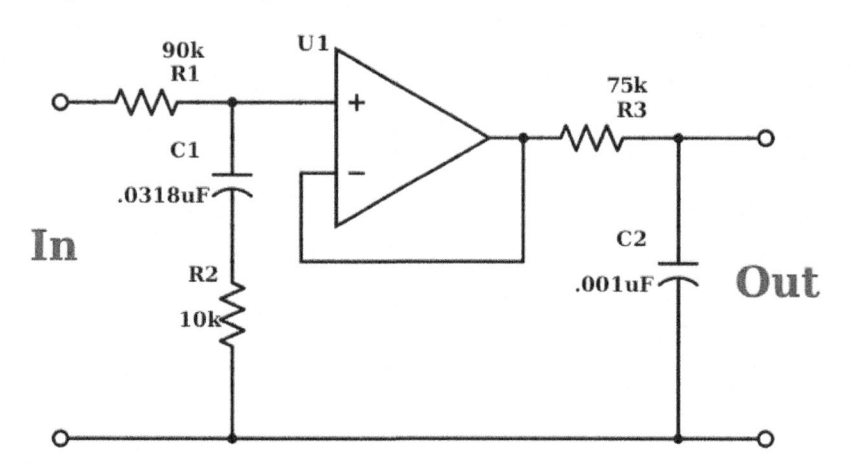

Components	Analyze their function… (e.g. HPF, inverting voltage follower buffer)
U1	
R1, R2, C1	
R3, C2	

Calculate the time constant for the latter filter (consisting of R3, C2):

Calculate one of the time constant for the first filter (based on R2, C1):

Are the results consistent with RIAA equalization time constants?

14. Analyze the following circuit:

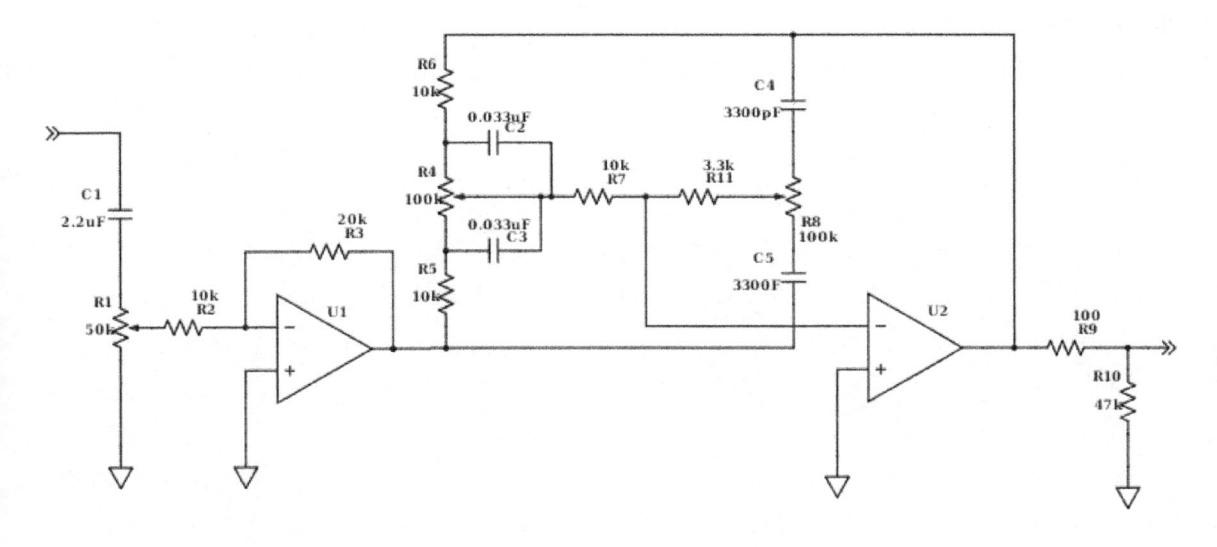

Component(s)	Analyze the function (e.g., non-inverting amp)	Calculate requested
C1		
R1		
R2, R3, U1		voltage gain =
R4-7, C2-3		
R8, R11, C4-5		
U2		
R9, R10		voltage gain =

Identify what the potentiometers control:

Potentiometer	control
R1	
R4	
R8	

(this page deliberately left blank)

Made in United States
North Haven, CT
09 October 2025